Signals and Communication Technology

More information about this series at http://www.springer.com/series/4748

Mikio Tohyama

Sound in the Time Domain

Springer

Mikio Tohyama
Research Consulting Firm
Wave Science Study
Kanagawa
Japan

ISSN 1860-4862 ISSN 1860-4870 (electronic)
Signals and Communication Technology
ISBN 978-981-10-5887-5 ISBN 978-981-10-5889-9 (eBook)
https://doi.org/10.1007/978-981-10-5889-9

Library of Congress Control Number: 2017952503

Printed on acid-free paper

This Springer imprint is published by Springer Nature
The registered company is Springer Nature Singapore Pte Ltd.
The registered company address is: 152 Beach Road, #21-01/04 Gateway East, Singapore 189721, Singapore

Preface

Sounds create an evolving signal of the events occurring around us in our daily lives. The field of acoustics is that branch of natural science concerned with the physical and psychological properties of such sounds. As with other areas of natural sciences, the properties of sound can be described by mathematics. As mathematics is an abstract science, sound is also analyzed in terms of abstract spaces. However, the primitive structures contained with the sounds encountered in our real lives are left without one's notice. The motivation of this book is the return to the nature of sound from the perspective of time structures.

Fourier transforming a function into its image function has been the fundamental and historical approach to the science of waves and particularly the field of acoustics. Given a function that depends on the time variable, the Fourier transformation provides a means to extract the spectral properties of the function. The spectral properties are normally represented by the frequency variable. Thus, the spectral function can be characterized by relating each spectral component to a sinusoidal function. Any function of the time variable or waveform can be represented as a superposition of sinusoidal functions. This approach by decomposition in treating the waveform as a sum of sinusoidal functions has been heavily exploited in signal analysis of waves and particularly sound in the frequency domain using Fourier transformation theory. Consequently, the behavior of sound waves in the time domain has had less attention although sound waveforms can be normally recorded in the time domain.

This book focuses on the characteristics of sound waves in the time domain. This time domain approach provides an informative and intuitively understandable description of various acoustic topics such as sound waves traveling in an acoustic tube or in other media where spectral or modal analysis can be intensively performed. The effects of reflection from the boundary on propagating waves are explicitly manifest, for example.

The envelope or the signal dynamics is representative of sound signatures in the time domain. The global behavior of the waveform defined by the envelope is determined by local structures related to the spectral characteristics including the phase of the waveform. The signal dynamics as represented by envelopes provide the main cues in speech intelligibility. This fact has oriented the author's research to the time domain approach and reestablished the importance of the phase of the wave, which has been given less attention in the frequency domain.

The periodic nature of sound is another typical signature associated with the time domain that relates to the harmonic and nonharmonic structures in the global spectral properties that are independent of the phase. Sensation of pitch that mostly relates to the period of sound is a historical topic in sound perception; however, it seems still a very attractive topic to the author. Sound generated by a musical instrument such as a piano looks nonharmonic. Such sound waveforms may be interpreted through period analyses in the time domain to be subject to the missing fundamental condition where the fundamental is out of auditory range.

Alternatively, sound waves travel in media basically governed by the theory of linear system as described by linear equations. Here the impulse responses and convolution are key concepts. The solutions of the linear equations are superposition of the particular and general solutions. This decomposition reminds the author of the decomposition of sound waves into direct and reflected (or reverberation) sound waves in the time domain. This type of decomposition would be informative not only for room acoustics but the study of musical sounds such as those from piano string vibrations, where the waveform can be understood as a superposition of the direct wave resulting from the excitation by the hammer and the resonant modal waves generated in the string vibration.

Sound propagation in a medium or radiation from a sound source can also be formulated in accordance with convolution theory. This formulation has also motivated the author to take up the time domain approach. In addition to that, the analysis of room reverberation in the time domain has historical roots in the field of acoustics.

This book deals with the signature analysis of sound from the view point of the time domain. The author has tried to make this book self-contained and useful as a textbook for graduate and undergraduate courses offered at universities. The book starts from the very introductory topic of sinusoidal waves basing the material on the formal relationship between the time and frequency domains. For that purpose, the fundamental notions of Fourier or z-transformations and linear systems theory are summarized along with interesting examples from acoustical research. Research engineers and scientists might also have interest in this book for its different approach. In particular, the expressions concerning waveforms including the impulse responses might be informative to audio engineers who are familiar with digital signal analysis.

Exercises have been prepared for every chapter. These exercises are very simple and designed to be solved straightforwardly by hand without the need for a computer. Thus, they assist in reconfirming the fundamental ideas and notions present in every chapter. However, the author has had to assume some prior knowledge on acoustics and signal processing so that from a practical point of view the finished book remained concise. For readers who are not familiar with acoustics in general including signal analysis, the references below have been found to be useful.

W.M.Hartmann, Signals, sound, and sensation, Springer (1997)
J. Blauert and N. Xiang, Acoustics for engineers, Springer (2008)
M. Tohyama, Sound and signals, Springer (2011)
M. Tohyama, Waveform analysis of sound, Springer (2015)

The author thanks Tomomi Hasegawa, Miyabi Konishi, Yoshifumi Hara, Satoru Gotoh, Yoshinori Takahashi, Takatoshi Okuno, Mitsuo Matsumoto, and Michiko Kazama (Waseda University, Kogakuin University), Prof. Manabu Fukushima (Nippon Bunri University), Prof. Hirofumi Nakajima and Prof. Kazunori Miyoshi (Kogakuin University), Prof. Yoshio Yamasaki, and Prof. Katsuhiko Shirai (Waseda University) for their research collaborations that have oriented the author towards the issues of sound synthesis in the time domain described in this book. The author also thanks Hirofumi Onitsuka and his research colleagues (Yamaha Corporation), Hiroyuki Satoh and his group members (Ono Sokki Co. Ltd), and Youske Tanabe and his colleagues (Hitachi Research Corporation) for their fruitful and intensive discussions. The author also thanks Edanz Group Japan Co. Ltd for checking the author's written English.

The author acknowledges that this book is motivated by the very kind and responsive editorship of Dr. Christoph Baumann (Springer). The author also acknowledges the very kind guidance provided by Dr. Kouji Maruyama (Wolfram Research Asia Ltd) in using Mathematica. The author expresses his great appreciation to Prof. Tammo Houtgast (Amsterdam Free University) and Dr. Yoshimutsu Hirata (SV Research Associates) for the long-term research cooperation spanning many years and the very fruitful suggestions and discussions. The author sincerely thanks Prof. Yoichi Ando (Kobe University) for the inspiration and motivation to take up the time domain approach. Finally, the author extends his appreciation to all authors of research articles referred to in this book.

Kanagawa, Japan
May 2017

Mikio Tohyama

Contents

About the Author

Mikio Tohyama holds a Doctorate of Engineering from Waseda University, Tokyo, Japan.

For the next 18 years beginning in 1975, Dr. Tohyama was involved in research projects in acoustics and signal processing at the Nippon Telegraph and Telephone (NTT) Research Laboratories. He was a professor at Kogakuin University (1993–2003) and Waseda University (2003–2012) in sound and perception. Since 2012 he has operated his research consulting firm, Wave Science Study (WSS). His present interest is sound signature expression in the time domain oriented to sound perception. Dr. Tohyama enjoys playing the piano every day.

e-mail address: m_tohyama@fuji.waseda.jp home page: http://wavesciencestudy.com

Chapter 1
Signal Dynamics as Superpositions of Sinusoidal Waves in the Time Domain

Abstract The basis of waveform representation is a sinusoidal function. This chapter introduces the basic nature of compound signals, that is, superpositions of sinusoidal waves. The time envelopes of the compound signals represent the signal dynamics arising from temporal changes in magnitude. The local spectral characteristics like the group delay determines the time-shift of the envelope as well as the dynamic range for the envelope given by the spectral characteristics of the magnitude of the sinusoidal components. Auto-correlation functions provide means to estimate the period of the compound signal even under the missing fundamental condition, in which the component corresponding to the fundamental frequency is lost. The auto-correlation analysis shows that a nonharmonic signal is, from a practical point of view, an example of a missing fundamental condition in which the fundamental is lower than the audible range. Cross-correlation analysis for the normal and time-reversed signals is a possible way to evaluate symmetry in the waveform in the time domain.

Keywords Sinusoidal wave · Exponential function · Analytic signal · Envelope · Beat · Modulation · Phase speed · Group speed · Group delay · Period · Fundamental frequency · Missing fundamental · Harmonics · Almost periodic function · Interference · Cross-correlation · Correlation coefficient · Auto-correlation

1.1 Sinusoidal Waves

1.1.1 Representation of Sinusoidal Waves

A sinusoidal wave is expressible mathematically in the form

$$
\begin{aligned}
x(t) &= A \sin \phi(t) \qquad (1.1)\\
&= A \sin(\omega t + \theta)\\
\phi(t) &= \omega t + \theta \quad \text{(rad)}, \qquad (1.2)
\end{aligned}
$$

M. Tohyama, *Sound in the Time Domain*, Signals and Communication Technology,
https //doi.org/10.1007/978-981-10-5889-9_1

where $\phi(t)$ is called the instantaneous phase or simply phase, and θ gives the initial phase, determining the amplitude (or magnitude) of the wave at $t = 0$, i.e., $x(0) = A \sin \theta$. Its angular frequency can be also defined as the first derivative of the instantaneous phase, that is,

$$\omega = \frac{d}{dt}\phi(t) \quad (\text{rad/s}). \tag{1.3}$$

The angular frequency is called the instantaneous angular frequency if it is a function of t. A sinusoidal wave can be interpreted as a periodic wave with a single frequency $f = \omega/2\pi$ (Hz). In the field of acoustics, a sinusoidal sound is called a pure tone.

Using the angle-sum trigonometric formula, the sinusoidal wave can be also rewritten as

$$\begin{aligned} x(t) &= A \sin(\omega t + \theta) \\ &= B \sin \omega t + C \cos \omega t, \end{aligned} \tag{1.4}$$

that is, as a sum of sinusoidal and co-sinusoidal functions. Here

$$\begin{aligned} A &= \sqrt{B^2 + C^2} \\ \theta &= \tan^{-1} \frac{C}{B}. \end{aligned} \tag{1.5}$$

A sinusoidal wave or function is defined by its magnitude, angular frequency (rad/s), and initial phase (rad). The above expression indicates that a sinusoidal wave can be expressed as a sum of sinusoidal and co-sinusoidal functions with the same angular frequency ω(rad/s) and the two magnitudes B and C for the sinusoidal and co-sinusoidal functions are determined by the magnitude A and the initial phase θ.

1.1.2 Complex Exponential Function

Both the sinusoidal and co-sinusoidal waves can be expanded as complex exponential functions. From the standard formula

$$A\mathrm{e}^{\mathrm{i}\phi} = A(\cos \phi + \mathrm{i} \sin \phi), \tag{1.6}$$

and substituting $\phi = \omega t + \theta$, the complex exponential function is expanded as a complex sum of co-sinusoidal and sinusoidal functions

$$A\mathrm{e}^{\mathrm{i}(\omega t + \theta)} = A_c(\cos \omega t + \mathrm{i} \sin \omega t), \tag{1.7}$$

where A_c is called the complex magnitude obtained from A and θ,

$$A_c = A e^{i\theta} \tag{1.8}$$

$$|A_c| = A \tag{1.9}$$

$$\angle A_c = \theta. \tag{1.10}$$

Inversely, note that the complex magnitude A_c determines both the magnitude A and the initial phase θ modulo additions of 2π.

1.1.3 Complex Variables and Logarithmic Functions

A complex variable $z = x + iy$ can be expressed as a magnitude and an angle such that

$$z = x + iy = |z|e^{i\theta} = re^{i\theta}, \tag{1.11}$$

where $|z| = r$ is called the magnitude (or absolute) and θ denotes the angle. Graphically, a complex variable is displayed on the complex plane, as shown in Fig. 1.1.

The absolute value and the polar angle are given by

$$|z| = \sqrt{x^2 + y^2} \tag{1.12}$$

$$\theta = \angle z = \tan^{-1}\frac{y}{x}. \tag{1.13}$$

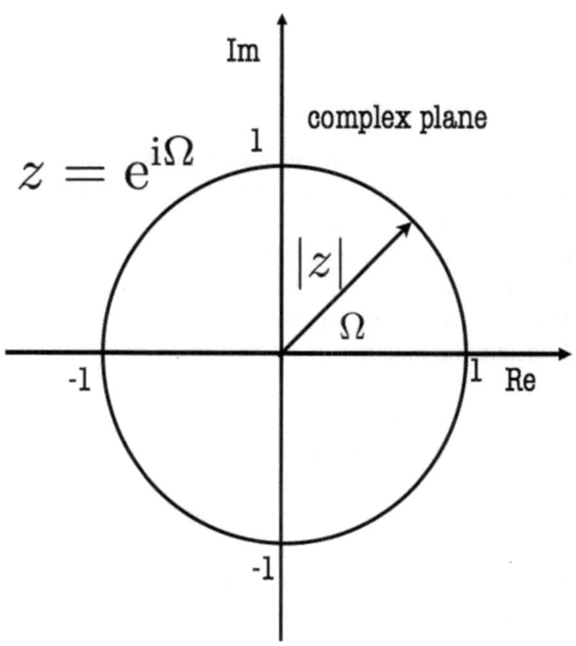

Fig. 1.1 Representation of a complex variable on the complex plane

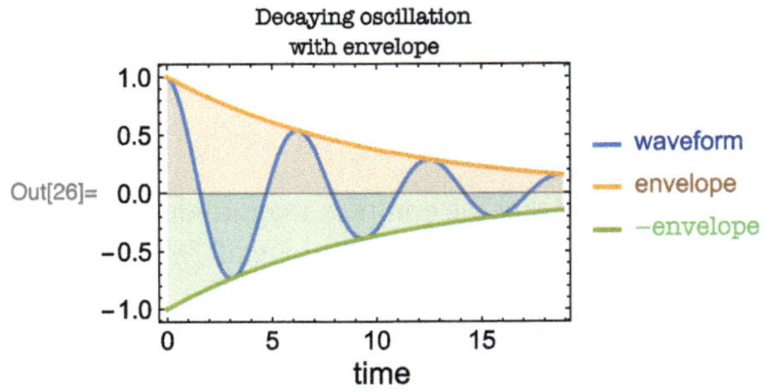

Fig. 1.2 Example of decaying sinusoid

Introducing the complex variable ω^c such that

$$\omega^c = \omega_r^c + \mathrm{i}\omega_i^c = \omega + \mathrm{i}\delta \tag{1.14}$$

where

$$\omega_r^c = \omega \tag{1.15}$$

into ω in the complex exponential function such that

$$\begin{aligned} \mathrm{e}^{\mathrm{i}\omega^c t} &= \mathrm{e}^{\mathrm{i}(\omega+\mathrm{i}\delta)t} = \mathrm{e}^{-\delta t}\mathrm{e}^{\mathrm{i}\omega t} \\ &= r(t)\mathrm{e}^{\mathrm{i}\theta(t)}, \end{aligned} \tag{1.16}$$

where

$$r(t) = \mathrm{e}^{-\delta t} \tag{1.17}$$

$$\theta(t) = \omega t, \tag{1.18}$$

and $r(t)$ is called the instantaneous magnitude or envelope. By allowing the angular frequency to be a complex variable, a temporally decaying sinusoidal function can be expressed as a complex exponential function. Figure 1.2 presents an example of decaying sinusoid with temporally decaying instantaneous magnitude.

Analogously, by introducing complex variables as arguments to the logarithmic function, the phase of a sinusoidal function can be expressed in an alternative way. Consider the function

$$y = a^x, \tag{1.19}$$

where a denotes a positive real variable. Taking the logarithmic function to base e, i.e., the natural logarithm,

$$\log_e y = x \log_e a. \tag{1.20}$$

If $a = \mathrm{e}$, $\log_e y = x$. In extending x to a complex variable, $z = x + \mathrm{i}y$, the logarithmic function of a complex variable can be written as

$$\begin{aligned}\log_e z &= \log_e(x + iy) \\ &= \log_e |z|e^{i\theta} \\ &= \log_e |z| + i\theta.\end{aligned} \tag{1.21}$$

The real (imaginary) part of the logarithmic function of a complex variable z gives us the logarithm of the magnitude (angle) of the complex variable. Using the logarithmic function, a variable can be decomposed into its magnitude and angular parts. For example, setting

$$a^i = re^{i\theta}, \tag{1.22}$$

where $a > 0$ (a real number), then

$$i \log_e a = \log_e r + i\theta. \tag{1.23}$$

Consequently,

$$a^i = 1 \cdot e^{i \log_e a}, \tag{1.24}$$

where $r = 1$ and $\theta = \log_e a$.

The logarithmic and exponential functions are helpful in manipulating sinusoidal functions. Following the properties of the exponential functions such that

$$f(x) = e^x \tag{1.25}$$

$$f(y) = e^y \tag{1.26}$$

$$\begin{aligned}f(x + y) &= f(x)f(y) \\ &= e^{x+y},\end{aligned} \tag{1.27}$$

the angle-sum formula for sinusoidal functions,

$$\sin(x + y) = \sin x \cos y + \cos x \sin y, \tag{1.28}$$

can be derived.

1.2 Temporal Fluctuations of Sinusoidal Waves in the Time Domain

1.2.1 Analytic Signals and Envelopes

The signal dynamics arising from temporal changes in magnitude is represented by the envelope in the time domain. The envelope is defined by the magnitude or the absolute of the analytic representation of the signal. Consider a sinusoidal signal such as

$$x(t) = A \cos(\omega t + \theta). \tag{1.29}$$

The sinusoidal signal can be expressed as

$$A\cos(\omega t+\theta)=\frac{A}{2}(\mathrm{e}^{\mathrm{i}(\omega t+\theta)}+\mathrm{e}^{-\mathrm{i}(\omega t+\theta)}). \tag{1.30}$$

Taking twice the positive frequency term only, then the signal can be expressed as

$$\begin{aligned} z(t) &= A\mathrm{e}^{\mathrm{i}(\omega t+\theta)}=A\mathrm{e}^{\mathrm{i}\phi(t)} \qquad (1.31)\\ &= A(\cos(\omega t+\theta)+\mathrm{i}\sin(\omega t+\theta))\\ &= x(t)+\mathrm{i}y(t)\\ &= |z(t)|(\cos\phi(t)+\mathrm{i}\sin\phi(t))\\ x(t) &= |z(t)|\cos\phi(t)=A\cos(\omega t+\theta)=\Re[z(t)] \qquad (1.32)\\ y(t) &= |z(t)|\sin\phi(t)=A\sin(\omega t+\theta)=\Im[z(t)] \qquad (1.33)\end{aligned}$$

as the complex function. A signal extended as a complex function is called an analytic representation of the signal where the real part comprises the original real function. The absolute value of the analytic signal is called the envelope (or Hilbert envelope). Whereas the real part of the complex function contains the original signal, the imaginary part, specifically

$$y(t)=A\cos\left(\omega t+\theta-\frac{\pi}{2}\right), \tag{1.34}$$

is shifted by a phase $-\pi/2$ from the original signal.

The definition of the envelope can be extended to a signal represented as superposition of sinusoidal functions. Consider a signal given by a sum of two sinusoidal signals,

$$x(t)=A\cos\omega_1 t+B\cos\omega_2 t. \tag{1.35}$$

Introducing the analytic representation of the signal,

$$\begin{aligned} z(t) &= A\cos\omega_1 t+B\cos\omega_2 t \qquad (1.36)\\ &\quad +\mathrm{i}\left[A\cos\left(\omega_1 t-\frac{\pi}{2}\right)+B\cos\left(\omega_2 t-\frac{\pi}{2}\right)\right]\end{aligned}$$

is derived. The square of the magnitude of the analytic representation above is

$$|z(t)|^2=A^2+B^2+2AB\cos(\omega_1-\omega_2)t. \tag{1.37}$$

The envelope is a time-dependent function different from that for each single sinusoidal wave.

Figure 1.3 presents the square of the envelope and the square of the original signal. The square of the original function, is

$$x^2(t)=(A\cos\omega_1 t+B\cos\omega_2 t)^2. \tag{1.38}$$

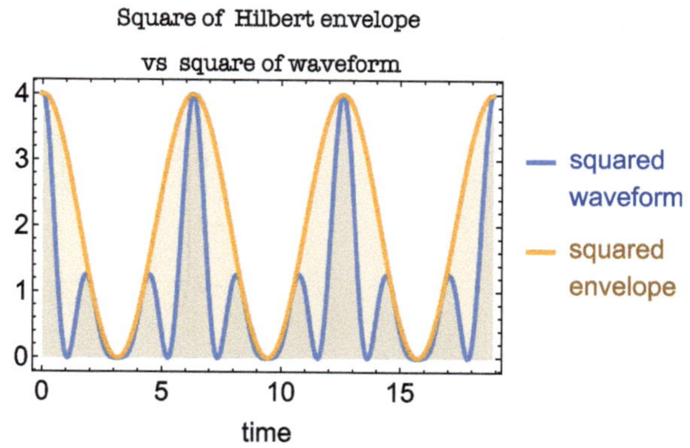

Fig. 1.3 Example of the square of the envelope by Eq. 1.37 and the value of Eq. 1.38

The magnitude of the analytic representation or the envelope displays a smoothing of the original function.

1.2.2 Beats

Superpositions of sinusoidal waves are fundamental in acoustics. In particular, a sum of two sinusoidal components for which the frequencies are very similar creates an effect called beats. Consider a pair of sinusoidal functions

$$x_1(t) = \sin(\omega - \Delta\omega)t \tag{1.39}$$
$$x_2(t) = \sin(\omega + \Delta\omega)t, \tag{1.40}$$

where $\omega > \Delta\omega$. The sum of the two components is

$$\begin{aligned} y(t) &= x_1(t) + x_2(t) & (1.41)\\ &= 2\cos\Delta\omega t \sin\omega t \\ &= A(t)\sin\omega t \\ A(t) &= 2\cos\Delta\omega t. & (1.42) \end{aligned}$$

Figure 1.4 depicts the temporal change in its magnitude $|A(t)|$, referred to as the instantaneous magnitude or envelope. Note that the period of |A(t)| is $\pi/\Delta\omega$(s). The beats generated by two sinusoidal components can be understood as a sinusoidal wave with a frequency that is the average frequency of the two sinusoids and with a temporally changing magnitude (envelope) of period $\pi/\Delta\omega$(s).

Beats afford two manners of signal representation. One is

$$y(t) = A(t)\sin\omega t, \tag{1.43}$$

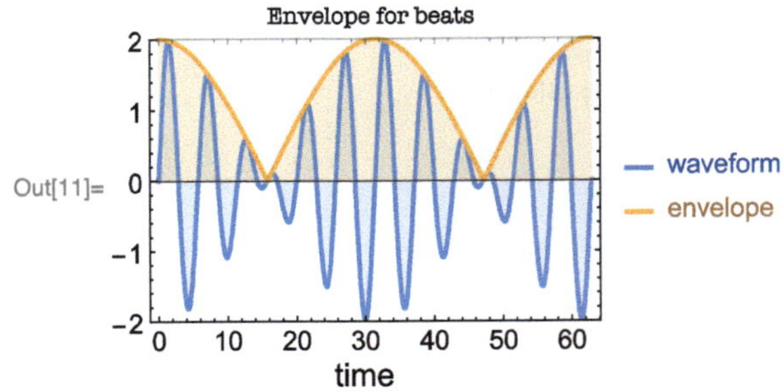

Fig. 1.4 Illustration of instantaneous magnitude (envelope)

which is helpful in representing the signal dynamics in the time domain in terms of the envelope or instantaneous magnitude. In general, when a representation is rewritten in the form

$$y(t) = A(t) \sin \omega t = A(t) x_c(t), \tag{1.44}$$

$x_c(t)$ is called the carrier. The other representation is

$$y(t) = \sin(\omega - \Delta\omega)t + \sin(\omega + \Delta\omega)t, \tag{1.45}$$

which is helpful in the frequency domain as each component has been separated and uniquely identified by the sinusoidal frequency.

If $\Delta\omega$ is small enough such that the beats are audible or detectable, i.e., the period of the envelope can be determined, then the time domain representation would be suitable. In contrast, if $\Delta\omega$ is broad, then the two distinct sinusoidal components might be detectable, and hence the frequency domain representation would be more appropriate.

1.2.3 Modulation

A superposition of three sinusoidal waves is often called modulation when all three frequencies of the sinusoidal components are close. Consider the function

$$y(t) = \frac{1}{2} \sin(\omega - \Delta\omega)t + \sin \omega t + \frac{1}{2} \sin(\omega + \Delta\omega)t. \tag{1.46}$$

Using the expressions developed for beats, the equation above can be rewritten as

$$\begin{aligned} y(t) &= 2\cos^2 \frac{\Delta\omega}{2} t \cdot \sin \omega t \qquad (1.47) \\ &= (1 + \cos \Delta\omega t) \sin \omega t \\ &= A(t) \sin \omega t. \end{aligned}$$

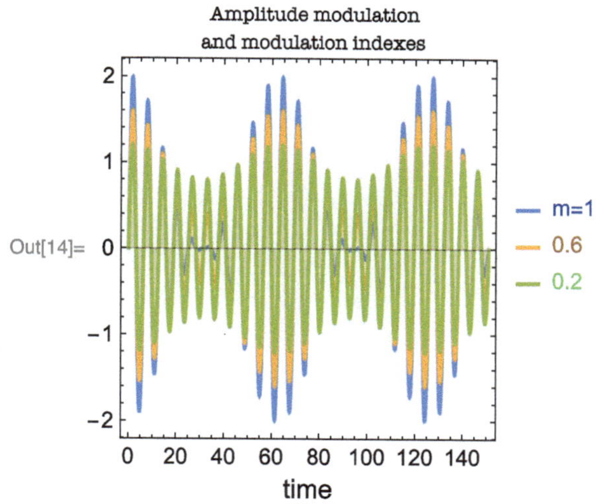

Fig. 1.5 Sampling of different modulations

Note $A(t) \geq 0$, and the instantaneous magnitude or the envelope is a smooth function of time different from that for beats. In addition, the period of $A(t)$, $2\pi/\Delta\omega$, is different from that for beats.

In general, the modulation is given by

$$y(t) = (1 + m\cos\Delta\omega t)\sin\omega t, \tag{1.48}$$

where $0 \leq m \leq 1$ is called the modulation index. If $m = 1$, the modulation is referred to as perfect. In contrast, no modulation is produced if $m = 0$. Figure 1.5 shows samples of different modulations.

Both modulation and beats are significant features of sound in the time domain from a point of view of signal envelopes (the instantaneous magnitude) and temporal dynamics of signals.

A modulation is sensitive to the initial phase conditions in the sinusoidal components. Consider a three-wave superposition that includes a phase effect,

$$\begin{aligned} y(t) &= \frac{1}{2}\sin[(\omega - \Delta\omega)t - \Delta\theta] + \sin\omega t + \frac{1}{2}\sin[(\omega + \Delta\omega)t + \Delta\theta] \\ &= [1 + \cos\Delta\omega(t - \tau)]\sin\omega t, \end{aligned} \tag{1.49}$$

where

$$\tau = -\frac{\Delta\theta}{\Delta\omega} \quad \text{(s)} \tag{1.50}$$

is the group delay (s). Whereas there is no phase effect on the carrier, the group delay is observed as a delay in the envelope. Figure 1.6 presents examples of such modulations.

Fig. 1.6 Group delay in the envelopes with different magnitudes

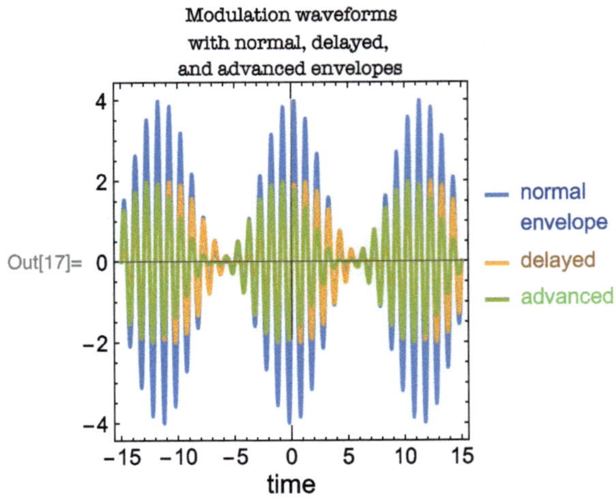

1.2.4 *Phase and Group Speed*

Wave propagation produces a time delay between observations at different locations A and B, because of the finite speed of the wave advancing from A to B. Given a wave that is sinusoidal with a single angular frequency ω, a progressive wave is observed by setting $\tau = x/c$ and $k = \omega/c$ such that

$$y_A = P \sin \omega t \tag{1.51}$$

$$\begin{aligned} y_B &= P \sin \omega(t - \tau) = P \sin \omega(t - x/c) \\ &= P \sin(\omega t - \phi) = P \sin(\omega t - kx), \end{aligned} \tag{1.52}$$

where positions A and B are taken as $x = 0$ and $x = x$, respectively. The difference in the phase between A and B is given by

$$\phi = kx. \qquad \text{(rad)}. \tag{1.53}$$

The ratio

$$\frac{\phi}{\omega} = \frac{kx}{\omega} = \frac{x}{\omega/k} = \frac{x}{c_p}, \tag{1.54}$$

where $c_p = \omega/k$ defines the phase velocity c_p(m/s) which is identical to the speed of sound, when the speed of sound is independent of the sinusoidal frequency. The medium is called non-dispersive (dispersive) if the speed of sound is independent of (depends on) the frequency. The speed of sound, however, is not always independent of frequency. The vibration of an elastic plate propagates with a speed proportional to $\sqrt{\omega}$ [1]. Consider a sinusoidal wave propagating with speed depending on frequency ω, $c(\omega)$. Assume it is a compound wave composed of three sinusoidal components with frequencies $\omega_0 - \Delta\omega$, ω_0, and $\omega_0 + \Delta\omega$, as for example

$$y_A = \frac{1}{2}\sin(\omega_0 - \Delta\omega)t + \sin\omega_0 t + \frac{1}{2}\sin(\omega_0 + \Delta\omega)t \tag{1.55}$$

$$= (1 + \cos\Delta\omega t)\sin\omega_0 t$$

at the point of observation A ($x = 0$). If $\omega_0 \gg \Delta\omega$ holds, the speed of sound can be approximated as

$$c(\omega_0 \pm \Delta\omega) \cong c(\omega_0) \pm \Delta c \tag{1.56}$$

for the given medium.

At observation position B, that is x(m) away from position A, the phase must change through the propagation of sound in the medium. Introducing the frequency-dependent speed of sound as given in Eq. 1.56, the phase can be expressed as

$$\phi(\omega_0 \pm \Delta\omega) \cong \pm\Delta\phi \tag{1.57}$$

by setting

$$\phi(\omega_0) = 0. \tag{1.58}$$

The waveform observed at point B can be written as

$$y_B = (1 + \cos\Delta\omega(t + \Delta\phi/\Delta\omega))\sin\omega_0 t. \tag{1.59}$$

By comparing Eqs. 1.55 and 1.59, the local property of this frequency-dependent phase function yields a time advance in the envelope of the compound sinusoidal function producing the magnitude (or amplitude) modulation form as illustrated in Fig. 1.6. Recalling phase velocity $c_p = \omega/k$, the local behavior of the phase velocity,

$$\frac{\Delta\omega}{\Delta k} = c_g \quad \text{(m/s)} \tag{1.60}$$

gives the group velocity. Introducing $\phi = kx$ once more,

$$\frac{\Delta\phi}{\Delta\omega} = x\frac{\Delta k}{\Delta\omega} = \frac{x}{c_g} = -\tau \quad \text{(s)} \tag{1.61}$$

is derived. Here $-\Delta\phi/\Delta\omega$ is often called the group delay, which gives the time delay of the envelope. Note the sign of the derivative of the phase, which is the local property of the phase function.

In contrast to Eq. 1.57, assume a phase change such that

$$\phi(\omega_0 \mp \Delta\omega) \cong \pm\Delta\phi \tag{1.62}$$

by setting $\phi(\omega_0) = 0$. The observed waveform at position B can be rewritten as

$$y_B = (1 + \cos \Delta\omega(t - (-\Delta\phi/\Delta\omega))) \sin \omega_0 t. \tag{1.63}$$

The local slope of the phase at $\omega = \omega_0$ is negative in accordance with Eq. 1.62. Consequently, $-\Delta\phi/\Delta\omega$ in Eq. 1.63, which is positive here, yields the envelope delay, in contrast to the previous example where $\Delta\phi/\Delta\omega$ is positive, as shown in Fig. 1.6. This explains why $-\Delta\phi/\Delta\omega$ is called the group delay, when $\Delta\phi/\Delta\omega$ is negative.

1.2.5 Period and Fundamental Frequency

In the time domain periodicity is often a basic feature of a compound signal composed of sinusoids. If, for a nonzero integer p, the relation

$$x(t) = x(t \pm pT) \tag{1.64}$$

holds, the signal is said to be periodic with period T(s), with T indicating a shift along the time axis. That is, a periodic signal is identical, when its waveform is shifted by T. The amount of shift determines the period.

Consider a signal is composed of N sinusoidal components,

$$y(t) = \sum_{k=0}^{N-1} A_k \sin(\omega_k t + \theta_k), \tag{1.65}$$

where A_k, ω_k, and θ_k denote the magnitude, the angular frequency (rad/s), and the initial phase (rad), respectively, for the sinusoidal components. If the signal $y(t)$ is periodic, then the period is determined from the fundamental angular frequency which is the great common divisor (GCD) of ω_k $(k = 0, \dots, N-1)$. The angular frequency ω_k can be written as $\omega_k = \alpha\omega_0$ where ω_0 is the GCD and α_k is a positive integer. The ratios of the angular frequencies are given by rational numbers for a periodic wave.

1.2.6 Missing Fundamental

The period or fundamental frequency can be estimated by the GCD of the frequencies of the sinusoidal components that comprise the compound signal. However, the sinusoid of fundamental frequency is not always necessary in generating signal periodicity. Suppose that a compound signal has three sinusoidal components of frequencies 100, 200, and 300 (Hz). The fundamental frequency is then 100 (Hz). Note that even if the signal contains only the 200 and 300 (Hz) components without the 100 (Hz) component, the fundamental frequency is still 100 (Hz).

The fundamental frequency, in general, gives the sense of pitch in hearing a tone. In terms of hearing, this sense of pitch for a periodic signal if there is no fundamental component is referred to as pitch with missing fundamental [2]. The discrimination of pitch between male and female telephone voices exemplifies this feature of missing fundamental. In general, the telephone network limits the transmission frequency band to mostly 300–4000 (Hz), despite the fact that the fundamental frequencies are located under 300 (Hz) for both male and female voices. Nevertheless, voice communications over telephones hold well in general.

1.2.7 Harmonic, Nonharmonic, and Almost Periodic Functions

The fundamental angular frequency ω_0 determines the periodicity of a periodic wave. The harmonics are the other sinusoidal components in the periodic wave. All their angular frequencies are $p\omega_0$, p a positive integer, in a periodic (or harmonic) signal. The harmonics determine the waveform in a single cycle (period) of a periodic signal.

If the ratios of frequencies to the fundamental one are not integers, the signal is called nonharmonic. However the definition of nonharmonic seems a little too idealized, that is, not realistic. Even if the ratio can be given by a rational number, it can in practice be nonharmonic. Suppose that the fundamental angular frequency satisfies $\omega_0 = \omega_1$, and the second component has angular frequency $\omega_2 = l\omega_1$. Here if $l = 2$, then the second component is really the second harmonic, and the total waveform for the superposition of the two components is periodic with period $2\pi/\omega_1$ (s). Now set $l = 2.1$. The ratio of frequencies becomes

$$\frac{\omega_2}{\omega_1} = \frac{21}{10}, \tag{1.66}$$

which is a rational number. However the period lengthens to 10 times $2\pi/\omega_1$. Often this type of change in the period would create a tonal change from a listener's perspective rather than a sense of pitch, because this longer period does not make any sense in terms of pitch. With slight differences in frequencies, the variations in ratios between the fundamental and higher harmonics, which are referred to as nonharmonicities, are important issues in tuning or designing musical instruments.

Figure 1.7 presents the variation in waveform by slightly changing the frequency ratio between the fundamental and the second harmonic. The superposition of sinusoidal components having slight differences in frequencies creates beats or modulation represented by the envelope or slowly changing signal dynamics. Slight differences in the frequency ratios from the regular harmonic relations generate fluctuations in the waveform itself.

If the ratios become irrational numbers, then the waveform is no longer periodic in the strict sense. Such signals are called almost periodic. Fig. 1.8 shows an almost periodic waveform [3] such that

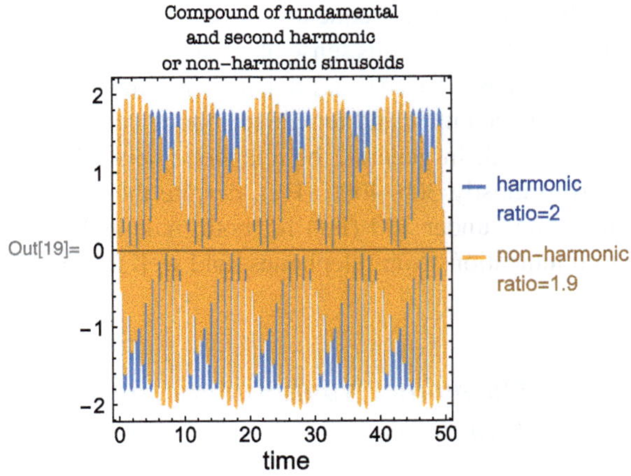

Fig. 1.7 Variation of frequency ratio between fundamental and second harmonic

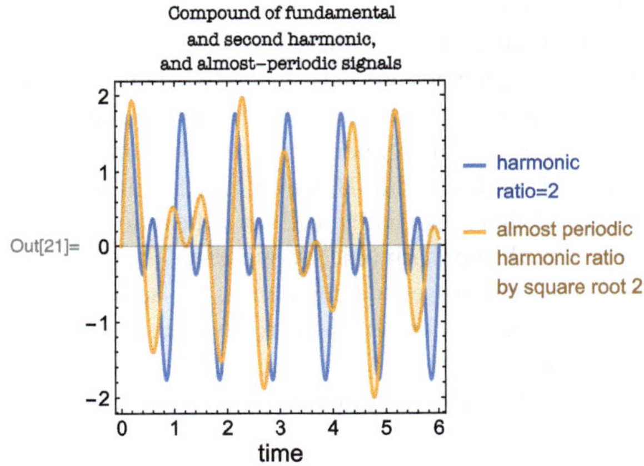

Fig. 1.8 Example of an almost periodic function

$$y(t) = \sin \omega_0 t + \sin \sqrt{2}\omega_0 t. \tag{1.67}$$

The period of the function above is not finite from a theoretical point of view. The envelope, which is determined from the magnitude of the analytic expression of a signal, would represent the difference between the harmonic and nonharmonic signals in the time domain. The nonharmonic relation in the sinusoidal components can be interpreted as a missing-fundamental condition from a practical point of view. Take the compound signal

$$y(t) = A \cos \omega_1 t + B \cos 2.1 \cdot \omega_1 t, \tag{1.68}$$

which has a missing fundamental of frequency $0.1\omega_1$. The square of the envelope becomes

$$|z(t)|^2 = A^2 + B^2 + 2AB \cos 1.1\omega_1 t, \tag{1.69}$$

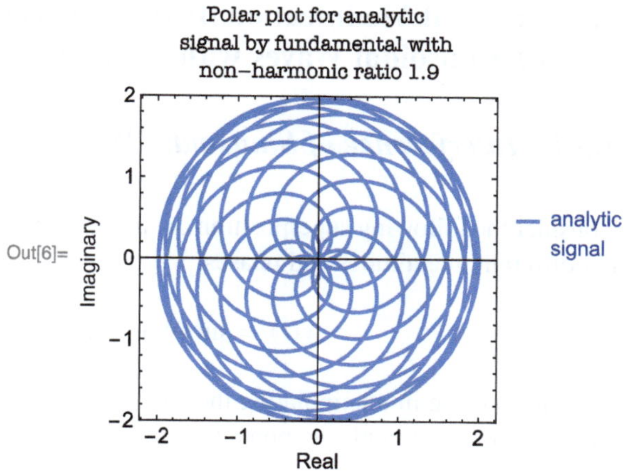

Fig. 1.9 Polar plot of the analytic representation of a signal where the fundamental and second harmonic are nonharmonic, equivalently a missing-fundamental condition

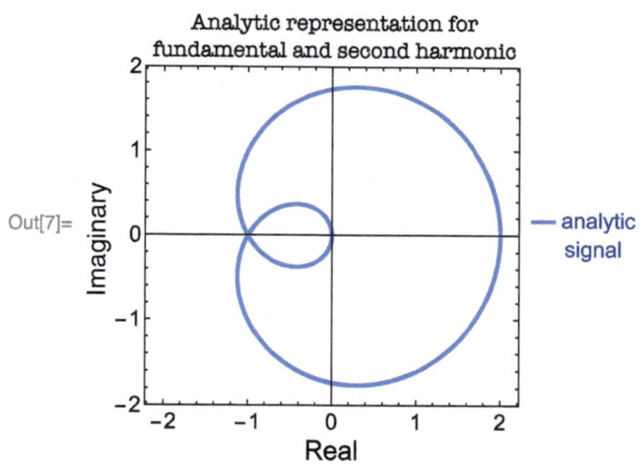

Fig. 1.10 Polar plot of the analytic representation of a harmonic signal composed of fundamental and second harmonic

where the analytic representation is given by

$$z(t) = A\cos\omega_1 t + B\cos 2.1\omega_1 t + \mathrm{i}\left[A\cos\left(\omega_1 t - \frac{\pi}{2}\right) + B\cos\left(2.1\omega_1 t - \frac{\pi}{2}\right)\right]. \tag{1.70}$$

Figure 1.9 presents a polar plot of the analytic representation of the complex function with real (horizontal axis) and imaginary parts (vertical axis). The distance from the origin represents the envelope. Comparing the harmonic relation as shown in Fig. 1.10, then the difference in the signal dynamics can be seen. In the time domain, nonharmonicities would be natural in musical instruments from a view point of signal dynamics.

1.3 Spectral Modification by the Superposition of Sinusoidal Waves with Different Phases

1.3.1 Interference of Sinusoidal Waves

Consider the following superposition of sinusoids with identical angular frequencies but different magnitudes and phases:

$$y(t) = \sin \omega t + a \sin(\omega t + \theta). \tag{1.71}$$

For convenience in manipulating the superposition, rewrite the equation above as a superposition of complex exponential functions, that is,

$$z(t) = \mathrm{e}^{\mathrm{i}\omega t} + a\mathrm{e}^{\mathrm{i}(\omega t+\theta)} = (1 + a\mathrm{e}^{\mathrm{i}\theta})\mathrm{e}^{\mathrm{i}\omega t}, \tag{1.72}$$

where the function $y(t)$ can be obtained by taking the imaginary part of $z(t)$. The equation above indicates that the magnitude of $z(t)$ varies by not only the magnitude but the phase of the second sinusoid. The variation in the magnitude by the superposition with different phases for the sinusoids is called interference.

The squared magnitude of $z(t)$ is found to be

$$|z(t)|^2 = 1 + 2a \cos \theta + a^2, \tag{1.73}$$

that is, a function of the magnitude and phase of the second one. The time average of the square of the real sinusoid $x(t)$ is defined by

$$\overline{|x(t)|^2} = \frac{1}{2} z(t) \cdot z^*(t) = \frac{1}{2} |z(t)|^2, \tag{1.74}$$

where

$$x(t) = A \cos(\omega t + \theta) \tag{1.75}$$

$$z(t) = A\mathrm{e}^{\mathrm{i}(\omega t+\theta)}, \tag{1.76}$$

and $\overline{*}$ denotes the average over one period. If $\theta = 0$, the squared magnitude in Eq. 1.73 becomes

$$|z(t)|^2 = (1 + a)^2. \tag{1.77}$$

The in-phase superposition $\theta = 0$ makes the superposition of magnitudes. In contrast, if $\theta = \pi/2$, the squared magnitude is given by

$$|z(t)|^2 = 1 + a^2 \tag{1.78}$$

yielding an out-of-phase superposition and thus renders the superposition of squared magnitudes. The out-of-phase superposition is equivalent to the expectation of the squared magnitude under the condition in which the phase is random so that any phase might be equally likely between 0–2π. In contrast, the magnitude could be zero when $\theta = \pi$. An anti-phase superposition makes the magnitude vanish if the magnitude of the second component is identically equal to the main component.

1.3.2 Superposition with Different Phases in the Time Domain

The superposition with different phases developed in the previous subsection can be interpreted as the superposition of sinusoids with time delays in the time domain. The time-delayed sinusoids are generated by reflections of sound in a space with reflective boundaries. An example of a time-delayed superposition would be a single reflection added to the direct wave. The waveform by the superposition can be expressed as

$$y^c(t) = \mathrm{e}^{\mathrm{i}\omega t} + a\mathrm{e}^{\mathrm{i}\omega(t-\tau)} = (1 + a\mathrm{e}^{-\mathrm{i}\omega\tau})\mathrm{e}^{\mathrm{i}\omega t}, \tag{1.79}$$

where τ denotes the time delay required for the sound of reflection to arrive at the observation point after the arrival of the direct sound wave. The equation takes the same form as in Eq. 1.72 by replacing the phase θ with $-\omega\tau$. The phase θ or $-\omega\tau$ is a function of the frequency and the time delay. Figure 1.11 shows the squared magnitude as a function of $\omega\tau$ by setting a as a parameter. When a approaches unity, the squared response introduces a periodic sequence of zeros at $\omega\tau = (2n + 1)\pi$ for an integer $n(\geq 0)$.

If the time delay is independent of ω, then zeros are created at

$$\omega_{0n} = (2n + 1)\pi/\tau = (2n + 1)2\pi F_0. \quad (\mathrm{rad/s}). \tag{1.80}$$

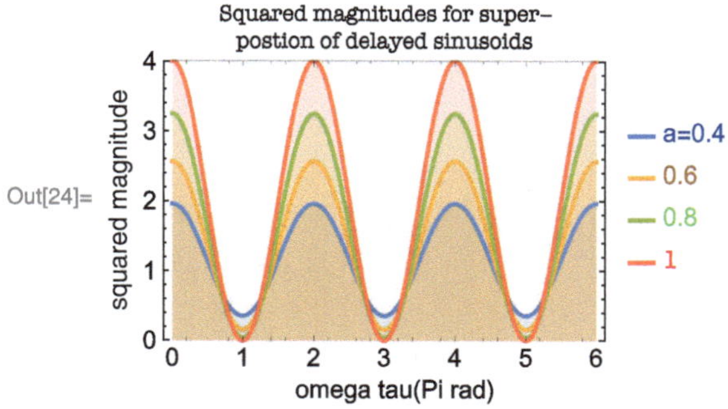

Fig. 1.11 Squared magnitude of the superposition of the direct and delayed signals

Here $F_0 = 1/2\tau$ means the lowest frequency of the sinusoidal wave for which the anti-phase relationship obtains between the direct and reflected sound signal at the receiving position. The tonal character of the compound sound might be greatly changed even by a single reflection, because the sinusoidal components with the angular frequencies ω_{0n} might cancel through interference between the direct and reflected sound waves.

As shown in Fig. 1.11, the troughs in the squared magnitude become shallower as a decreases from unity. Why the zeros disappear is an interesting question. Let us recall the complex angular frequency introduced in Eq. 1.14,

$$\omega^c = \omega_r^c + i\omega_i^c = \omega + i\delta. \tag{1.81}$$

Set

$$e^{-i\omega^c\tau} = z^{-1}, \tag{1.82}$$

where

$$z^{-1} = |z|^{-1}e^{-i\Omega} \tag{1.83}$$
$$|z|^{-1} = e^{\delta\tau} \tag{1.84}$$
$$\Omega = \omega\tau, \tag{1.85}$$

and Ω is called the normalized angular frequency(rad). The magnitude of the superposed waveform can be rewritten as

$$\begin{aligned} |y^c(t)| &= |1 + ae^{-i\omega^c\tau}| \\ &= |1 + az^{-1}|. \end{aligned} \tag{1.86}$$

Here a is a real positive number$(0 < a \leq 1)$. The zero is defined by the root of the equation

$$1 + az_0^{-1} = 0. \tag{1.87}$$

Setting

$$z_0 = |z_0|e^{i\Omega_0} = |z_0|e^{i\omega_0\tau}, \tag{1.88}$$

then

$$|z_0| = |a| = a \tag{1.89}$$
$$e^{i\Omega_0} = e^{i\omega_0\tau} = -1. \tag{1.90}$$

Consequently, the angular frequency ω_0 is found to be

$$\omega_0\tau = (2n+1)\pi \quad \text{(rad)}, \tag{1.91}$$

where $n \geq 0$ is an integer. The angular frequency $\omega_0(\text{rad/s})$ is defined independent of a, and the zeros are located inside the unit circle in the complex domain following a.

The angular frequency is the same as that when $a = 1$, however, note that

$$1 + a\mathrm{e}^{-\mathrm{i}\omega_0\tau} = 1 - a \neq 0, \tag{1.92}$$

where $0 < a < 1$. This explains why troughs are shallower when $|a|$ becomes smaller than unity. A zero inside the unit circle, for example, at

$$z_0 = \mathrm{e}^{-\delta_0\tau}\mathrm{e}^{\mathrm{i}\omega_0\tau} \tag{1.93}$$

is possible if the single reflected wave was attenuated by

$$a = \mathrm{e}^{-\delta_0\tau} \tag{1.94}$$

such that

$$\begin{aligned} 1 + az_0^{-1} &= 1 + a\mathrm{e}^{\delta_0\tau}\mathrm{e}^{-\mathrm{i}\omega_0\tau} \\ &= 1 - 1 \\ &= 0, \end{aligned} \tag{1.95}$$

where

$$\mathrm{e}^{-\mathrm{i}\omega_0\tau} = -1. \tag{1.96}$$

1.3.3 Cross-Correlation of Sinusoidal Waves

Suppose once more that

$$x(t) = A\sin(\omega t + \theta) \tag{1.97}$$

and

$$y(t) = B\sin(\omega t + \phi). \tag{1.98}$$

The average with respect to t of the square of the sum becomes

$$\begin{aligned} e &= \overline{(x(t) + y(t))^2} \\ &= \overline{x^2(t)} + 2\sqrt{\overline{x^2(t)}}\sqrt{\overline{y^2(t)}}R_{xy} + \overline{y^2(t)}, \end{aligned} \tag{1.99}$$

where $\overline{*}$ means the average in a single period with respect to t, and

$$R_{xy} = \frac{\overline{x(t) \cdot y(t)}}{\sqrt{\overline{x^2(t)}}\sqrt{\overline{y^2(t)}}}, \tag{1.100}$$

which is referred to as the cross-correlation coefficient between $x(t)$ and $y(t)$. Here

$$\overline{x(t)} = \overline{y(t)} = 0 \tag{1.101}$$

and

$$-1 \leq R_{xy} \leq 1. \tag{1.102}$$

When $R_{xy} = 0$, the functions $x(t)$ and $y(t)$ are mutually uncorrelated, and then

$$e = \overline{x^2(t)} + \overline{y^2(t)} = \frac{A^2 + B^2}{2}, \tag{1.103}$$

that is, an average of the sum of the squares. In contrast, when $R_{xy} = \pm 1$, the average of the squares becomes

$$e = \frac{(A \pm B)^2}{2}. \tag{1.104}$$

Specifically, the average of the square of the sum depends on the cross-correlation coefficient between the components.

Taking the cross-correlation coefficient between $x(t)$ and $y(t)$ in Eqs. 1.97 and 1.98

$$R_{xy} = \frac{2}{T} \int_0^T \sin(\omega t + \theta) \sin(\omega t + \phi) dt = \cos(\theta - \phi), \tag{1.105}$$

where T(s) denotes the period. The cross-correlation coefficient is a function of the phase difference, and it becomes zero if the sinusoidal waves have different frequencies. In addition, R_{xy} could be zero when

$$\theta - \phi = \frac{\pi}{2} \tag{1.106}$$

even if the frequencies are mutually identical because

$$\int_0^T \sin \omega t \cos \omega t dt = 0. \tag{1.107}$$

The cross-correlation coefficient and the phase difference would be understood in terms of the shift in the waveform. Suppose that a sinusoidal wave has a phase difference $\Delta\phi$ with respect to another with the same frequency. The two waves can be made identical by shifting the waveform by phase $\Delta\phi$ when the magnitudes are normalized.

1.3.4 Correlation Functions

The correlation coefficient can be extended into the correlation function. Take a time-shift $\Delta t = \tau$ instead of the phase shift. Consider a sinusoidal function

$$x(t) = A\cos\omega t. \tag{1.108}$$

Applying the time-shift τ to the function yields

$$x(t-\tau) = A\cos\omega(t-\tau), \tag{1.109}$$

where the waveform is delayed by τ in the time domain. The periodic nature of the waveform may be observed by shifting τ so that the original waveform overlays the delayed one. Such periodic characteristics can be represented by the auto-correlation function such that

$$r(\tau) = \frac{1}{T}\int_0^T x(t)x(t-\tau)dt \tag{1.110}$$

$$= \frac{A^2}{T}\int_0^T \cos\omega t \cdot \cos\omega(t-\tau)dt$$
$$= \frac{A^2}{2}\cos\omega\tau.$$

The auto-correlation function is periodic with respect to the time-shift τ, because the waveform is periodic. The auto-correlation function is independent of the initial phase of the original waveform. Indeed, the auto-correlation function for

$$x(t) = A\cos(\omega t + \theta) \tag{1.111}$$

is the same as that in Eq. 1.110. The auto-correlation function indicates that the period of the original waveform is independent of the phase, if the original waveform is periodic.

The auto-correlation function depends on the squared magnitude. Suppose the superposition of sinusoidal waves

$$x(t) = A\cos\omega t + B\cos 2\omega t \tag{1.112}$$

is given. Taking the auto-correlation function yields

$$\begin{aligned} r(\tau) &= \frac{1}{T}\int_0^T x(t)x(t-\tau)dt \\ &= \frac{A^2}{2}\cos\omega\tau + \frac{B^2}{2}\cos 2\omega\tau. \end{aligned} \tag{1.113}$$

Like the original function, the auto-correlation function is periodic as well; however, it is a function of the average of the squared magnitude for each component. The magnitudes of the sinusoidal components, A and B, are called the magnitude spectrum of the waveform in the frequency domain. Similarly, $A^2/2$ and $B^2/2$ are called the power spectrum. The auto-correlation function exhibits its periodic nature in the time domain by discarding the phase characteristics, and it determines the power spectral property without the phase in the frequency domain.

Consider a pair of sinusoidal functions

$$x(t) = A\cos(\omega t + \theta) \tag{1.114}$$

$$y(t) = B\cos(\omega t + \phi). \tag{1.115}$$

By analogy to the cross-correlation coefficient given in Eq. 1.100, the cross-correlation function $r_{xy}(\tau)$ can be defined as

$$\begin{aligned} r_{xy}(\tau) &= \frac{1}{T}\int_0^T x(t)y(t-\tau)dt \\ &= \frac{AB}{2}\cos(\omega\tau + \theta - \phi), \end{aligned} \tag{1.116}$$

where

$$r_{xy}(\tau) = r_{yx}(-\tau) \tag{1.117}$$

showing the difference between the time delay and time advance. Introducing a time delay to $x(t)$ is equivalent to applying a corresponding time advance shift to $y(t)$, where the cross-correlation is defined for the pair $x(t)$ and $y(t)$.

The cross-correlation functions manifest the symmetry of the pair of waveforms within the time delay that could be interpreted as compensating the phase difference between the paired functions. An interesting example is the effect of time reversal on the correlation function [4]. If a function $x(t)$ is an even function, then

$$x(t) = x_e(t) = x_e(-t). \tag{1.118}$$

In contrast, for an odd function,

$$x(t) = x_o(t) = -x_o(-t). \tag{1.119}$$

Taking a sinusoidal function

$$x(t) = A\cos(\omega t + \theta) \tag{1.120}$$

$$\begin{aligned} x^-(t) &= x(-t) = A\cos(-\omega t + \theta) \\ &= A\cos(\omega t - \theta), \end{aligned} \tag{1.121}$$

where the auto-correlation function for $x(-t)$ is the same as that for $x(t)$.

Now take the cross-correlation function for the pair of functions $x(t)$ and $x(-t)$, then

$$r_{xx^-}(\tau) = \frac{A^2}{2}\cos(\omega\tau + 2\theta) \tag{1.122}$$

$$\begin{aligned} r_{xx^-}(0) &= \frac{A^2}{2}\cos 2\theta \\ &= \hat{R}_{xx^-}, \end{aligned} \tag{1.123}$$

where $r_{xx^-}(0)$ indicates the cross-correlation coefficient between $x(t)$ and $x(-t)$ by normalizing the magnitude. Here

$$\begin{aligned} x(t) &= A\cos(\omega t + \theta) \\ &= A(\cos\omega t\cos\theta - \sin\omega t\sin\theta) \\ &= x_e(t) + x_o(t). \end{aligned} \tag{1.124}$$

The expression above represents the decomposition of $x(t)$ into even and odd functions of t. Taking the cross-correlation coefficient for the even parts, that is, $x_e(t)$ and $x_e(-t)$, then

$$\hat{R}_{ee^-} = \frac{A^2}{2}\cos^2\theta, \tag{1.125}$$

where the magnitude has not been normalized. Similarly, for the odd part

$$\hat{R}_{oo^-} = -\frac{A^2}{2}\sin^2\theta. \tag{1.126}$$

Consequently, the cross-correlation coefficients between $x(t)$ and $x(-t)$ become

$$\begin{aligned} R_{xx^-} &= \frac{1\cdot\cos^2\theta + (-1)\cdot\sin^2\theta}{\cos^2\theta + \sin^2\theta} \\ &= \cos 2\theta, \end{aligned} \tag{1.127}$$

depending on the decomposition into the even and odd functions. In other words, the cross-correlation between $x(t)$ and $x(-t)$ exhibits the symmetry of the waveform in the time domain. The correlation coefficient can be rewritten as

$$R_{xx^-} = \frac{1 \cdot \overline{x_e(t)^2} + (-1) \cdot \overline{x_o(t)^2}}{\overline{x_e(t)^2} + \overline{x_o(t)^2}}, \tag{1.128}$$

where $\overline{x_e(t)} = 0$.

1.4 Exercises

1. Sketch the complex function

$$z = \cos \omega t + \mathrm{i} \sin \omega t$$

in the complex plane with t varying from $\omega t = 0$ to 2π. What does ω signify?
2. Suppose a is a real number. Draw

$$z = a^{\mathrm{i}\theta}$$

on the complex plane varying θ from 0 to 2π.
3. Consider once more that

$$z = \cos \omega t + \mathrm{i} \sin \omega t.$$

Draw the functions

$$z^{-1}, \quad z^*, \quad z + z^*,$$
$$z - z^*, \quad z \cdot z^*, \quad z^n,$$

where z^* denotes the complex conjugate of z such that $z = x + \mathrm{i}y, \quad z^* = x - \mathrm{i}y$, n denotes an integer.
4. Suppose

$$z = |z|\mathrm{e}^{\mathrm{i}\theta} = x + \mathrm{i}y.$$

Indicate in sketches the magnitude and angular parts of the functions

$$-z, \quad z^2, \quad 1/z, \quad 1/z^*.$$

5. Derive the angle-sum and angle-difference formulae

$$\sin(x \pm y) = \sin x \cos y \pm \cos x \sin y$$
$$\cos(x \pm y) = \cos x \cos y \mp \sin x \sin y.$$

6. Examine the difference between the two sinusoidal functions of angular frequencies $\omega - \Delta\omega$ and $\omega + \Delta\omega$. The magnitude of the difference is a function of time, and

the difference also produces beats in the time domain. If the difference is very small, the beat period is very long. When two sinusoidal functions are almost identical, it takes a long time to perceive the difference, because the difference in the initial portion of a long period is so small.

7. Obtain the envelope and carrier of the compound signal

$$y(t) = \frac{1}{2}\sin(\omega - \Delta\omega)t + \cos\omega t + \frac{1}{2}\sin(\omega + \Delta\omega)t.$$

The envelope or the instantaneous magnitude with $A(t) \geq 1$ indicates there is no possibility in producing perfect modulation. Instead, the instantaneous phase is modulated [5, 6].

8. Derive the instantaneous phase and angular frequency for the signal in the exercise above. This signal indicates that the modulation is sensitive to the phase condition between the sinusoidal components.

9. Obtain the fundamental angular frequencies ω_0 and the periods of the compound signals

$$y(t) = A_1\cos(\omega_1 t + \theta_1) + A_2\cos(\omega_2 t + \theta_2) + A_3\cos(\omega_3 t + \theta_3),$$

when

$$\begin{aligned}
&\text{(a)} \quad \omega_2/\omega_1 = 2/1, \quad \omega_3/\omega_1 = 3/1\\
&\text{(b)} \quad \omega_2/\omega_1 = 2/1, \quad \omega_3/\omega_1 = 4/1\\
&\text{(c)} \quad \omega_2/\omega_1 = 5/4, \quad \omega_3/\omega_1 = 6/4\\
&\text{(d)} \quad \omega_2/\omega_1 = 6/5, \quad \omega_3/\omega_1 = 7/5,
\end{aligned}$$

and $\omega_1 = 2\pi \cdot 100$ (rad/s). What do you think is the effect of the initial phase on the period or fundamental frequency? The fundamental frequency could be much lower than the lowest frequency component. The major chord C is composed of three frequencies in which the frequency ratios are 4:5:6. The fundamental frequency is lower than that for the root of the chord. The sense of pitch might be more complicated than perceived just on the fundamental frequency given by the GCD of the frequencies.

10. Decompose the function

$$y(t) = \sin\omega t + \sin\frac{21}{10}\omega t$$

into a superposition of harmonics.

11. Calculate the period of the function

$$y(t) = \sin(2\pi \cdot 100t) + \sin(2\pi \cdot 141.42t).$$

12. Obtain the time average of squared functions $y^2(t)$ over a single period for the following

$$
\begin{aligned}
y(t) &= A\sin(\omega t+\theta)\\
y(t) &= \sin\omega t + A\sin(\omega t+\theta)\\
y(t) &= \sin\omega t + A\cos(\omega t+\theta)\\
y(t) &= A\sin(\omega t+\theta_1) + B\sin(2\omega t+\theta_2)\\
y(t) &= A\sin(\omega t+\theta_1)\cdot B\sin(2\omega t+\theta_2)\\
y(t) &= A\sin(\omega t+\theta_1)\cdot B\sin(\omega t+\theta_2).
\end{aligned}
$$

13. Derive the cross-correlation coefficients between the following pairs:

$$
\begin{aligned}
x(t) &= A\cos\omega t\\
y(t) &= B\cos(\omega t+\phi)
\end{aligned}
$$

$$
\begin{aligned}
x(t) &= A\sin\omega t\\
y(t) &= B\sin(\omega t+\phi)
\end{aligned}
$$

$$
\begin{aligned}
x(t) &= A\cos\omega t\\
y(t) &= B\cos 2\omega t
\end{aligned}
$$

$$
\begin{aligned}
x(t) &= A\sin\omega t\\
y(t) &= B\sin 2\omega t
\end{aligned}
$$

$$
\begin{aligned}
x(t) &= A\cos\omega t + B\sin\omega t\\
y(t) &= A\cos\omega t - B\sin\omega t
\end{aligned}
$$

$$
\begin{aligned}
x(t) &= A\cos\omega t + B\cos 2\omega t\\
y(t) &= A\cos\omega t - B\cos 2\omega t
\end{aligned}
$$

$$
\begin{aligned}
x(t) &= A\cos\omega t + B\cos(\omega t+\theta)\\
y(t) &= A\cos\omega t - B\cos(\omega t+\theta)
\end{aligned}
$$

$$
\begin{aligned}
x(t) &= A\sin\omega t + B\sin(\omega t+\theta)\\
y(t) &= A\sin\omega t - B\sin(\omega t+\theta).
\end{aligned}
$$

14. Derive the auto-correlation functions for the functions such that

$$\begin{aligned} x(t) &= A\cos\omega t \\ x(t) &= B\cos(\omega t+\theta) \\ x(t) &= A\sin\omega t \\ x(t) &= B\sin(\omega t+\theta) \\ y(t) &= A\cos\omega t + B\cos(\omega t+\theta) \\ y(t) &= A\sin\omega t + B\sin(\omega t+\theta) \\ y(t) &= A\cos\omega t + B\sin 2\omega t \\ y(t) &= A\cos\omega t - B\sin 2\omega t. \end{aligned}$$

15. Derive the cross-correlation functions for the following pairs of functions:

$$\begin{aligned} &\text{(a)} \quad & x(t) &= A\cos\omega t \\ && y(t) &= B\cos(\omega t+\theta) \\ &\text{(b)} \quad & x(t) &= A\sin\omega t \\ && y(t) &= B\sin(\omega t+\phi) \\ &\text{(c)} \quad & x(t) &= A\cos\omega t \\ && y(t) &= B\cos(-\omega t+\phi) \\ &\text{(d)} \quad & x(t) &= A\sin\omega t \\ && y(t) &= B\sin(-\omega t+\phi) \\ &\text{(e)} \quad & x(t) &= A\cos\omega t + B\cos 2\omega t \\ && y(t) &= A\cos(\omega t+\phi) - B\cos(2\omega t+\theta) \\ &\text{(f)} \quad & x(t) &= A\sin\omega t + B\sin 2\omega t \\ && y(t) &= A\sin(\omega t+\phi) - B\sin(2\omega t+\theta) \\ &\text{(g)} \quad & x(t) &= A\cos\omega t + B\cos(\omega t+\theta) \\ && y(t) &= A\cos(\omega t+\phi) + B\cos\omega t \\ &\text{(h)} \quad & x(t) &= A\sin\omega t + B\sin(\omega t+\theta) \\ && y(t) &= A\sin(\omega t+\phi) + B\sin\omega t. \end{aligned}$$

16. Find the average of the squared values for the following quartets of functions

$$\begin{aligned} &\sin x + \sin 2x \\ &\sin x + \sin(x+\theta) \\ &\cos x + \cos 2x \\ &\cos x + \cos(x+\theta) \end{aligned}$$

$$\sin x + \cos 2x$$
$$\sin x + \cos(2x + \theta)$$
$$\sin 2x + \cos x$$
$$\sin 2x + \cos(x + \theta)$$

$$\sin(-x) + \cos x$$
$$\sin(-x + \theta) + \cos(x + \theta)$$
$$\cos(-x) + \sin x$$
$$\cos(-x + \theta) + \sin(x + \theta)$$

17. Determine the zeros and the corresponding frequencies for these monomials

$$1 + z = 0, \quad 1 + z^2 = 0$$
$$1 + z^3 = 0, \quad 1 + z^4 = 0$$
$$1/2 + z = 0, \quad 1/2 + z^2 = 0$$
$$1/2 + z^3 = 0, \quad 1/2 + z^4 = 0$$
$$2 + z = 0, \quad 2 + z^2 = 0$$
$$2 + z^3 = 0, \quad 2 + z^4 = 0$$

given

$$z = \mathrm{e}^{\mathrm{i}\Omega}$$
$$\Omega = \omega T$$
$$T = 1.$$

18. Construct the multiplication tables for the solutions (zeros) of these equations,

$$1 + z = 0, \quad 1 + z^2 = 0$$
$$1 + z^3 = 0, \quad 1 + z^4 = 0$$

given

$$z = \mathrm{e}^{\mathrm{i}\Omega}$$
$$\Omega = \omega T$$
$$T = 1.$$

References

1. R.H. Lyon, *Machinery Noise and Diagnostics* (Butterworth, London, 2000)
2. R. Meddis, L. O'Mard, A unitary model of pitch perception. J. Acoust. Soc. Am. **102**(3), 1811–1820 (1997)
3. N. Wiener, *The Fourier Integral and Certain of its Applications* (Dover, New York, 1958)
4. S. Gotoh, M. Tohyama, T. Houtgast, The effect of permutations of time samples in the speech waveform on intelligibility. J. Acoust. Soc. Am. **142**(1), 249–255 (2018)
5. M. Tohyama, *Sound and Signals* (Springer, Berlin, 2011)
6. W.M. Hartmann, *Signals, Sound, and Sensation* (Springer, Berlin, 1997)

Chapter 2
Sinusoidal Waves as Random Variables

Abstract The probability distribution associated with a random variable is a basic notion on which signal analysis of sounds is based. The probability distribution for a sum of a pair of independent random variables underlies the notion of convolution that is basic to signal analysis from the point of view of the linear system theory. The correlation between a pair of random variables is a standard metric to characterize pairs of random variables as well as the independence of random variables. The difference between the uncorrelated (or orthogonality) and statistically independent pairs of random variables renders typical auditory effects on sound image sensations. The perception of sound image experienced during binaural listening to an independent pair of random noise is called subjective diffuseness. An example of the probability density function for a sinusoidal wave is given to help understand orthogonality as being different from the statistical independence. The square correlation is a measure that represents the energy ratio of the uncorrelated pair of variables. Direct sound and its reverberation is a typical example of an uncorrelated pair in reverberant sound fields.

Keywords Probability · Random variables · Expectation · Convolution · Statistical independence · Orthogonality · Uncorrelation · Correlation function · Probability density function · Probability density function for sinusoidal function · Binaural listening · Subjective diffuseness · Sound image perception

2.1 Random Variables

2.1.1 Probability Distribution and Expectation

The concept of correlation is now introduced into random variable analysis. The expressions for correlation in the previous subsections have seemingly not been developed in terms of non-deterministic events; however, the time average used in the various expressions can be interpreted according to the ensemble average or expectation of random variables. By random variable, here denoted as X, is meant a variable that can have any of a range of values that occur randomly, i.e., occurs

M. Tohyama, *Sound in the Time Domain*, Signals and Communication Technology,
https://doi.org/10.1007/978-981-10-5889-9_2

non-deterministically, but can be described with an associated probability. To each random variable individually, this probability, denoted P, is given deterministically; it is a real number in the range $0 \leq P \leq 1$ that provides an estimate of how likely an event of interest occurs [1].

The probability associated with a random variable X taking value x is denoted by $P_X(X = x)$. Suppose that a random variable X takes one of two values, x_0 or x_1, with probability p and q, respectively, where $p + q = 1$. The expectation of the random variable X is defined as

$$E[X] = p \cdot x_0 + q \cdot x_1. \tag{2.1}$$

Consider a pair of random variables X and Y. One of the main interests in random variable analysis would be how likely the event Y (or X) occurs if X (or Y) occurs. If some statistical relevance can be attached to the pair occurring, then it might be informative and helpful in a probabilistic prediction of two random events. Take a pair of random variables X and Y with probabilities $\mathbf{p}_X$ and $\mathbf{p}_Y$, respectively, such that

$$S_{xy} = \begin{pmatrix} \mathbf{x}^T \\ \mathbf{y}^T \end{pmatrix} = \begin{pmatrix} x_0 & x_1 \\ y_0 & y_1 \end{pmatrix} \tag{2.2}$$

and

$$P_{XY} = \begin{pmatrix} \mathbf{p}_x^T \\ \mathbf{p}_y^T \end{pmatrix} = \begin{pmatrix} p_{x_0} & p_{x_1} \\ p_{y_0} & p_{y_1} \end{pmatrix} \tag{2.3}$$

where

$$p_{x_0} + p_{x_1} = 1 \tag{2.4}$$
$$p_{y_0} + p_{y_1} = 1. \tag{2.5}$$

The distribution of probabilities for the random events is rewritten as a matrix,

$$P_{XY} = \begin{pmatrix} p_{00} & p_{01} \\ p_{10} & p_{11} \end{pmatrix} \tag{2.6}$$

where p_{01} denotes the probability that the outcome $(X = x_0, Y = y_1)$ occurs, and likewise for the other entries. Recall that to each random variable, there is associated a probability distribution. The probability $p_X(X = x)$ in which the event X takes x is determined unchanged, irrespective of whether the actual random outcome of the event is observed for the random variable. This condition or assumption that the probability assigned to a random variable is independent of the outcome seems unlikely in a practical situation. However, a stationary random process or a series of random events could be fundamental in understanding theoretically the random nature of events or signals from a probabilistic point of view.

In the matrix P_{XY}, if

$$p_{ij} = p_X(X = x_i) \cdot p_Y(Y = y_j) \tag{2.7}$$

holds, then the random variables X and Y are called statistically independent to each other. With this notion of independence

$$\begin{aligned} p_{00} + p_{01} &= p_X(X = x_0) \cdot p_Y(Y = y_0) + p_X(X = x_0) \cdot p_Y(Y = y_1) \quad (2.8) \\ &= p_X(X = x_0)(p_Y(Y = y_0) + p_Y(Y = y_1)) \\ &= p_X(X = x_0) \\ p_{10} + p_{11} &= p_X(X = x_1) \cdot p_Y(Y = y_0) + p_X(X = x_1) \cdot p_Y(Y = y_1) \quad (2.9) \\ &= p_X(X = x_1), \end{aligned}$$

where

$$p_Y(Y = y_0) + p_Y(Y = y_1) = 1. \tag{2.10}$$

Similarly,

$$p_{00} + p_{10} = p_Y(Y = y_0) \tag{2.11}$$
$$p_{01} + p_{11} = p_Y(Y = y_1), \tag{2.12}$$

where

$$p_X(X = x_0) + p_X(X = x_1) = 1 \tag{2.13}$$

and $p_X(X = x_0)$ indicates that the random variable X takes $X = x_0$ in the random events solely for the random variable X.

Going one step further now toward a practical situation or an instance with no statistical independence, the probability is rewritten as

$$\begin{aligned} p_{ij} &= p_{XY}(X = x_i, Y = y_j) \quad (2.14) \\ &= p_{X|Y}(X = x_i | Y = y_j) \cdot p_Y(Y = y_j), \end{aligned}$$

where

$$p_{X|Y}(X = x_i | Y = y_j) = \frac{p_{XY}(X = x_i, Y = y_j)}{p_Y(Y = y_j)} \tag{2.15}$$

denotes the conditional probability of the event $X = x_i$, subject to the occurrence of the event $Y = y_j$ with probability $p_Y(Y = y_j)$, which is assigned to the random variable Y. Even if the conditional probability is introduced in regard to random variables, Eqs. 2.10 and 2.13 still hold. Specifically

$$p_{X|Y}(X = x_0|Y = y_0) + p_{X|Y}(X = x_1|Y = y_0) = 1 \tag{2.16}$$
$$p_{X|Y}(X = x_0|Y = y_1) + p_{X|Y}(X = x_1|Y = y_1) = 1 \tag{2.17}$$
$$p_{Y|X}(Y = y_0|X = x_0) + p_{Y|X}(Y = y_1|X = x_0) = 1 \tag{2.18}$$
$$p_{Y|X}(Y = y_0|X = x_1) + p_{Y|X}(Y = y_1|X = x_1) = 1. \tag{2.19}$$

If the random variables X and Y are not statistically independent, the probability associated with both X and Y occurring could be obtained from the probabilities for the joint random events. The probabilities given by the matrix P_{XY} is called the joint probability distribution for a pair of random variables.

One of the main features of importance regarding the joint distribution would be the expectation or ensemble average for a sum of the random variables $Z = X + Y$. The expectation can be formally defined as

$$E[Z] = E[X + Y] \tag{2.20}$$

Following the relations of Eqs. 2.16–2.19

$$E[Z] = E[X] + E[Y], \tag{2.21}$$

which indicates the taking an ensemble average is a linear operation. The relation above holds regardless of the interdependence between the random variables X and Y. It is not necessary to take into account the joint probabilities for random variables for estimating the expectation of the sum of the random variables.

2.1.2 Sum of Independent Random Variables and Convolution

Consider again a pair of independent random variables X and Y. The probability distribution for $Z = X + Y$ can be written as

$$p(Z = z_i) = \sum_i p(X = x_i)p(Y = y_i = z_i - x_i) \tag{2.22}$$

where $z_i = x_i + y_i$. This equation defines the operation referred to as convolution.

Consider a pair of identical dice with faces numbered from 1 to 6. The roll of the two dice is assumed independent one from the other. Now sum the numbers showing on the upper surface after each roll. The sum takes random integer values between 2 and 12. The probability distribution associated with the sum can be expressed as

$$p(Z = z) = \sum_i p_X(X = x_i)p_Y(Y = z - x_i) \tag{2.23}$$
$$= \frac{1}{36}\sum_i N(x_i, z - x_i)$$
$$= \frac{1}{36}M(z),$$

where $M(z)$ denotes the number of the combinations for x_i and y_i that sum to make z. These combinations are given in matrix form by

$$C \cdot \mathbf{d} = \mathbf{m}, \tag{2.24}$$

where C indicates the combination or convolution matrix counting the combinations, **d** denotes the 6−dimensional vector for which all entries are unity, and **m** is the 11−dimensional vector giving the number of combinations using the entry $M(z)$ for $Z = z$; that is,

$$\begin{pmatrix} 1\,0\,0\,0\,0\,0 \\ 1\,1\,0\,0\,0\,0 \\ 1\,1\,1\,0\,0\,0 \\ 1\,1\,1\,1\,0\,0 \\ 1\,1\,1\,1\,1\,0 \\ 1\,1\,1\,1\,1\,1 \\ 0\,1\,1\,1\,1\,1 \\ 0\,0\,1\,1\,1\,1 \\ 0\,0\,0\,1\,1\,1 \\ 0\,0\,0\,0\,1\,1 \\ 0\,0\,0\,0\,0\,1 \end{pmatrix} \begin{pmatrix} 1 \\ 1 \\ 1 \\ 1 \\ 1 \\ 1 \end{pmatrix} = \begin{pmatrix} 1 \\ 2 \\ 3 \\ 4 \\ 5 \\ 6 \\ 5 \\ 4 \\ 3 \\ 2 \\ 1 \end{pmatrix}. \tag{2.25}$$

2.2 Correlations

2.2.1 Correlation Functions for Random Variables

According to the definition of the expectation, the cross-correlation coefficient between two random variables X and Y can be rewritten as

$$R_{XY} = \frac{E[(X - \mu_X)(Y - \mu_Y)]}{\sqrt{E[X^2]} \cdot \sqrt{E[Y^2]}}, \tag{2.26}$$

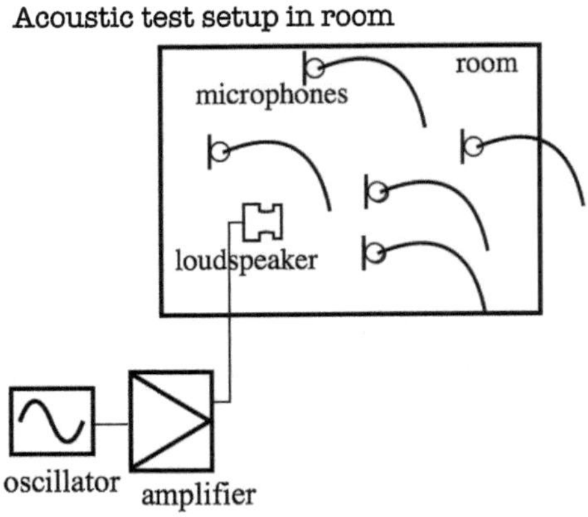

Fig. 2.1 Schematic of an acoustic test setup in which an ensemble average is taken

where

$$\mu_X = E[X] \tag{2.27}$$

$$\mu_Y = E[Y]. \tag{2.28}$$

Similarly, the cross-correlation function is defined as

$$r_{XY}(\tau) = E[X(t)Y(t-\tau)] \tag{2.29}$$

and in setting X = Y, the auto-correlation function becomes

$$r_{XX}(\tau) = E[X(t)X(t-\tau)]. \tag{2.30}$$

The ensemble average is taken for a set in which many pairs of observed variables X and Y are available at every time and the ensemble average is assumed to be independent of the observation time t. Figure 2.1 presents a sketch of a typical setup in which many pairs of microphones are arranged to record the sounds in a closed space.

This situation exemplifies an event when many people are listening to sounds in the room. Assuming a pair of microphones represents the ears of a listener, taking the ensemble average of pairs of random variables may be the audio signal perceived.

2.2.2 *Correlation and Square Correlation*

Correlation as mathematically encoded by the cross-correlation coefficients is a fundamental concept in establishing the mutual relationship or connection between pairs of random variables.

Assuming a pair of two data (such as vectors)

$$\mathrm{L} = \mathrm{X} \tag{2.31}$$
$$\mathrm{R} = a\mathrm{X} + \mathrm{Y} \tag{2.32}$$
$$\mathrm{X} \cdot \mathrm{Y} = 0. \tag{2.33}$$

The cross-correlation coefficient between the two is formally written as

$$\begin{aligned} R_{\mathrm{LR}} &= \frac{\mathrm{X} \cdot (a\mathrm{X} + \mathrm{Y})}{|\mathrm{X}||(a\mathrm{X} + \mathrm{Y})|} \\ &= \frac{a|\mathrm{X}|^2}{|\mathrm{X}||(a\mathrm{X} + \mathrm{Y})|} \\ &= \frac{a|\mathrm{X}|}{|(a\mathrm{X} + \mathrm{Y})|} \to \pm 1. \qquad (\mathrm{Y} \to 0) \end{aligned} \tag{2.34}$$

Then R_{LR}^2 becomes

$$R_{\mathrm{LR}}^2 = \frac{|a|^2|\mathrm{X}|^2}{|(a\mathrm{X} + \mathrm{Y})|^2}. \tag{2.35}$$

Consequently

$$R_{\mathrm{LR}}^2 |(a\mathrm{X} + \mathrm{Y})|^2 = |a|^2|\mathrm{X}|^2. \tag{2.36}$$

This expression hints at the meaning of the square correlation. It reflects the ratio of the direct and reverberant sound energy that is very frequently used in describing features involved in the acoustics of a room. Simple examples entail instances when $|\mathrm{X}| = |\mathrm{Y}| = 1$, then

$$|\mathrm{X} + \mathrm{Y}| = \sqrt{|\mathrm{X}|^2 + |\mathrm{Y}|^2} = \sqrt{2}. \tag{2.37}$$

The cross-correlation coefficient becomes $1/\sqrt{2}$ assuming $a = 1$, and square correlation is 1/2.

2.3 Probability Distribution for Sinusoidal Waves

2.3.1 Probability Density Function

Given the probability distribution P_{X} for when an arbitrary random variable X takes a value in the range x and $x + dx$, expressly,

$$P_{\mathrm{X}}(x \le \mathrm{X} < x + dx) = p(x)dx, \tag{2.38}$$

$p(x)$ is then called the probability density function and satisfies the normalizing condition [1]

$$\int_{X} p(x)dx = 1. \tag{2.39}$$

The range of integration is taken over all values associated with random variable X. A graphical display of the probability distribution from a practical view point is customarily called the histogram.

The expectation or the ensemble average for the random variable X is defined by

$$E[X] = \int_{X} xp(x)dx. \tag{2.40}$$

Assuming the observable variable in X^2 instead of X, then the expectation for X^2 is given by

$$E[X^2] = \int_{X} x^2 p(x)dx \tag{2.41}$$

and generally

$$E[f(X)] = \int_{X} f(x)p(x)dx. \tag{2.42}$$

is defined for the observed value $f(X)$ of interest.

The expectation or ensemble average for a random variable X is a linear operation. Consider now the random variable

$$Z = X + Y, \tag{2.43}$$

where X and Y are random variables. The joint probability density function denoting the probability density function for a pair of the random variables X and Y is given by

$$p_{(X,Y)}(x, y) = p_{X|Y}(x|y) \cdot p_{(\overline{X},Y)}(y), \tag{2.44}$$

where $p_{X|Y}(x|y)$ denotes the conditional probability density function, and $p_{(\overline{X},Y)}(y)$ is called the marginal probability density function defined by

$$p_{(\overline{X},Y)}(y) = \int_{X} p_{(X,Y)}(x, y)dx. \tag{2.45}$$

Following the definitions of the density functions, the expectation of the random variable X + Y becomes

$$\begin{aligned} E[X + Y] &= \int_{X} x \cdot p_{(X,\overline{Y})}(x) + \int_{Y} y \cdot p_{(\overline{X},Y)}(y) \\ &= E_{\overline{Y}}[X] + E_{\overline{X}}[Y] = E[X] + E[Y]. \end{aligned} \tag{2.46}$$

If

$$p_{X|Y}(x|y) = p_X(x) \tag{2.47}$$

$$p_{Y|X}(y|x) = p_Y(y) \tag{2.48}$$

hold, then the random variables X and Y are statistically independent of each other. Consequently, the joint probability density function for a statistically independent pair is written as

$$p_{(X,Y)}(x, y) = p_X(x) \cdot p_Y(y), \tag{2.49}$$

where $p_X(x)$ and $p_Y(y)$ denote the probability density functions for the random variables X and Y, respectively. Hence the expectation of a product of a statistically independent pair becomes

$$E[X \cdot Y] = E[X] \cdot E[Y]. \tag{2.50}$$

2.3.2 *Probability Density Function for Sinusoidal Wave*

A sinusoidal wave can be viewed as a random event when it is randomly sampled along its time axis. Figure 2.2 presents an example of such a case where a sample shows a random variable distributed between ± 1.

Consider a probability density for a sinusoidal wave function for which the function is randomly sampled. For that purpose consider the inverse function $x = f^{-1}(y)$ of a function $y = f(x)$, as depicted in Fig. 2.3.

The probability for $y < Y \leq y + dy$ can be interpreted as

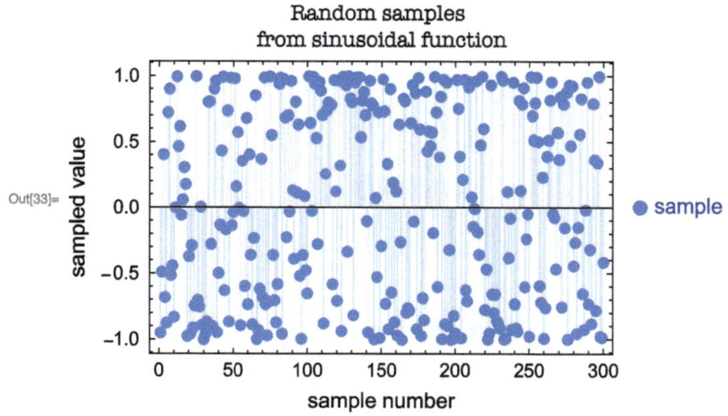

Fig. 2.2 Sinusoidal wave randomly sampled

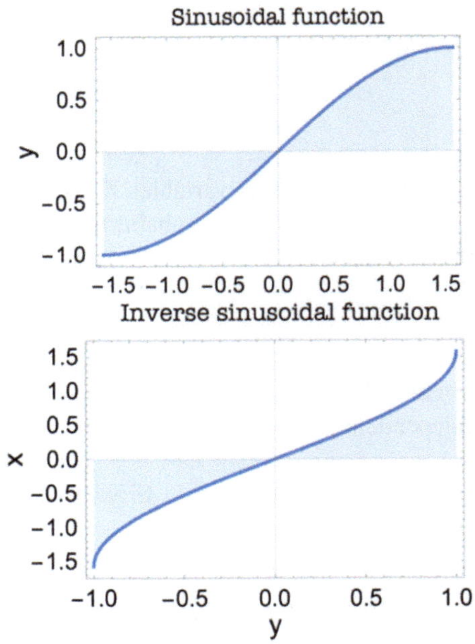

Fig. 2.3 Schematic of the inverse function

$$\begin{aligned} & P_{\mathrm{Y}}(y < \mathrm{Y} \le y + dy) \\ & = P_{\mathrm{X}}(f^{-1}(y) < \mathrm{X} \le f^{-1}(y + dy)) \\ & = p_{\mathrm{X}}(x) \cdot |f^{-1}(y + dy) - f^{-1}(y)|, \end{aligned} \tag{2.51}$$

where $p_{\mathrm{X}}(x)$ denotes the probability density function for the random variable X. Introducing

$$f^{-1}(y + dy) - f^{-1}(y) = \frac{dx}{dy} dy, \tag{2.52}$$

where $x = f(y)$, then

$$\begin{aligned} & P_{\mathrm{Y}}(y < \mathrm{Y} \le y + dy) \\ & = p_{\mathrm{Y}}(y) dy \end{aligned} \tag{2.53}$$

$$= p_{\mathrm{X}}(x) \left| \frac{dx}{dy} \right| dy \tag{2.54}$$

$$p_{\mathrm{Y}}(y) = p_{\mathrm{X}}(x) \left| \frac{dx}{dy} \right| .$$

As an example, take the sinusoidal function

$$y = A \sin \frac{2\pi}{L} x. \tag{2.55}$$

Its inverse function is

$$x = \frac{L}{2\pi} \sin^{-1} \frac{y}{A} \tag{2.56}$$

and therefore its derivative is

$$\left| \frac{dx}{dy} \right| = \frac{L}{2\pi} \frac{1}{\sqrt{A^2 - y^2}}. \tag{2.57}$$

Assuming X is a random variable in the interval $0 < \mathrm{X} < L/2$ where L is the period and

$$p_{\mathrm{X}}(x) = \frac{2}{L}, \tag{2.58}$$

then the probability density function for Y becomes

$$p_{\mathrm{Y}}(y) = \frac{1}{\pi \sqrt{A^2 - y^2}}. \tag{2.59}$$

The probability density function for a sinusoidal wave satisfies

$$\int_{-A}^{A} p_{\mathrm{Y}}(y) dy = 1 \tag{2.60}$$

$$\mathrm{E[Y]} = \int_{-A}^{A} y \cdot p_{\mathrm{Y}}(y) dy = 0 \tag{2.61}$$

$$\mathrm{E[Y^2]} = \int_{-A}^{A} y^2 \cdot p_{\mathrm{Y}}(y) dy = \frac{A^2}{2}. \tag{2.62}$$

The corresponding results for the sinusoidal wave of Eq. 2.55 is

$$\mathrm{E[Y]} = \frac{1}{T} \int_0^T A \sin \frac{2\pi}{T} t dt = 0 \tag{2.63}$$

$$\mathrm{E[Y^2]} = \frac{1}{T} \int_0^T A^2 \sin^2 \frac{2\pi}{T} t dt = \frac{A^2}{2}. \tag{2.64}$$

A random variable X that depends on time t is called ergodic if its expectation or ensemble average is equal to its cycle average.

2.3.3 Uncorrelation and Independence of Random Variables

Consider a pair of random variables X and Y. When the expectation for $\mathrm{Z} = \mathrm{XY}$

$$\mathrm{E[Z]} = \mathrm{E[XY]} = 0 \tag{2.65}$$

the two random variables are referred to as being uncorrelated. In contrast, when the probability density function for $Z = X \cdot Y$ becomes

$$p_Z(z) = p_X(x)p_Y(y) \tag{2.66}$$

the pair of random variables are statistically independent of each other. If two random variables X and Y are independent, then

$$E[Z] = E[XY] = E[X] \cdot E[Y] = 0 \tag{2.67}$$

given that

$$E[X] = E[Y] = 0. \tag{2.68}$$

Conversely, a pair of random variables is not always independent, even if

$$E[Z] = E[XY] = 0 \tag{2.69}$$

holds. Take the pair of random variables

$$X = A \sin \frac{2\pi}{L} U \tag{2.70}$$

$$Y = A \cos \frac{2\pi}{L} U, \tag{2.71}$$

where the probability density functions are identical to each other and the expectations are

$$E[X] = E[Y] = 0. \tag{2.72}$$

Recalling that the product of random variables $Z = XY$ becomes

$$\begin{aligned} Z &= XY \\ &= A^2 \sin \frac{2\pi}{L} U \cdot \cos \frac{2\pi}{L} U \\ &= \frac{A^2}{2} \sin \frac{2\pi}{L/2} U \end{aligned} \tag{2.73}$$

the probability density function for the random variable Z can be expressed as for the sinusoidal function

$$p_Z(z) = \frac{1}{\pi\sqrt{(A^2/2)^2 - z^2}} \tag{2.74}$$

$$E[Z] = 0. \tag{2.75}$$

The probability density function for the random variable Z is not identical to the product of the probability density functions; that is,

$$p_Z(z) \neq p_X(x) \cdot p_Y(y). \tag{2.76}$$

The pair of random variables may have no cross-correlation, even if the variables are not a statistical independent pair. When the cross-correlation or inner product is zero for a pair of random variables that are not statistically independent, the pair is called orthogonal or $\pi/2-$phase shifted to each other rather than statistically independent.

2.3.4 Binaural Merit in Listening to Pairs of Signals

Binaural listening refers to the difference between statistical independence and a $\pi/2-$phase shift for pairs of signals. For example, a binaural pair of independent random noise samples produces a so-called subjective diffuseness [2–5]. In contrast, a pair of $\pi/2$-phase-shifted random noise stimuli does not produce any diffuseness even under binaural listening conditions, although just a non-focal sound image may be produced instead.

Such differences can be displayed using a parametric plot. Figure 2.4 shows samples of pairs for random variables X and Y where

$$x = \cos\phi, \tag{2.77}$$

$$y = \cos(\phi + \theta), \tag{2.78}$$

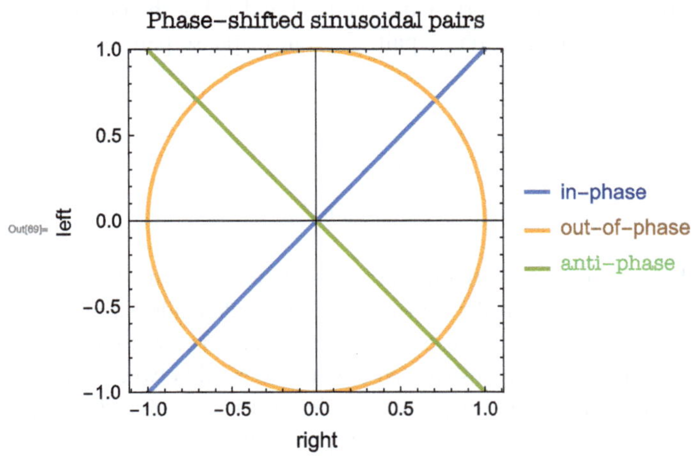

Fig. 2.4 Parametric plots for pairs of phase-shifted sinusoidal functions

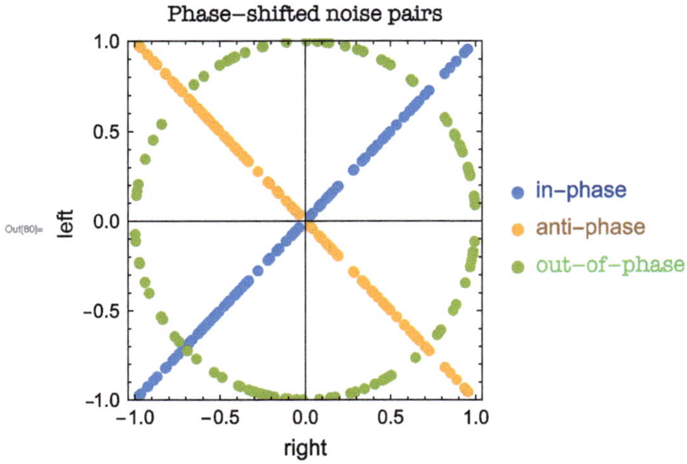

Fig. 2.5 Parametric plots for pairs of out-of-phase random variables

and ϕ ranges between $0 \leq \phi < 2\pi$, and θ is the phase shift. If $\theta = 0$, the pair is said to be in phase; if $\theta = \pi/2$ or $\theta = \pi$, the pair is said to be orthogonal (out-of-phase) or in anti-phase, respectively. For an in-phase pair, the sound image is clearly focused in the head between the pair of ears. In contrast, for an anti-phase pair, the sound image is unnaturally shifted to the back of the head. For the out-of-phase pair no focal sound image is produced. The images are different from that for a pair of independent random noise samples. A subjective diffuseness does not appear for the sinusoidal pair. This sense of subjective diffuseness can be made through the randomness in the time domain of two independent noises.

Figure 2.5 presents examples of random variables X and Y where the random variables are distributed uniformly between -1 and 1. For the in-phase pair $\mathrm{X} = \mathrm{Y}$, where $\mathrm{X} = -\mathrm{Y}$ for the anti-phase pair. The out-of-phase pair was generated using distributions

$$\mathrm{X} = \cos \Phi \tag{2.79}$$

$$\mathrm{Y} = \sin \Phi, \tag{2.80}$$

where Φ denotes a random variable that is distributed uniformly between $-\pi \leq \Phi < \pi$. The difference between the out-of-phase and independent random variables is clearly displayed. Figure 2.6, presents sample plots for a pair of independent random variables that are uniformly distributed over the interval between -1 and 1.

A comparison of the out-of-phase samples in Figs. 2.5 and 2.6 emphasizes the difference. The differences between the in-phase and out-of-phase pairs can also be displayed using the cross-correlation functions for the pair of random variables. Take again the pair of sinusoidal functions. The pair of functions

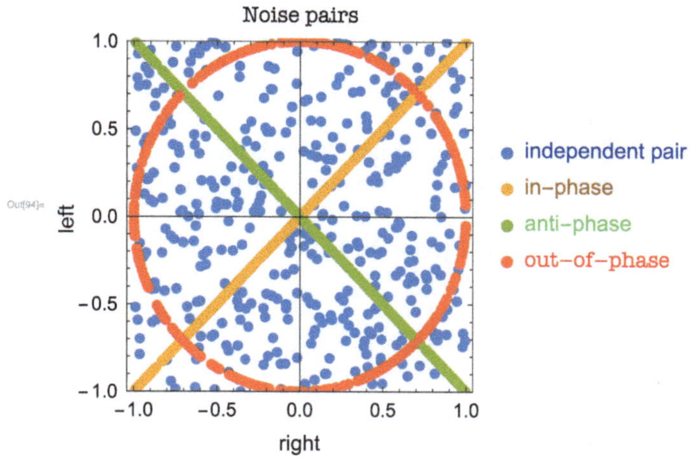

Fig. 2.6 Parametric plots for a pair of independent random variables

$$X = \sin \Phi \tag{2.81}$$

$$Y = \sin \Phi \tag{2.82}$$

yields cross-correlation function

$$\begin{aligned} r(\theta) &= \frac{1}{2\pi}\int_0^{2\pi} \sin\phi \sin(\phi - \theta) d\phi \\ &= \frac{1}{2}\cos\theta. \end{aligned} \tag{2.83}$$

This function is periodic and even with respect to θ; a peak forms at $r(0)$. In contrast, for the out-of-phase pair

$$X = \sin \Phi \tag{2.84}$$

and

$$Y = \cos \Phi \tag{2.85}$$

the cross-correlation function is given by

$$\begin{aligned} r(\theta) &= \frac{1}{2\pi}\int_0^{2\pi} \sin\phi \cos(\phi - \theta) d\phi \\ &= \frac{1}{2}\sin\theta \end{aligned} \tag{2.86}$$

which is a periodic function that is odd with respect to the variable θ and $r(0) = 0$.

Take a random variable X; its auto-correlation function is

$$r(n) = \mathrm{E}[\mathrm{R}(m, n)] = \mathrm{E}[\mathrm{E}[\mathrm{X}(m)\mathrm{X}(m-n)]], \tag{2.87}$$

where $\mathrm{X}(m)$ denotes the random variable that is sampled at time m. Assuming the random variable X corresponds to white noise, then the cross-correlation function is expected to be

$$r(n) = \mathrm{E}[\mathrm{X}(m)\mathrm{X}(m-n)] = u_0\delta(n), \tag{2.88}$$

where

$$u_0 = \mathrm{E}[\mathrm{X}^2(m)]. \tag{2.89}$$

The result above indicates that the cross-correlation function becomes

$$r(n) = u_0\delta(n \pm N) \tag{2.90}$$

for a pair of variables such that

$$\mathrm{X} = \mathrm{X} \tag{2.91}$$

$$\mathrm{Y} = \mathrm{X}(n \mp N). \tag{2.92}$$

The cross-correlation function is not periodic, and has peaks at $n = \pm N$.

The sound image is clearly focused in binaural listening of the pair of random variables stated above. The focal point is situated just between both ears when $N = 0$. As $|N|$ increases, the image point is shifted to the left or right side depending on the sign of N. Nevertheless, within the some range, which is determined by the inter-aural distance, the sound image is focused in consequence of the cross-correlation function being even with peaks located at $n = \pm N$.

For an independent pair of variables X and Y, the cross-correlation function becomes

$$\begin{aligned} r(n) &= \mathrm{E}[\mathrm{X}(m)\mathrm{Y}(m-n)] \\ &= \mathrm{E}[\mathrm{X}(m)]\mathrm{E}[\mathrm{Y}(m-n)], \\ &= 0, \end{aligned} \tag{2.93}$$

where the peaks have disappeared. The binaural sense from listening to this pair of independent random variables is called subjective diffuseness in which the sound image is not focused but randomly changes with time, provided that no differences between the variables are detected in monaural listening.

Taking an example of an out-of-phase pair described by Eqs. 2.84 and 2.85, the cross-correlation function is interpreted as an instance when the variable Φ is randomly sampled after being uniformly distributed over the interval between $-\pi$ and π. The result highlights the differences in the cross-correlation functions between out-of-phase and independent pairs, where for both instances $r(0) = 0$. However, the cross-correlation for the out-of-phase pair is an odd function for which a pair of peaks of opposite sign is observed as indicated in Eq. 2.86. In contrast, the cross-correlation for the independent pair is formally an even function and $r(n) = 0$ independent of n. These differences are characteristic of the out-of-phase (or orthogonal) and independent random variable pairs.

2.4 Exercises

1. Consider a probability density function of random variable X given by

$$p(x) = 1/2. \quad -1 \leq x \leq 1$$

Calculate

$$\begin{aligned} &\mathrm{E[X]} \\ &\mathrm{E[X^2]} \\ &\mathrm{E[X^3]}. \end{aligned}$$

2. Consider once more the probability density function of random variable X defined above. Evaluate the cross-correlation coefficient between X and Y for

$$\begin{aligned} \mathrm{Y} &= a\mathrm{X} + b \\ \mathrm{Y} &= a\mathrm{X}^2 + b\mathrm{X} + c \\ \mathrm{Y} &= a\mathrm{X}^3 + b\mathrm{X}^2 + c\mathrm{X} + d. \end{aligned}$$

3. Suppose

$$\begin{aligned} \mathrm{E[X]} &= 0 \\ \mathrm{E[Y]} &= 0 \\ \mathrm{E[XY]} &= r \\ \mathrm{E[X^2]} &= \mathrm{E[Y^2]} = 1. \end{aligned}$$

Calculate the expectation value for the random variables

$$\begin{aligned} Z &= X + Y \\ Z &= X - Y \\ Z &= (X + Y)^2 \\ Z &= (X - Y)^2 \\ K &= \frac{E(X - Y)^2}{E(X + Y)^2}. \end{aligned} \tag{2.94}$$

The cross-correlation coefficient can be obtained using the power ratio between the difference and sum for the pair of random variables.
4. Suppose once more

$$\begin{aligned} E[X] &= 0 \\ E[Y] &= 0 \\ E[XY] &= r \\ E[X^2] &= E[Y^2] = 1. \end{aligned}$$

Obtain the expectation values for random variables

$$\begin{aligned} Z &= X \\ Z &= rX \\ Z &= \sqrt{1 - r^2}Y \\ Z &= rX + \sqrt{1 - r^2}Y \\ Z &= X(rX + \sqrt{1 - r^2}Y). \end{aligned}$$

The cross-correlation coefficient between the pair $\mathbf{Z} = r\mathbf{X} + \sqrt{1 - r^2}\mathbf{Y}$ and $\mathbf{X}$ with the same power is r [6].
5. The convolution of a pair of sequences $a(n)$ and $b(n)$ is written as

$$c(n) = a * b(n) = \sum_m a(m)b(n - m).$$

Make a pair of polynomials for $a(n)$ and $b(n)$ such that

$$\begin{aligned} A(X) &= \sum_n a(n)X^n \\ B(X) &= \sum_n b(n)X^n. \end{aligned} \tag{2.95}$$

Suppose the corresponding polynomial for $c(n)$ is defined as

Fig. 2.7 Schematic for the convolution of Exercise 6

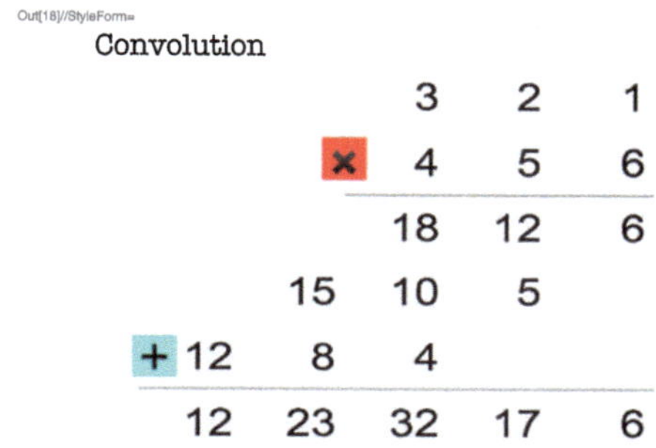

$$C(X) = A(X)B(X) = \sum_n \hat{c}(n)X^n.$$

Show

$$\hat{c}(n) = a * b(n) = c(n).$$

6. Derive the matrix form of the convolution for

$$\begin{aligned} a(n) &= 1, 2, 3 \\ b(n) &= 6, 5, 4 \\ c(n) &= a * b(n). \end{aligned} \tag{2.96}$$

Confirm $c(n) = a * b(n) = b * a(n)$. Interestingly, the convolution can be calculated as shown in Fig. 2.7 [7].

References

1. R. Nelson, *Probability, Stochastic Processes, and Queueing Theory* (Springer, Berlin, 1995)
2. Y. Ando, *Auditory and Visual Sensation* (Springer, Berlin, 2009)
3. P. Damaske, Subjective Untersuchung von Schallfeldern, Acustica 19 199–213 (1967/68)
4. M. Tohyama, A. Suzuki, Interaural cross-correlation coefficients in stereo-reproduced sound fields. J. Acoust. Soc. Am. **85**(2), 780–786 (1989)
5. D.H. Cooper, Notions of conventional stereo interaural cross-correlation coefficients in stereo-reproduced sound fields; Interaural cross-correlation coefficients in stereo-reproduced sound fields J. Acoust. Soc. Am. 85 780–786 (1989). J. Acoust. Soc. Am. **86**(6), 2452–2454 (1989)
6. K.J. Gabriel, H.S. Colburn, Interaural correlation discrimination I. Bandwidth and level dependence. J. Acoust. Soc. Am. **69**(5), 1394–1401 (1981)
7. S. Hirobayashi, *Private communication* (2003)

Chapter 3
Fourier Transform and Superposition of Sinusoidal Functions

Abstract The Fourier transformation theory provides the theoretical basis for understanding the representation of any signal as a superposition of sinusoidal functions in time and frequency planes. This chapter deals with the Fourier and inverse Fourier transforms of sequences in the time and frequency domains. A single pulse in the time (frequency) domain is transformed to a sinusoidal function in the frequency (time) domain. Thus, a sequence composed of pulses is represented as a superposition of sinusoidal functions with magnitude and phase. The auto-correlation of a sequence in the time domain, whose Fourier transform discarding phase information determines the power spectral function in the frequency domain, gives an estimate of the period of the sequence. In contrast, the auto-correlation of the spectral function, including the magnitude and phase in the frequency domain, yields the time sequence of the squared original samples that show the change in the signal power or signal dynamics in the time domain. The change in the signal power is formally defined by the envelope for the analytic signal. As well as the envelope the periodic property represents the signature of a waveform in the time domain. The periodicity is lost as the spectral function is spread over the line spectral components. This loss of periodicity can be represented by the envelope of the auto-correlation function of the waveform. The envelope as a limit is a sinc function when the power spectral function is uniformly spread around the discrete spectral components.

Keywords Generating function · Combination · Fourier transform · Convolution · Auto-correlation · Auto-convolution · Analytic sequence · Envelope · Analytic function · Sinc function · Low-pass filtering · High-pass filtering · Windowing · Triangular windowing · Group delay · Fourier series · Dirichlet function

M. Tohyama, *Sound in the Time Domain*, Signals and Communication Technology,
https://doi.org/10.1007/978-981-10-5889-9_3

3.1 Convolution, Generating Functions, and Fourier Transform

3.1.1 Generating Function and Combination

Consider two independent random variables U and V, for which U takes integer values 0, 1, 2 randomly and equally likely, and similarly V takes integers 1, 3, 4; then sum the two integers, S = U + V. The probability distribution for S is constructed from the generating functions defined for the two random variables, specifically [1]

$$U(X) = X^0 + X^1 + X^2 \tag{3.1}$$

$$V(X) = X^1 + X^3 + X^4. \tag{3.2}$$

Recall that the combination U and V that determines S can be expressed as a convolution. Taking the product of the generating functions $U(X)$ and $V(X)$ gives

$$\begin{aligned} S(X) &= U(X) \cdot V(X) \\ &= (X^0 + X^1 + X^2)(X^1 + X^3 + X^4) \\ &= X^1 + X^2 + 2X^3 + 2X^4 + 2X^5 + X^6 \\ &= \sum_n s_n X^n. \end{aligned} \tag{3.3}$$

The result derived from the product of generating functions indicates that after normalization the probability distribution is

$$P(\mathrm{S} = n) = \frac{s_n}{\sum_{k=0}^{6} s_k}. \tag{3.4}$$

The combination can be expressed in matrix form

$$\begin{pmatrix} 1\,0\,0\,0\,0 \\ 1\,1\,0\,0\,0 \\ 1\,1\,1\,0\,0 \\ 0\,1\,1\,1\,0 \\ 0\,0\,1\,1\,1 \\ 0\,0\,0\,1\,1 \\ 0\,0\,0\,0\,1 \end{pmatrix} \begin{pmatrix} 0 \\ 1 \\ 0 \\ 1 \\ 1 \end{pmatrix} = \begin{pmatrix} 0 \\ 1 \\ 1 \\ 2 \\ 2 \\ 2 \\ 1 \end{pmatrix} = \mathbf{m} \tag{3.5}$$

where $\mathbf{m}$ is the vector giving the number of combinations from S = 0 to S = 6. The product of the generating functions and the matrix formulation for convolution can be expressed in the closed form

$$\begin{aligned} m(n) &= u * v(n) = v * u(n) \\ &= \sum_l u(n-l)v(l) \\ &= \sum_l u(l)v(n-l) \end{aligned} \tag{3.6}$$

where $n = n - l + l = l + n - l$, and $u(n)$ and $v(n)$ denote sequences with values

$$u(0) = u(1) = u(2) = 1 \tag{3.7}$$
$$v(0) = v(2) = 0 \tag{3.8}$$
$$v(1) = v(3) = v(4) = 1 \tag{3.9}$$

and $m(n)$ gives the number of combinations for $\mathrm{S} = n$.

3.1.2 Fourier Transform and Convolution

The generating function introduced for the sequence composed of 0 and 1s above is a polynomial of a formal variable X. Extending the sequences to real numbers $a(n)$ and substituting a complex exponential function for X, then

$$A(X)|_{\mathrm{e}^{-\mathrm{i}\Omega}} = A(\mathrm{e}^{-\mathrm{i}\Omega}) = \sum_n a_n \mathrm{e}^{-\mathrm{i}\Omega n} \tag{3.10}$$

is called the Fourier transform of the sequence a_n. Suppose that a pair of sequences $a(n) = a_n$ and $b(n) = b_n$ is given. The sequence produced through the convolution of a_n and b_n is defined as

$$\begin{aligned} c_n &= c(n) = a * b(n) = b * a(n) \\ &= \sum_m a_m \cdot b_{n-m} = \sum_m a_{n-m} \cdot b_m. \end{aligned} \tag{3.11}$$

The Fourier transform of the sequence $c(n)$ is written

$$\begin{aligned} C(\mathrm{e}^{-\mathrm{i}\Omega}) &= \sum_n c_n \mathrm{e}^{-\mathrm{i}\Omega n} \\ &= \sum_n a_n \mathrm{e}^{-\mathrm{i}\Omega n} \cdot \sum_n b_n \mathrm{e}^{-\mathrm{i}\Omega n} \\ &= A(\mathrm{e}^{-\mathrm{i}\Omega}) \cdot B(\mathrm{e}^{-\mathrm{i}\Omega}), \end{aligned} \tag{3.12}$$

that is, the product of the Fourier transforms of the sequences a_n and b_n. The relationship between convolution and the Fourier transform is just the same as that for combination and the generating function.

3.1.3 Periodicity of the Fourier Transform

Fourier transforming a sequence a_n ($n = 0, \ldots, N-1$) involves creating a harmonic superposition of the sinusoidal functions of angular frequencies Ω normalized within the range 0 to 2π. Hence, the Fourier transform of a sequence is also referred to as the spectral or frequency characteristics of the sequence. The Fourier transform of the sequence a_n, denoted $A(\mathrm{e}^{-\mathrm{i}\Omega})$, is a periodic continuous function of $\mathrm{e}^{-\mathrm{i}\Omega}$ with period 2π. Moreover,

$$A(\mathrm{e}^{-\mathrm{i}\Omega}) = \sum_n a_n \mathrm{e}^{-\mathrm{i}\Omega n} \tag{3.13}$$

corresponds to the Fourier series expansion of the periodic function $A(\mathrm{e}^{-\mathrm{i}\Omega})$. This periodicity is fundamental to the Fourier transform of discrete sequences.

3.1.4 Inverse Fourier Transform

A Fourier transform of sequence a_n determines a periodic and continuous function of $\mathrm{e}^{-\mathrm{i}\Omega}$. The Fourier transform of a discrete sequence is referred to as the Fourier series expansion of the periodic and continuous function. Because the set of functions $\mathrm{e}^{-\mathrm{i}\Omega n}$ is orthogonal, the sequence a_n can be reconstructed in the time domain using the inverse Fourier transform;

$$\hat{a}_n = \frac{1}{2\pi}\int_0^{2\pi} A(\mathrm{e}^{-\mathrm{i}\Omega})\mathrm{e}^{\mathrm{i}\Omega n} d\Omega \tag{3.14}$$

where

$$A(\mathrm{e}^{-\mathrm{i}\Omega}) = \sum_n a_n \mathrm{e}^{-\mathrm{i}\Omega n} \tag{3.15}$$

and the orthonormality of the set of complex exponential functions is given by

$$\frac{1}{2\pi}\int_0^{2\pi} \mathrm{e}^{\mathrm{i}\Omega(n-m)} d\Omega = \delta(n-m) \tag{3.16}$$
$$= \begin{cases} 0 & n \neq m \\ 1 & n = m. \end{cases}$$

Here, the sequence of values a_n now assumes the role as coefficients of the Fourier series. The relationship between $\hat{a}_n$ and the original sequence a_n is of principal interest. Suppose that the original sequence is defined for $0 \leq n \leq N-1$. Substituting $A(\mathrm{e}^{-\mathrm{i}\Omega})$ into Eq. 3.14

$$\begin{aligned}\hat{a}_n &= \frac{1}{2\pi}\int_0^{2\pi}\left[\sum_{m=0}^{N-1} a_m \mathrm{e}^{-\mathrm{i}\Omega m}\right]\mathrm{e}^{\mathrm{i}\Omega n} d\Omega \qquad (3.17)\\ &= \frac{1}{2\pi}\sum_{m=0}^{N-1} a_m \int_0^{2\pi} \mathrm{e}^{\mathrm{i}(n-m)\Omega} d\Omega\\ &= \sum_{m=0}^{N-1} a_m \delta(n-m)\\ &= \begin{cases} a_n & 0 \leq n \leq N-1\\ 0 & \text{otherwise.}\end{cases}\end{aligned}$$

The inverse Fourier transform, $\hat{a}_n$, reproduces the original sequence a_n for $0 \leq n \leq N-1$ but vanishes for all other values of n. The sequence is not periodically expanded at all in the time domain. The periodicity lies instead in the spectral or frequency characteristics of the Fourier-transformed a_n, $A(\mathrm{e}^{-\mathrm{i}\Omega})$. This periodicity produces a discrete sequence, but does not create a periodic sequence in the time domain.

In brief, the sequence a_n is represented by the inverse Fourier transform. Regarding the inverse Fourier transform as a superposition of the sinusoidal sequence, the Fourier transform gives the magnitude and phase of the sinusoidal sequence that depends on a continuous variable Ω. The representation of a sequence by the Fourier transform is called the signal representation by a continuous spectral function.

3.1.5 *Auto-Correlation and Auto-Convolution for Signal Dynamics in Time Domain*

The auto-correlation sequence can be defined for sequences as well as for functions in the time domain. Consider the Fourier transform of a sequence a_n such that

$$A(\mathrm{e}^{-\mathrm{i}\Omega}) = \sum_{n=0}^{N-1} a_n \mathrm{e}^{-\mathrm{i}\Omega n}, \qquad (3.18)$$

where a_n is a series of real numbers. Taking the square of the absolute for the Fourier-transformed sequence,

$$P(\mathrm{e}^{-\mathrm{i}\Omega}) = |A(\mathrm{e}^{-\mathrm{i}\Omega})|^2 = A(\mathrm{e}^{-\mathrm{i}\Omega})A(\mathrm{e}^{\mathrm{i}\Omega}) \tag{3.19}$$

yields the power spectral density function of the time sequence. Taking the inverse Fourier transform of the power spectral density function then gives

$$\begin{aligned}
&\frac{1}{2\pi}\int_0^{2\pi} P(\mathrm{e}^{-\mathrm{i}\Omega})\mathrm{e}^{\mathrm{i}\Omega n} d\Omega \\
&= \frac{1}{2\pi}\int_0^{2\pi} |A(\mathrm{e}^{-\mathrm{i}\Omega})|^2 \mathrm{e}^{\mathrm{i}\Omega n} d\Omega \\
&= \frac{1}{2\pi}\int_0^{2\pi} \sum_{l=0}^{N-1} a_l \mathrm{e}^{-\mathrm{i}\Omega l} \sum_{m=0}^{N-1} a_m \mathrm{e}^{\mathrm{i}\Omega m} \mathrm{e}^{\mathrm{i}\Omega n} d\Omega \\
&= \sum_m a_m \cdot a_{m+n} = a \otimes a(n)
\end{aligned} \tag{3.20}$$

where orthonormality of the sinusoidal functions in the form

$$\frac{1}{2\pi}\int_0^{2\pi} \mathrm{e}^{-\mathrm{i}\Omega(l-m-n)} d\Omega = \begin{cases} 1 & l-m-n=0 \\ 0 & \text{otherwise} \end{cases} \tag{3.21}$$

has been used. The sequence

$$r(n) = \sum_m a_m \cdot a_{m+n} = \sum_m a_m \cdot a_{m-n} = a \otimes a(n) \tag{3.22}$$

is called the auto-correlation sequence, which is defined as the inverse Fourier transform of the power spectral density function for the sequence a_n. The auto-correlation of a time sequence can be defined independently of the phase function.

The auto-correlation sequence is an even sequence with regard to the integer n, which includes positive, negative, and zero. The Fourier transform of an even sequence is a real function; in addition, the function is nonnegative if the even sequence is the auto-correlation sequence of a real sequence. The Fourier transform for the auto-correlation sequence is the zero-phase function. Figure 3.1 shows examples of a sequence, its power spectral function, and the auto-correlation sequence.

Moreover, the correlation between a_n and a_{-n} where $\hat{a}_n = a_{-n}$ determines the auto-convolution r_n^* for a_n. Specifically,

$$\begin{aligned}
r_n^* = r^*(n) &= \sum_m a_m \cdot a_{n-m} \\
&= a * a(n) = a \otimes \hat{a}(n).
\end{aligned} \tag{3.23}$$

The auto-convolution can be rewritten as

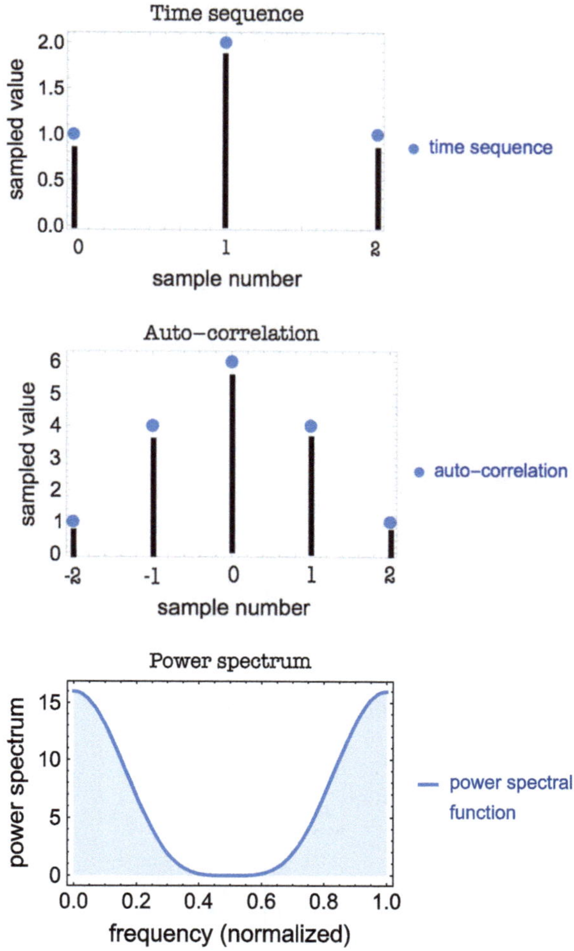

Fig. 3.1 Examples of a sequence, its power spectral function, and auto-correlation sequence

$$r^*(n) = \frac{1}{2\pi}\int_0^{2\pi} A^2(\mathrm{e}^{-\mathrm{i}\Omega})\mathrm{e}^{\mathrm{i}\Omega n}d\Omega \tag{3.24}$$

but is not given by the power spectral function; rather it depends on the phase spectral function where

$$A(\mathrm{e}^{-\mathrm{i}\Omega}) = |A(\mathrm{e}^{-\mathrm{i}\Omega})|\mathrm{e}^{\mathrm{i}\theta(\mathrm{e}^{-\mathrm{i}\Omega})} \tag{3.25}$$

$$A^2(\mathrm{e}^{-\mathrm{i}\Omega}) = |A(\mathrm{e}^{-\mathrm{i}\Omega})|^2\mathrm{e}^{\mathrm{i}2\theta(\mathrm{e}^{-\mathrm{i}\Omega})}. \tag{3.26}$$

The auto-correlation sequence can be rewritten as

$$
\begin{aligned}
r_n &= \frac{1}{2\pi}\int_0^{2\pi} |A(\mathrm{e}^{-\mathrm{i}\Omega})|^2 \mathrm{e}^{\mathrm{i}\Omega n} d\Omega \\
&= \sum_m a_m \cdot a_{m+n} = \sum_m a_m \cdot a_{m-n} \\
&= a \otimes a(n) = a * \hat{a}(n).
\end{aligned}
\tag{3.27}
$$

The phase spectral information is discarded.

Signal dynamics of a sequence a_n in the time domain can be represented basically by the squared sequence a_n^2 as a time-varying envelope. Taking the Fourier transform of a_n^2 gives

$$
\begin{aligned}
&\sum_{n=-\infty}^{\infty} a_n^2 \mathrm{e}^{-\mathrm{i}\Omega n} \\
&= \frac{1}{2\pi}\int_0^{2\pi} A(\mathrm{e}^{-\mathrm{i}\Omega_1}) d\Omega_1 \frac{1}{2\pi}\int_0^{2\pi} A(\mathrm{e}^{-\mathrm{i}\Omega_2}) d\Omega_2 \sum_{n=-\infty}^{\infty} \mathrm{e}^{\mathrm{i}(\Omega_1+\Omega_2-\Omega)n} \\
&= \frac{1}{2\pi}\int_0^{2\pi} A(\mathrm{e}^{-\mathrm{i}\Omega_1}) A(\mathrm{e}^{-\mathrm{i}(\Omega-\Omega_1)}) d\Omega_1 \\
&= A * A(\mathrm{e}^{-\mathrm{i}\Omega})
\end{aligned}
\tag{3.28}
$$

subject to the application of the formula

$$
\begin{aligned}
&\sum_{n=-\infty}^{\infty} \mathrm{e}^{\mathrm{i}(\Omega_1+\Omega_2-\Omega)n} \\
&= \lim_{r\to 1}\left[\sum_{n=-\infty}^{0} + \sum_{n=0}^{\infty} r \mathrm{e}^{\mathrm{i}(\Omega_1+\Omega_2-\Omega)n}\right] - 1 \\
&= \delta(\Omega_1 + \Omega_2 - \Omega).
\end{aligned}
\tag{3.29}
$$

The signal dynamics represented by the square of a sequence in the time domain can be expressed as the inverse Fourier transform of the auto-convolution of the spectral function of the sequence. The dynamics in the time domain depends on both magnitude and phase of the spectral functions. Recalling that the auto-correlation function represents the power spectral dynamics of the waveform, as described in Subsection 1.3.4, the spectral correlation (spectral auto-convolution) exhibits the signal dynamics in the time domain. Figure 3.2 shows the squared sequence along with the magnitude and phase parts of the spectral function, and similarly those for the original sequence. The linear phase (see Subsection 3.2.2) of the original sequence is reserved for the squared sequence.

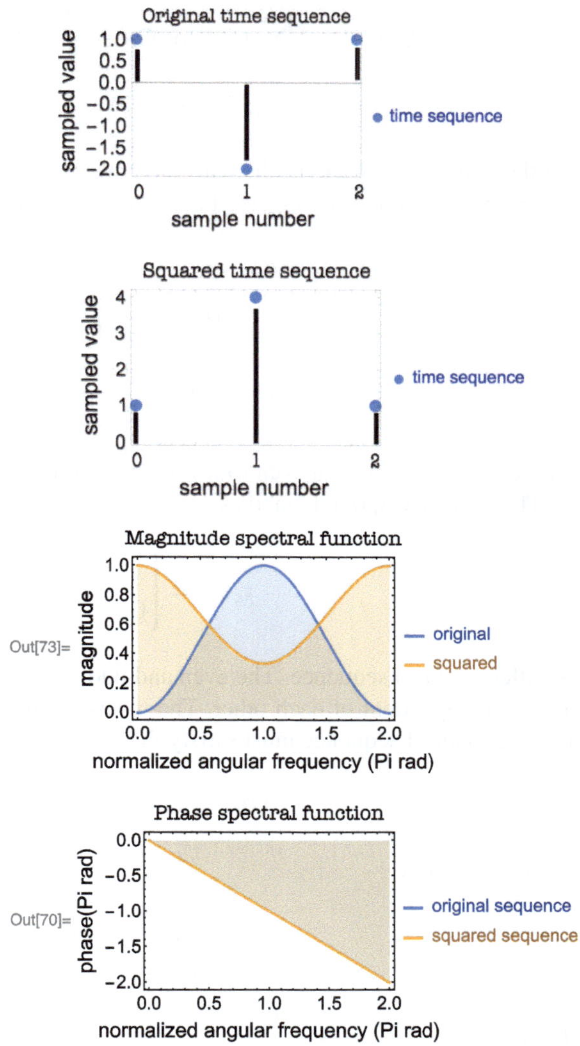

Fig. 3.2 Squared sequence with magnitude and phase parts of spectral function, and similarly the spectral functions of the original sequence

3.1.6 Decomposition of Sequence into Even and Odd Sequences

Even and odd sequences are defined in similar manner as for functions. Take the Fourier transform of a real sequence $x(n)$ such that

$$X(\mathrm{e}^{-\mathrm{i}\Omega}) = \sum_{n} x(n)\mathrm{e}^{-\mathrm{i}\Omega n}. \tag{3.30}$$

The Fourier-transformed function is decomposed into real and imaginary parts,

$$X(e^{-i\Omega}) = \sum_n x(n)\cos\Omega n + i\left(-\sum_n x(n)\sin\Omega n\right) \tag{3.31}$$
$$= X_r(\Omega) + iX_i(\Omega)$$

for the real sequence $x(n)$. The real (imaginary) part is an even (odd) function with respect to Ω. Taking the inverse Fourier transform of the real and imaginary parts yields

$$x_r(n) = x_e(n) = \frac{1}{2\pi}\int_0^{2\pi} X_r e^{i\Omega n} d\Omega \tag{3.32}$$
$$x_i(n) = x_o(n) = i\frac{1}{2\pi}\int_0^{2\pi} X_i e^{i\Omega n} d\Omega, \tag{3.33}$$

where $x_e(n)(x_o(n))$ is an even (odd) sequence for n.

The sequence $x_c(n)$ defined as

$$x_c(n) = \begin{cases} x(n) & n \geq 0 \\ 0 & n < 0 \end{cases} \tag{3.34}$$

is called a causal sequence. The even and odd components for this causal sequence are not independent of each other. The inverse Fourier transforms $x_r(n)$ and $x_i(n)$ for a real causal sequence must satisfy the relation

$$x(n) = x_r(n) + x_i(n) = 0 \tag{3.35}$$

for $n < 0$, where

$$x_r(n) = x_r(-n) \tag{3.36}$$
$$x_i(n) = -x_i(-n). \tag{3.37}$$

Then

$$x_r(n) = \begin{cases} x_i(n) & n \geq 0 \\ -x_i(n) & n < 0. \end{cases} \tag{3.38}$$

Figure 3.3 presents a schematic of the decomposition of a causal sequence into even and odd components.

A real causal sequence can be reconstructed using only the Fourier transform of the real (or imaginary) part. Taking only the real or imaginary part does not mean that under the Fourier transform the phase is discarded. This is because even if only the real part is taken, it still contains the sign changes corresponding to changes in the phase. Taking the real (or imaginary) part can be interpreted as constructing the even (or odd) part. The value $x(0)$ is required for reconstructing $x(n)$ from the imaginary part of its Fourier transform because $x_i(0) = 0$.

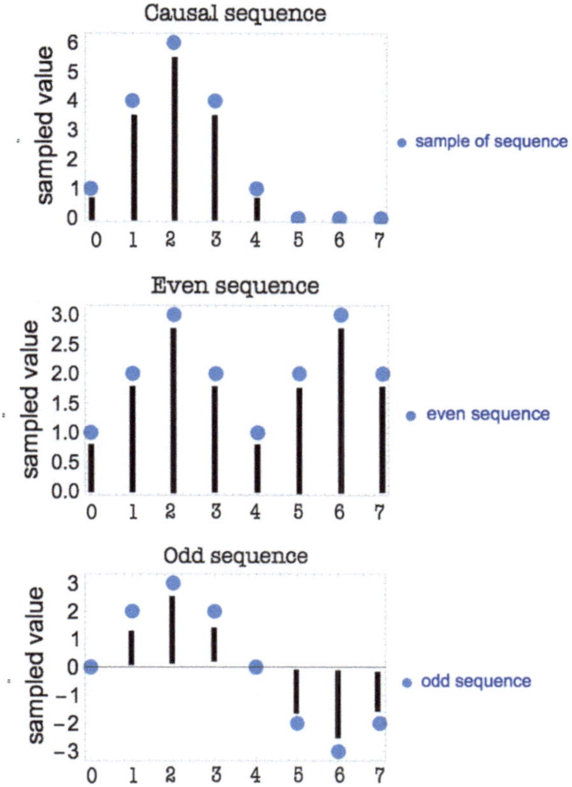

Fig. 3.3 Schematic of the decomposition of a causal sequence into even and odd sequences

3.1.7 Analytic Sequence and Envelope

A causal spectral function can be defined for the function as well as the time sequence. Consider the Fourier transform for a real sequence $x(n)$ according to Eq. 3.31. The real and imaginary parts of the Fourier transform are even and odd functions in the frequency domain. If the Fourier transform of $\hat{x}(n)$ is defined as

$$\hat{X}(\mathrm{e}^{-\mathrm{i}\Omega}) = \begin{cases} 2X(\mathrm{e}^{-\mathrm{i}\Omega}) & 0 < \Omega < \pi \\ X(\mathrm{e}^{-\mathrm{i}\Omega}) & \Omega = 0, \pi \\ 0 & \pi < \Omega < 2\pi, \end{cases} \tag{3.39}$$

the spectral function $\hat{X}(\mathrm{e}^{-\mathrm{i}\Omega})$ is called the causal spectral function. Taking the inverse Fourier transform of the causal spectral function then gives

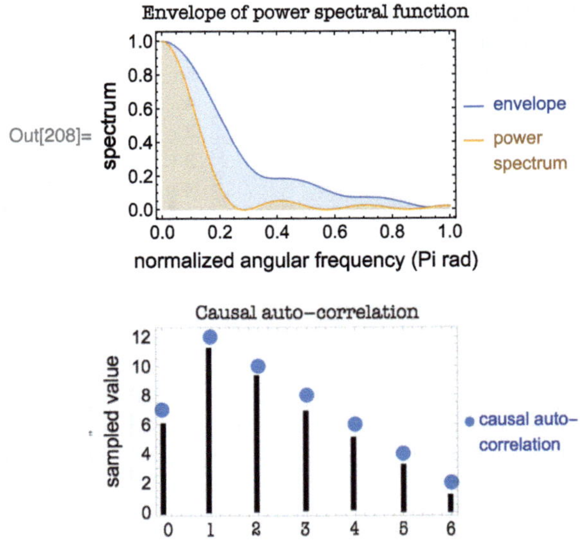

Fig. 3.4 Envelope of a power spectral density function obtained using causal auto-correlation sequence

$$z(n) = \hat{x}(n) \tag{3.40}$$
$$= \frac{1}{2\pi}\int_0^{2\pi} \hat{X}e^{i\Omega n}d\Omega$$
$$= x(n) + iy(n) = |z(n)|e^{i\theta(n)}$$
$$x(n) = \frac{1}{\pi}\int_0^{\pi}(X_r \cos\Omega n - X_i \sin\Omega n)d\Omega \tag{3.41}$$
$$y(n) = \frac{1}{\pi}\int_0^{\pi}(X_r \sin\Omega n + X_i \cos\Omega n)d\Omega, \tag{3.42}$$

where the real part of $z(n)$ is equal to the original sequence $x(n)$. Here, $|z(n)|$ is called the (Hilbert) envelope and $\cos\theta(n)$ is called the carrier, and the original sequence can be represented as

$$x(n) = |z(n)|\cos\theta(n). \tag{3.43}$$

The newly created imaginary part obtained from the original sequence corresponds to the $-\pi/2$ phase-shifted original sequence.

The power spectral density function shows a kind of squared envelope of the spectral function, because it is obtained from the squared magnitude of the spectral function. The magnitude of the spectral function however does not correspond to the Hilbert envelope in accordance with the causality of the spectral function, because in general the real and imaginary parts could be independent. The power spectral density function is defined as the Fourier transform of the auto-correlation sequence, which is an even sequence in the time domain. Introducing the causal auto-correlation sequence

$$\hat{r}(n) = \begin{cases} 2r(n) & n > 0 \\ r(0) & n = 0 \\ 0 & n < 0, \end{cases} \tag{3.44}$$

the Fourier transform of the causal auto-correlation sequence becomes the complex function that can be interpreted as the complex power spectral function. The absolute of the complex power spectral function produces the envelope of the power spectral density function, following the definition of the Hilbert envelope. Causality underpins the dynamics in both the time and frequency domains. Figure 3.4 presents an example of a power spectral envelope.

3.2 Symbolic Expression of Time Delay and Convolution

A time sequence might not be observed in sound in daily life. However, it could be interpreted as a symbolic expression of a time delay of an event such as a reflected sound.

3.2.1 *Magnitude Spectral Modification by Superposition of Direct and Delayed Sound*

As described in the previous Sect. 1.3, consider a sinusoidal wave followed by a reflected sound with a time delay of τ. The superposition of the sinusoidal and the delayed waves can be expressed as a convolution of the sinusoidal function and symbolic sequence $h(n)$ where $h(0) = 1$ and $h(1) = 1$, such that

$$\begin{aligned} y(t) &= x * h(t) \\ &= Ae^{i\omega t} \cdot h(0) + Ae^{i\omega(t-\tau)} \cdot h(1) \\ &= Ae^{i\omega t}(1 + e^{-i\omega\tau}) \\ &= Ae^{i\omega t} H(e^{-i\omega\tau}) \end{aligned} \tag{3.45}$$

$$H(e^{-i\omega\tau}) = 1 + e^{-i\omega\tau}, \tag{3.46}$$

where the sequence $h(n)$ is formally read as

$$\begin{aligned} h(0) &= h(t)|_{t=0} = 1 \\ h(1) &= h(t)|_{t=\tau} = 1. \end{aligned} \tag{3.47}$$

An interesting feature is how the delayed sound interferes with the direct sound. The effect of the delayed sound on the direct sinusoidal wave produces a spectral

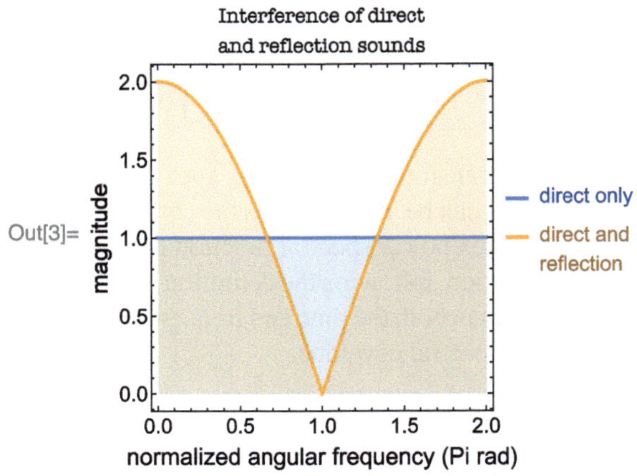

Fig. 3.5 Example of the magnitude of sinusoidal sound followed by single reflection as a function of $\omega\tau = \Omega$ where ω denotes the angular frequency of sinusoidal sound and τ is the delay time for a single reflection, and Ω is normalized angular frequency

modification in the frequency domain in which the Fourier transform of the sequence is defined. Taking the Fourier transform of $h(n)$

$$H(\mathrm{e}^{-\mathrm{i}\Omega}) = \sum_{n=0}^{1} h(n)\mathrm{e}^{-\mathrm{i}\Omega n} = 1 + \mathrm{e}^{-\mathrm{i}\Omega}, \tag{3.48}$$

which corresponds to Eq. 3.46 with $\omega\tau$ substituted by Ω. The equation above represents the spectral characteristics for the direct sinusoidal sound followed by a single reflection with time delay τ.

Figure 3.5 shows an example of the magnitude of the sinusoidal sound followed by a single reflection as a function of $\omega\tau$ where ω denotes the angular frequency of the sinusoidal sound and τ is the delay time for the single reflection. The reflected sound wave periodically modifies the magnitude of the direct sound wave as a sinusoidal function of $\omega\tau = \Omega$ in the frequency domain.

3.2.2 Effect of Phase on the Direct Sound from the Time Delay

A time delay in the reflected wave produces a change in the phase of the direct sound. Suppose a sequence $h(n)$ is composed of $h(1)$ only. Taking a complex sinusoidal wave, a convolution of the signal and the sequence $h(n)$ becomes

$$y(t) = x * h(t) = A\mathrm{e}^{\mathrm{i}\omega(t-\tau)} \tag{3.49}$$

$$H(\mathrm{e}^{-\mathrm{i}\omega\tau}) = \mathrm{e}^{-\mathrm{i}\omega\tau} \tag{3.50}$$

where $H(\mathrm{e}^{-\mathrm{i}\omega\tau})$ corresponds to the Fourier transform of the sequence $h(n)$ setting the unit interval of the sequence as τ. The magnitude of the original sinusoidal sound is maintained without changes even after the time delay; however, the phase changes by $\omega\tau$. A phase change that is proportional to the frequency is termed linear phase.

A delay introduces a linear phase that is proportional to the delay time. This linear phase does not, theoretically, create a change in the original waveform, but the origin of the time axis is shifted by the time delay so that $x(t-\tau)$. The time delay, which is required in the propagation of a wave, is determined to be $\tau = r/c$(s) where r(m) denotes the distance in space and c(m/s) gives the propagation speed of the wave. If the speed of sound is independent of the frequency, the time-delayed wave is expressed as $x(t-\tau)$ for a waveform $x(t)$.

Interestingly, the linear phase effect can be produced by other combinations of time delays. If the Fourier transform of $h(n)$ is written

$$H(\mathrm{e}^{-\mathrm{i}\Omega}) = \sum_n h(n)\mathrm{e}^{-\mathrm{i}\Omega n} \tag{3.51}$$

$$= |H(\mathrm{e}^{-\mathrm{i}\Omega})|\mathrm{e}^{-\mathrm{i}\Omega m}, \tag{3.52}$$

then the sequence $h(n)$ produces a linear phase shift $\theta = -\Omega m$ for the direct sound. Note that the magnitude $|H(\mathrm{e}^{-\mathrm{i}\Omega})|$ is not always constant even for a given linear phase delay. Only the pure delay sequence has a constant magnitude independent of the frequency associated with the linear phase. Take as an example the sequence $h(n)$ for which

$$h(0) = 1 \tag{3.53}$$
$$h(1) = 2$$
$$h(2) = 1.$$

Its Fourier transform is

$$H(\mathrm{e}^{-\mathrm{i}\Omega}) = \sum_{n=0}^{2} h(n)\mathrm{e}^{-\mathrm{i}\Omega n} \tag{3.54}$$
$$= 2H_0(\mathrm{e}^{-\mathrm{i}\Omega})\mathrm{e}^{-\mathrm{i}\Omega},$$

where

$$H_0(\mathrm{e}^{-\mathrm{i}\Omega}) = 1 + \cos\Omega \geq 0. \tag{3.55}$$

The sequence $h(n)$ gives a unit sample of delay to a direct sinusoidal sound, although the magnitude depends on the frequency of the direct wave. If a delay sequence produces a linear phase, the sequence $h(n)$ is symmetric with respect to n. However, the inverse does not hold always. Take another example of $h(n)$ such that

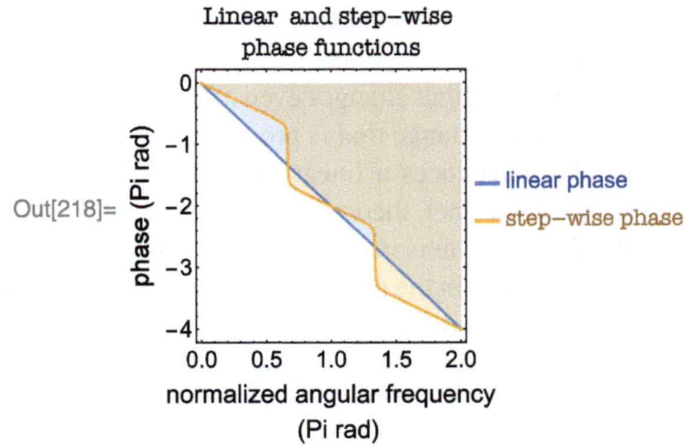

Fig. 3.6 Examples of phase functions; A (B) shows a linear phase (stepwise phase) function

$$h(0) = 1 \tag{3.56}$$
$$h(1) = \frac{1}{2}$$
$$h(2) = 1.$$

The sequence is symmetric with respect to n; however, it is not a linear phase. The Fourier transform of $h(n)$ becomes

$$H(\mathrm{e}^{-\mathrm{i}\Omega}) = \sum_{n=0}^{2} h(n)\mathrm{e}^{-\mathrm{i}\Omega n} \tag{3.57}$$
$$= \frac{1}{2}H_0(\mathrm{e}^{-\mathrm{i}\Omega})\mathrm{e}^{-\mathrm{i}\Omega}$$
$$H_0(\mathrm{e}^{-\mathrm{i}\Omega}) = 1 + 4\cos\Omega, \tag{3.58}$$

where $1 + 4\cos\Omega$ is both positive and negative. The change in sign produces a phase shift of π. The phase given by $\angle H$ is not a linear phase function of Ω, but the phase is a superposition of a linear function of Ω and a pulse sequence taking π only at Ω_0 and satisfies

$$1 + 4\cos\Omega_0 = 0. \tag{3.59}$$

Figure 3.6 shows examples of two sequences with their phase functions for a linear phase (A) and a stepwise phase (B). Recall that the auto-correlation sequence is an even sequence and has a zero-phase function. A time-shift is introduced into the auto-correlation sequence so that the auto-correlation sequence might be virtually given for $n \geq 0$ and the symmetric property of the sequence retained. Its phase function is then always linear.

3.2.3 Clustered Time Sequence and Its Spectral Effect

Given the time delay sequence $h(n)$, its Fourier transform is

$$H(\mathrm{e}^{-\mathrm{i}\Omega}) = \sum_{n=0}^{N-1} h(n)\mathrm{e}^{-\mathrm{i}\Omega n}, \tag{3.60}$$

which generally produces a spectral modification in the direct sinusoidal sound such as a change in the magnitude and phase, both of which may depend on the angular frequency. An example of a spectral change occurs with a time delay sequence composed of clustered samples. Consider again the time delay sequence $h(n)$

$$\begin{aligned} h(0) &= 1 \\ h(1) &= a \\ h(2) &= 1, \end{aligned} \tag{3.61}$$

where a is a real number. Its Fourier transform is

$$\begin{aligned} H(\mathrm{e}^{-\mathrm{i}\Omega}) &= \sum_{n=0}^{2} h(n)\mathrm{e}^{-\mathrm{i}\Omega n} \\ &= 2\mathrm{e}^{-\mathrm{i}\Omega}\left(\frac{a}{2} + \cos\Omega\right). \end{aligned} \tag{3.62}$$

Setting $a = 2$, this equation becomes

$$\begin{aligned} H(\mathrm{e}^{-\mathrm{i}\Omega}) &= 2\mathrm{e}^{-\mathrm{i}\Omega}(1 + \cos\Omega) \\ &= 4\mathrm{e}^{-\mathrm{i}\Omega}\cos^2\frac{\Omega}{2}. \end{aligned} \tag{3.63}$$

Figure 3.7 presents the magnitude of the spectral function for this sequence.

The Fourier transform of $h(n)$ is a periodic function of period $\Omega = 2\pi$; however, taking the interval $0 \le \Omega \le \pi$, the function exhibits a cosine-type behavior. The magnitude decreases as the angular frequency increases in the interval $0 \le \Omega \le \pi$. For this reason, the sequence $h(n)$ is called a low-pass filtering sequence. The superposition of the direct and delayed sinusoidal sounds following the time delay sequence $h(n)$ reduces the magnitude from 4 to 0 as Ω increases when the magnitude of the direct sound is unity.

In contrast, setting $a = -2$,

$$H(\mathrm{e}^{-\mathrm{i}\Omega}) = 4\mathrm{e}^{-\mathrm{i}\pi}\mathrm{e}^{-\mathrm{i}\Omega}\sin^2\frac{\Omega}{2}. \tag{3.64}$$

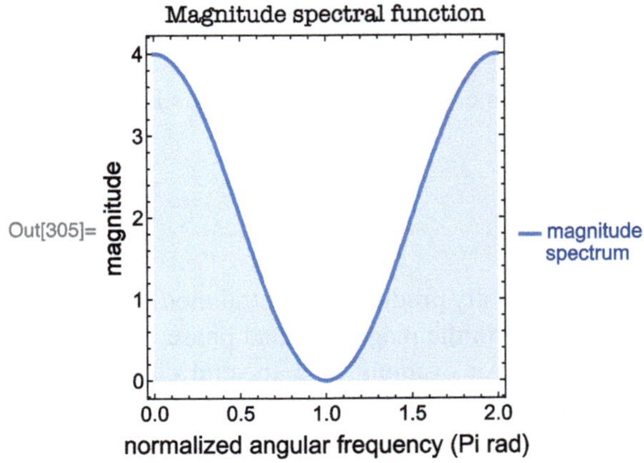

Fig. 3.7 Magnitude of the spectral function for the sequence given by Eq. 3.62 with $a = 2$

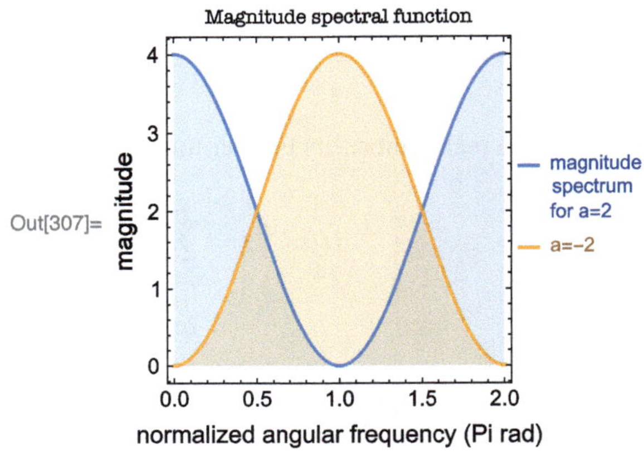

Fig. 3.8 Magnitude of the spectral function for the sequence given by Eq. 3.62 with $a = 2$ and $a = -2$

Again it is periodic; however over the interval $0 \leq \Omega \leq \pi$, the Fourier transform of the sequence becomes a sine-like function. Hence this sequence $h(n)$ is called a high-pass filtering sequence, which increases the magnitude of the superposed sinusoidal wave as angular frequency increases. Figure 3.8 presents the magnitude of the spectral function for the sequence given by Eq. 3.62 with $a = -2$. The phases of both spectral functions are linear, the only difference being the addition of $-\pi$ for high-pass filtering.

The periodic and cosine/sine-like behavior are typical of the Fourier transforms of sequences. Nonetheless, consider two further examples,

$$h_1(0) = 1, \quad h_1(1) = 2, \quad h_1(2) = 1 \tag{3.65}$$

$$h_2(0) = -1, \quad h_2(1) = 2, \quad h_2(2) = -1. \tag{3.66}$$

Taking the Fourier transforms yields

$$H_1(e^{-i\Omega}) = 4e^{-i\Omega}\cos^2\frac{\Omega}{2} \tag{3.67}$$

$$H_2(e^{-i\Omega}) = 4e^{-i\Omega}\sin^2\frac{\Omega}{2}. \tag{3.68}$$

The sum of these two functions is

$$H_1(e^{-i\Omega}) + H_2(e^{-i\Omega}) = 4e^{-i\Omega}, \tag{3.69}$$

which corresponds to the Fourier transform of the sum of the sequences

$$h_0(n) = h_1(n) + h_2(n), \tag{3.70}$$

where

$$h_0(0) = 0, \quad h_0(1) = 4, \quad h_0(2) = 0. \tag{3.71}$$

The sum of the low-pass and high-pass filtering sequences generates the impulse sequence with a delay.

3.2.4 *Sinc Function and Auto-Correlation*

In the previous subsection, examples were introduced of clustered time sequences composed of a finite number of clustered entries. However, from a theoretical point of view, sequences made of an infinite number of components can represent the signal property in the time domain. One example is the sinc function. Consider an even periodic spectral function defined in the interval $0 \le \Omega < \pi$ such that

$$|A(e^{-i\Omega})|^2 = \begin{cases} \frac{2\pi}{2\Delta\Omega} & \Omega_L \le \Omega \le \Omega_H \\ 0 & \text{otherwise} \end{cases} \tag{3.72}$$

and similarly defined for the negative Ω domain with interval $-\pi \le \Omega < 0$ and an even function. Here $\Delta\Omega = \Omega_H - \Omega_L$. The inverse Fourier transform of the spectral function gives the auto-correlation sequence

$$\begin{aligned} r(n) &= \frac{1}{2\pi}\left[\int_{\Omega_L}^{\Omega_H} + \int_{-\Omega_H}^{-\Omega_L}\right]\frac{2\pi}{2\Delta\Omega}e^{-i\Omega n}d\Omega \\ &= \frac{\sin\frac{\Delta\Omega}{2}n}{\frac{\Delta\Omega}{2}n}\cos\Omega_c n \\ &= \text{sinc}\left(\frac{\Delta\Omega}{2}n\right)\cos\Omega_c n \end{aligned} \tag{3.73}$$

where Ω_c denotes the central angular frequency, and

$$\mathrm{sinc}\left(\frac{\Delta\Omega}{2}n\right) = \frac{\sin\frac{\Delta\Omega}{2}n}{\frac{\Delta\Omega}{2}n} \tag{3.74}$$

is called the sinc function. Because $\Delta\Omega$ is nonzero, the periodicity of the sinusoidal function is lost by the continuous superposition of the sinusoidal components over the interval $\Delta\Omega$. Representing the time envelope of the auto-correlation sequence, the sinc function decreases to zero from unity as time increases. The trend follows a $1/t$ behavior where t can be interpreted as $t = nT_s$ with T_s the time interval of the sequence.

From another point of view, the sinc function can be interpreted as the time delay sequence or the impulse response in representing ideal low-pass filtering. The sinc function is a limit of the auto-correlation sequence when the central frequency approaches zero. When the frequency interval is confined to $-\Delta\Omega/2 < \Omega < \Delta\Omega/2$ (no response is obtained outside the interval), the spectral effect of the time delay sequence is called ideal low-pass filtering. Although a formally well-defined sequence, the sinc function is nevertheless not realized as a time sequence. This is because it is an even function that produces a sequence that includes both positive and negative integers n. Only the positive integer part of the sequence can be realized as a time delay sequence; however, the negative n part of the sequence imparts instead a time advance that is not possible in the real world. A practical virtual approach to obtain ideal low-pass filtering is establishing a time-shift amounting to some value M such that

$$\frac{\sin\frac{\Delta\Omega}{2}n}{\frac{\Delta\Omega}{2}n} \cong 0 \quad (\text{for} \quad n < -M). \tag{3.75}$$

The time-delayed sinc function

$$\frac{\sin\frac{\Delta\Omega}{2}(n-M)}{\frac{\Delta\Omega}{2}(n-M)} \tag{3.76}$$

suffices as a time delay sequence and affords virtual low-pass filtering from a practical point of view.

The reason why the time delay sequence is an even sequence is that the power spectral function is a real even function. Real and imaginary parts are necessary to obtain a real causal function from its Fourier transform.

3.3 Fourier Transform of Functions

3.3.1 Fourier Transform

The definition of the Fourier transform of a sequence extends formally over to continuous functions. Consider a continuous function $x(t)$; its Fourier transform is given by

$$X(\omega) = \int_{-\infty}^{\infty} x(t)\mathrm{e}^{-\mathrm{i}\omega t} dt. \tag{3.77}$$

Hence, the function $x(t)$ is the inverse Fourier transform of $X(\omega)$ expressed as

$$x(t) = \frac{1}{2\pi} \int_{-\infty}^{\infty} X(\omega)\mathrm{e}^{\mathrm{i}\omega t} d\omega \tag{3.78}$$

for which the following result has been used [2]

$$\begin{aligned} \frac{1}{2\pi} \int_{-\infty}^{\infty} \mathrm{e}^{\mathrm{i}\omega t} d\omega &= \frac{1}{2\pi} \int_{-\infty}^{\infty} \mathrm{i}\omega \frac{\mathrm{e}^{\mathrm{i}\omega t}}{\mathrm{i}\omega} d\omega \\ &= \frac{\partial}{\partial t} \frac{1}{2\pi} \int_{-\infty}^{\infty} \frac{\mathrm{e}^{\mathrm{i}\omega t}}{\mathrm{i}\omega} d\omega = \delta(t). \end{aligned} \tag{3.79}$$

Consequently, the function $x(t)$ can be viewed as a superposition or integration of complex sinusoidal functions over the continuous variable ω.

3.3.2 *Sinc Function*

A sequence that behaves like the sinc function has already been introduced. The sinc function itself can be defined as the inverse Fourier transform of the complex exponential function over a narrow interval,

$$\begin{aligned} h_0(t) &= \frac{1}{2\pi} \int_{-\Delta\omega/2}^{\Delta\omega/2} \frac{2\pi}{\Delta\omega} \mathrm{e}^{\mathrm{i}\omega t} d\omega \\ &= \frac{1}{\Delta\omega} \int_{-\Delta\omega/2}^{\Delta\omega/2} \mathrm{e}^{\mathrm{i}\omega t} d\omega \\ &= \frac{\sin \frac{\Delta\omega}{2} t}{\frac{\Delta\omega}{2} t}. \end{aligned} \tag{3.80}$$

The sinc function can be interpreted as the result of frequency band limiting or low-pass filtering of the delta function.

The delta function is obtained as a limit of a function that is compressed to a point $t = 0$ in the time domain; however, its Fourier transform is spread over the entire frequency domain. In contrast, the Fourier transform is limited to a finite area, whereas the function given by the inverse Fourier transform is spread over the entire time domain as for the sinc function. A sinusoidal wave is restricted in the frequency domain to only a single frequency. Consequently, a sinusoidal function is spread with periodicity over the entire time domain without any harmonics.

There is a formal relationship between the delta function $\delta(t)$ and a function such that

$$\int_{-\infty}^{\infty} x(t)\delta(t-\tau)dt = x(\tau) \tag{3.81}$$

where the delta function $\delta(t)$ is shifted to $\delta(t-\tau)$ on the time domain. The Fourier transform of $\delta(t-\tau)$ becomes

$$\int_{-\infty}^{\infty} \delta(t-\tau)\mathrm{e}^{-\mathrm{i}\omega t}d\omega = \mathrm{e}^{-\mathrm{i}\omega\tau}, \tag{3.82}$$

that is, a complex exponential function in the frequency domain. Similarly, taking a single unit pulse $\delta(n-m)$, where m signifies the amount of delay, its Fourier transform becomes

$$\sum_n \delta(n-m)\mathrm{e}^{-\mathrm{i}\Omega n} = \mathrm{e}^{-\mathrm{i}\Omega m}, \tag{3.83}$$

which likewise produces a complex exponential function in the frequency domain.

The change in signal (signal dynamics) in the time domain is obtained by superposing the complex exponential spectral function in the frequency domain. In contrast, a single unit pulse in the frequency domain such as $\delta(\Omega-\Omega_0)$ yields the single complex exponential function $\mathrm{e}^{\mathrm{i}\Omega_0 n}$ in the time domain. The change in signal within a frequency interval can be produced by a superposition of complex exponential functions in the time domain. For example, a sound heard by an ear can be understood as the dynamics in either the time or the frequency domain, and even both.

3.3.3 Spectral Weighting and Auto-Correlation Function

Instead of ideal low-pass filtering, consider a spectral function linear in frequency. Taking its inverse Fourier transform yields

$$h(t) = \frac{1}{2\pi}\frac{2\pi}{\Delta\omega}\int_{-\Delta\omega/2}^{\Delta\omega/2} \frac{\mathrm{i}\omega}{\Delta\omega}\mathrm{e}^{\mathrm{i}\omega t}d\omega, \tag{3.84}$$

which can be rewritten as

$$\begin{aligned} h(t) &= \frac{\partial}{\partial t}\left[\int_{-\Delta\omega/2}^{\Delta\omega/2} \frac{1}{(\Delta\omega)^2}\mathrm{e}^{\mathrm{i}\omega t}d\omega\right] \\ &= \frac{1}{\Delta\omega}\frac{\partial}{\partial t}\frac{\sin\frac{\Delta\omega}{2}t}{\frac{\Delta\omega}{2}t} = \frac{1}{\Delta\omega}\frac{d}{dt}h_0(t), \end{aligned} \tag{3.85}$$

where $h_0(t)$ denotes the sinc function. The equation above is analogous to the delta function

$$\delta(t) = \frac{1}{2\pi}\int_{-\infty}^{\infty} \mathrm{e}^{\mathrm{i}\omega t} d\omega \qquad (3.86)$$
$$= \frac{1}{2\pi}\int_{-\infty}^{\infty} \mathrm{i}\omega \frac{\mathrm{e}^{\mathrm{i}\omega t}}{\mathrm{i}\omega} d\omega$$
$$= \frac{d}{dt} s(t)$$

where $s(t)$ denotes Heaviside's step function, which replaces the sinc function [2]. Hence, high-pass filtering in a narrow frequency band for which the spectral function is $\mathrm{i}\omega$ and therefore proportional to ω, produces the first derivative of the sinc function with respect to the time.

Assuming the power spectral density function is given by $2\pi/\Delta\omega$ independent of the frequency within the frequency interval, the auto-correlation function is then determined by the sinc function. Random noise for which the ensemble average of the power spectral density function is constant and independent of frequency within the frequency band is called band-limited white noise. Recall that the inverse Fourier transform of the power spectral density function gives the auto-correlation function in the time domain. The auto-correlation function of a band-limited white noise is given by

$$r(t) = \frac{\sin \frac{\Delta\omega t}{2}}{\frac{\Delta\omega t}{2}} \cos \omega_c t \qquad (3.87)$$

where ω_c denotes the central angular frequency of the frequency band and $\Delta\omega$ denotes the band width. The periodicity represented by the central frequency diminishes inversely proportional to the band width, as the band width widens. This periodicity as represented by the cosine function of the central frequency is modulated by random fluctuations of the noise as the frequency band widens.

The effect of fluctuations in the random noise on the auto-correlation function appears in the envelope of the auto-correlation function. The envelope of the correlation [3] governs the signal source characteristics as fluctuations in periodicity. If there is no fluctuation in the periodicity, discriminating the various types of musical instruments would be difficult.

An interesting question to pose is whether the power spectral characteristics of random noise or fluctuations affect the auto-correlation function. Suppose that the power spectral function is proportional to $|\omega|$. The auto-correlation function can be expressed as

$$r(t) = \frac{8}{\Delta\omega^2}\int_{0}^{\Delta\omega/2} \omega \cos \omega t d\omega \qquad (3.88)$$
$$= 2\frac{\sin \frac{\Delta\omega t}{2}}{\frac{\Delta\omega t}{2}} - \left(\frac{\sin \frac{\Delta\omega t}{4}}{\frac{\Delta\omega t}{4}}\right)^2$$

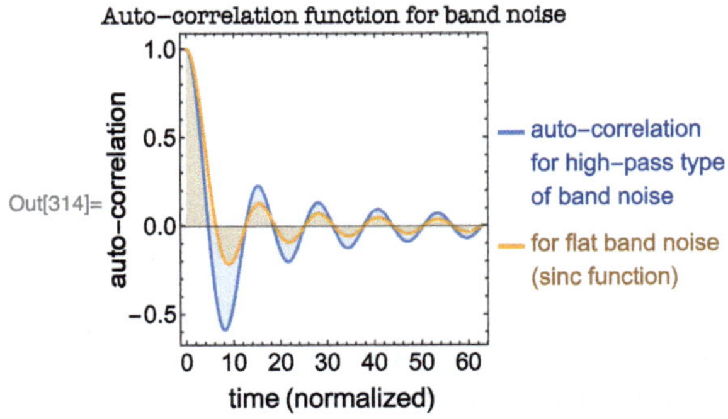

Fig. 3.9 Auto-correlation function for band noise for which the power spectral density function is proportional to frequency

and approaches the sinc function as time proceeds (Fig. 3.9). The power spectral characteristics for random noise show fluctuation in the frequency itself.

For a different perspective of fluctuations, consider random noise having power spectral characteristics that follow a $1/\omega$ dependence, i.e., pink noise. A typical characteristic of pink noise is a constant power density over logarithmic scales of the frequency. Setting a frequency range between say ω_1 and $N\omega_1$, the power spectrum in that range then becomes

$$P \sim \int_{\omega_1}^{N\omega_1} \frac{1}{\omega} d\omega = \log N. \tag{3.89}$$

In acoustic measurements, a frequency range over an octave is usually taken. This is because the ear can be modeled more or less in accordance with an octave-based frequency analysis. A constant power density over a logarithmic frequency scale is convenient for that purpose.

The term pink noise stems from an analogy with light waves. With $1/\omega$ power spectral characteristics, the power density increases as wavelength increases (the wave components are red shifted). Hence, in the field of acoustics, random noise with a $1/\omega$ trend in the power spectral density is called pink noise as a result of this comparison with white noise.

Another example of power spectral characteristics is provided by considering

$$P(\omega) = \frac{2}{\Delta\omega}\left(1 - \frac{2}{\Delta\omega}\omega\right), \tag{3.90}$$

which generates the sum

$$\omega \cdot \frac{4}{\Delta\omega^2} + \frac{2}{\Delta\omega}\left(1 - \frac{2}{\Delta\omega}\omega\right) = \frac{2}{\Delta\omega} \tag{3.91}$$

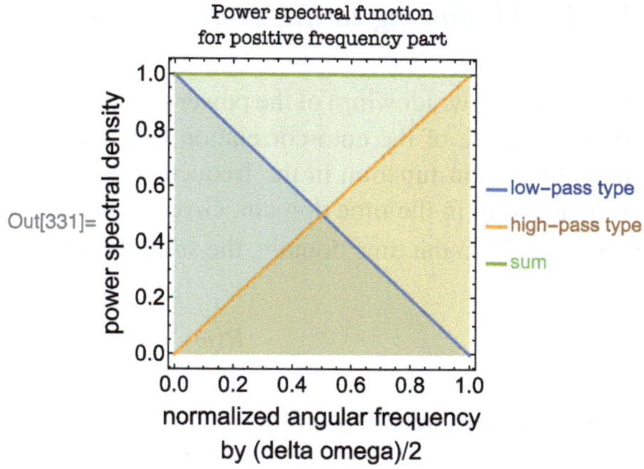

Fig. 3.10 Schematic of the power spectral functions

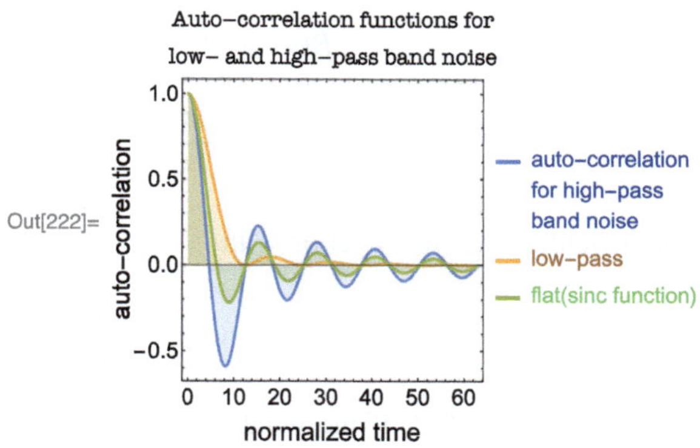

Fig. 3.11 Examples of the auto-correlation functions and power spectral characteristics

plotted Fig. 3.10 for the positive frequency part as even functions of ω. The power spectral characteristics $2/\Delta\omega$ produce twice the sinc function, whereas $4\omega/\Delta\omega^2$ produces Eq. 3.88. Hence, the auto-correlation function becomes

$$r(t) = \left(\frac{\sin \frac{\Delta\omega t}{4}}{\frac{\Delta\omega t}{4}} \right)^2 \tag{3.92}$$

for the power spectral function of Eq. 3.90. The rate of decrease in the envelope of the auto-correlation function becomes slow as the time proceeds. Figure 3.11 presents examples of the auto-correlation functions.

Change in signal in the time domain can be expressed from two viewpoints; the temporal envelope of the waveform such as modulation, beats, and missing fundamentals, and the temporal variation of the periodicity that may be displayed by the auto-correlation function.

3.3.4 Windowing in the Time Domain

Weighting (or windowing) of the power spectral function was introduced above from the perspective of the auto-correlation function in the time domain. Similarly, the auto-correlation function in the frequency domain can be related to the weighting or windowing in the time domain. Given a pair of delta functions $\frac{1}{2}\delta(t+T/2)$ and $\frac{1}{2}\delta(t-T/2)$ in the time domain, the sum of their Fourier transforms is then

$$R(\omega) = \cos\frac{\omega T}{2}. \tag{3.93}$$

Filling the time interval by $1/T$ between $t = \pm T/2$, the sum corresponding to Eq. 3.93 becomes

$$\begin{aligned} R(\omega) &= \int_{-T/2}^{T/2} w_0(t)\mathrm{e}^{-\mathrm{i}\omega t}dt \\ &= \frac{\sin\frac{\omega T}{2}}{\frac{\omega T}{2}} \end{aligned} \tag{3.94}$$

again yielding the sinc function with

$$w_0(t) = \frac{1}{T} \tag{3.95}$$

for $-T/2 < t < T/2$. If the window function $w_0(t)$ is applied to the range $t_c - (T/2) < t < t_c + T/2$, the time-shifted window function $w_1(t)$ can be written as

$$\hat{w}(t) = w_0 * f(t) \tag{3.96}$$

where

$$f(t) = \frac{1}{2}\delta(t+t_c) + \frac{1}{2}\delta(t-t_c). \tag{3.97}$$

The Fourier transform of $\hat{w}(t)$ produces a spectral auto-correlation function such that

$$\begin{aligned} \hat{R}(\omega) &= R_0(\omega)\cdot F(\omega) \\ &= \frac{\sin\frac{\omega T}{2}}{\frac{\omega T}{2}} cos\ \omega t_c \end{aligned} \tag{3.98}$$

where the spectral function for a single pulse in the time domain is a complex exponential function in the frequency domain.

Interpreting the function $w(t)$ as a distribution of time delays clustered between $t_c \pm T/2$ and centered at t_c, the spectral modification can then be expressed as a product of a complex exponential function corresponding to the central time delay t_c

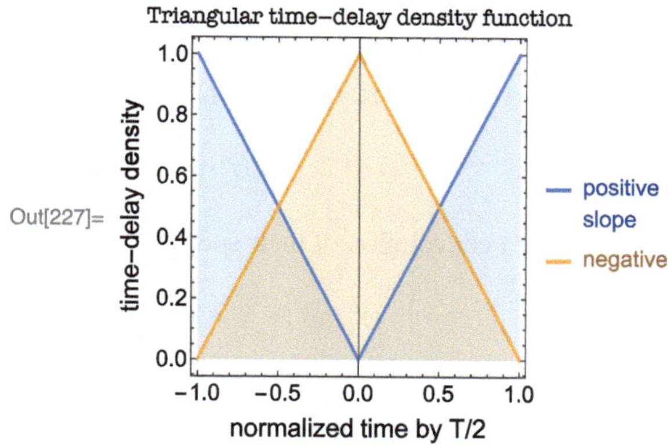

Fig. 3.12 Schematic of time delay distribution

without any fluctuations and the sinc function assuming that the time delay function governing the clustering or distribution of the delays is an even function in the time domain. The fluctuations in the time delays are modeled by the sinc function in the frequency domain.

Taking a similar example to that in the previous subsection for which the distribution of time delays is given by

$$w_1(t) = \frac{4}{T^2}|t| \tag{3.99}$$

for $-T/2 < t < T/2$, then

$$R_1(\omega) = 2\frac{\sin\frac{\omega T}{2}}{\frac{\omega T}{2}} - \left(\frac{\sin\frac{\omega T}{4}}{\frac{\omega T}{4}}\right)^2. \tag{3.100}$$

Again taking the time delay distribution,

$$w_2(t) = \frac{2}{T}\left(1 - \frac{2}{T}|t|\right) \tag{3.101}$$

as shown in Fig. 3.12, then accordingly,

$$\frac{2}{T}\left(1 - \frac{2}{T}|t|\right) + \frac{4}{T^2}|t| = \frac{2}{T} \tag{3.102}$$

and the Fourier transform of $w_2(t)$ is

$$R_2(\omega) = \left(\frac{\sin\frac{\omega T}{4}}{\frac{\omega T}{4}}\right)^2. \tag{3.103}$$

Recalling that the effect of windowing in the time domain can be interpreted as a convolution in the spectral domain, then the difference in the two types of window functions is well understood. If $w_1(t)$ is applied to the function $x(t)$, the Fourier transform of

$$y_1(t) = x(t) \cdot w_1(t) \tag{3.104}$$

is written as a convolution of their spectral functions such that

$$Y_1(\omega) = X * W_1(\omega) = X * R_1(\omega), \tag{3.105}$$

where $W_1(\omega)$ is given by Eq. 3.100. The convolution can be interpreted as a moving average or differentiation of $X(\omega)$ using weighting function $W_1(\omega)$, which takes both positive and negative signs. The weighting function reveals differential property of the spectral function.

In contrast, taking the window or time delay function $w_2(t)$, the spectral function $Y_2(\omega)$ is represented as a convolution for $X(\omega)$ and $W_2(\omega) = R_2(\omega)$, where $R_2(\omega)$ given by Eq. 3.103 takes only nonnegative values.

The window function can be interpreted as mimicking one feature of hearing, namely the forgetting function. The function $w_1(t)$ is an example of a forgetting function that in the time domain emphasizes the latest component of the signal rather than the previous ones. Such a forgetting function is interpreted as a function that emphasizes spectral differences or variations in the fine structure with respect to frequency. In contrast, certain auto-correlation functions can be used as window functions for smoothing the power spectral components without changing the signs of the power spectral function. This is because the Fourier transform of the auto-correlation function produces a power spectral density function that is nonnegative.

3.4 Triangular Windowing and Group Delay

3.4.1 *Phase and Group Delay Functions*

The phase function of the frequency characteristics $H(\mathrm{e}^{-\mathrm{i}\Omega})$ is defined as

$$\angle H = \tan^{-1} \frac{\Im[H]}{\Re[H]}. \tag{3.106}$$

Noting the derivative

$$\frac{d}{dx} \tan^{-1} x = \frac{1}{1+x^2} \tag{3.107}$$

and setting

$$X = \frac{\Im[H]}{\Re[H]} = \frac{H_\mathrm{I}}{H_\mathrm{R}}, \tag{3.108}$$

then

$$\tau_g = \frac{-d}{d\Omega} \tan^{-1} X \tag{3.109}$$
$$= \frac{-1}{1+X^2}\frac{dX}{d\Omega} = \frac{H_{\mathrm{I}}H'_{\mathrm{R}} - H'_{\mathrm{I}}H_{\mathrm{R}}}{|H|^2}$$

where the prime signifies differentiation with respect to Ω. Given that

$$H_{\mathrm{R}} = \sum_n h(n) \cos \Omega n \tag{3.110}$$

$$H_{\mathrm{I}} = -\sum_n h(n) \sin \Omega n \tag{3.111}$$

and assuming $h(n)$ is a real sequence, the first derivatives are then

$$H'_{\mathrm{R}} = -\sum_n nh(n) \sin \Omega n = G_{\mathrm{I}}, \tag{3.112}$$

$$H'_{\mathrm{I}} = -\sum_n nh(n) \cos \Omega n = -G_{\mathrm{R}}, \tag{3.113}$$

where

$$G = G_{\mathrm{R}} + \mathrm{i}G_{\mathrm{I}} \tag{3.114}$$
$$= -H'_{\mathrm{I}} + \mathrm{i}H'_{\mathrm{R}}$$

is the Fourier transform of the real sequence

$$g(n) = nh(n). \tag{3.115}$$

The sequence $g(n)$ corresponds to that obtained by multiplying the time response sequence $h(n)$ and the triangular window of positive slope. Consequently, the group delay can be obtained from

$$\tau_g = \frac{-d}{d\Omega} \tan^{-1} \frac{\Im[H(\mathrm{e}^{-\mathrm{i}\Omega})]}{\Re[H(\mathrm{e}^{-\mathrm{i}\Omega})]} = \frac{H_{\mathrm{I}}G_{\mathrm{I}} + G_{\mathrm{R}}H_{\mathrm{R}}}{|H|^2} \tag{3.116}$$

without requiring the derivative functions.

A superposition of direct and single reflection waves is an example in which the group delay depends on the frequency, even if the medium is nondispersive. Following the example

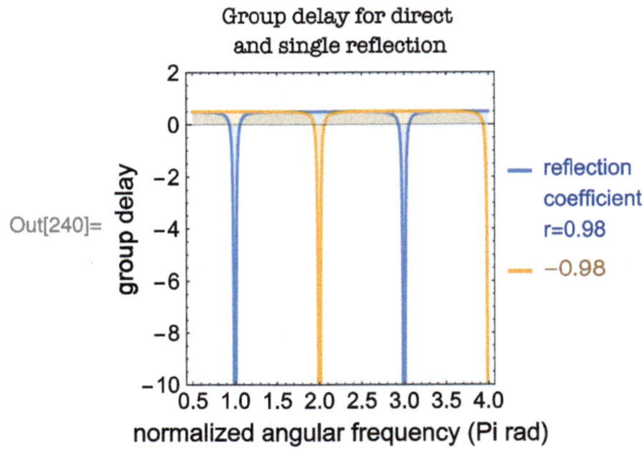

Fig. 3.13 Group delay function obtained from Eq. 3.124

$$H(\mathrm{e}^{-\mathrm{i}\Omega}) = \sum_{n=0}^{1} h(n)\mathrm{e}^{-\mathrm{i}\Omega n} = 1 + r\mathrm{e}^{-\mathrm{i}\Omega} \tag{3.117}$$

$$H_{\mathrm{R}} = \sum_{n=0}^{1} h(n) \cos \Omega n \tag{3.118}$$

$$H_{\mathrm{I}} = -\sum_{n=0}^{1} h(n) \sin \Omega n, \tag{3.119}$$

$$H_{\mathrm{R}} = 1 + r \cos \Omega \tag{3.120}$$

$$H_{\mathrm{I}} = -r \sin \Omega \tag{3.121}$$

$$G_{\mathrm{R}} = \Re\left[r\mathrm{e}^{-\mathrm{i}\Omega}\right] = r \cos \Omega \tag{3.122}$$

$$G_{\mathrm{I}} = \Im\left[r\mathrm{e}^{-\mathrm{i}\Omega}\right] = -r \sin \Omega \tag{3.123}$$

where r denotes the reflection coefficient for the single reflection wave with one sample of delay. In accordance with Eq. 3.116, the group delay of $H(\mathrm{e}^{-\mathrm{i}\Omega})$ can be written as

$$\tau_g = \frac{r(r + \cos \Omega)}{r^2 + 2r \cos \Omega + 1}. \tag{3.124}$$

Figure 3.13 presents the group delay functions for positive and negative reflection coefficients r. For a vibrating string, positive coefficients in general are not assumed at the end; however, for one-dimensional systems like acoustic tubes, the positive values are often assumed. The frequency characteristics fix zeros at

$$z_0 = r_0 e^{i\Omega_0} = -r = \begin{cases} re^{i\pi} & 0 < r \leq 1 \\ -r & -1 \leq r < 0. \end{cases} \tag{3.125}$$

As illustrated in Fig. 3.13, the group delay functions govern the locations of the angular frequencies yielding the zeros.

3.4.2 Group Delay Function for N−Sample Delay of Reflection

Suppose that the delay time is represented by N samples. The frequency response for the superposition of the direct and reflection sounds can be written as

$$H(e^{-i\Omega}) = 1 + re^{-iN\Omega}. \tag{3.126}$$

The zeros are determined by

$$z_0^N = |z_0|^N e^{iN\Omega_0} = -r \tag{3.127}$$

so that

$$1 + rz_0^{-N} = 0. \tag{3.128}$$

Accordingly,

$$|z_0| = |r|^{(1/N)} \tag{3.129}$$

$$e^{iN\Omega_0} = \begin{cases} -1 & 0 < r \leq 1 \\ 1 & -1 < r \leq 0. \end{cases} \tag{3.130}$$

The real and imaginary parts of Eq. 3.126 are written as

$$H_R = 1 + r\cos N\Omega \tag{3.131}$$

$$H_I = -r\sin N\Omega. \tag{3.132}$$

From

$$G_R = N \cdot r\cos N\Omega \tag{3.133}$$

$$G_I = -N \cdot r\sin N\Omega, \tag{3.134}$$

the group delay function can be expressed as

$$\tau_g = N\frac{r(r + \cos N\Omega)}{r^2 + 2r\cos N\Omega + 1}. \tag{3.135}$$

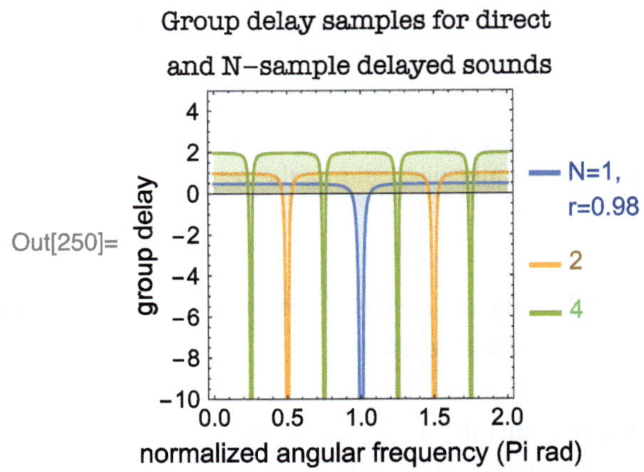

Fig. 3.14 Examples of group delay functions versus normalized angular frequency for N-sample delay

Figure 3.14 illustrates examples of the group delay functions for N-sample delay. The functions exhibit a scaling law under frequency scaling $\Omega \to \Omega/N$ where N indicates the number of zeros contained in the period 2π.

3.4.3 Group Delay Functions and Locations of Zeros

In linear systems, $|r| > 1$ can be assumed to be possible. Recall Eq. 3.124

$$\tau_g = \frac{r(r + \cos\Omega)}{r^2 + 2r\cos\Omega + 1}. \tag{3.136}$$

Assuming $r < 0$, then $\cos\Omega_0 = 1$ and

$$\tau_g(\Omega)\big|_{\Omega=\Omega_0} = \frac{r}{r+1}\begin{cases} < 0 & -1 < r < 0 \\ > 0 & r < -1. \end{cases} \tag{3.137}$$

In contrast, assuming $r > 0$, then $\cos\Omega_0 = -1$ and

$$\tau_g(\Omega)|_{\Omega=\Omega_0} = \frac{r}{r-1}. \tag{3.138}$$

Consequently, the group delay function can be expressed as

$$\tau_g(\Omega)|_{\Omega=\Omega_0}\begin{cases} < 0 & |r| < 1 \\ > 0 & |r| > 1. \end{cases} \tag{3.139}$$

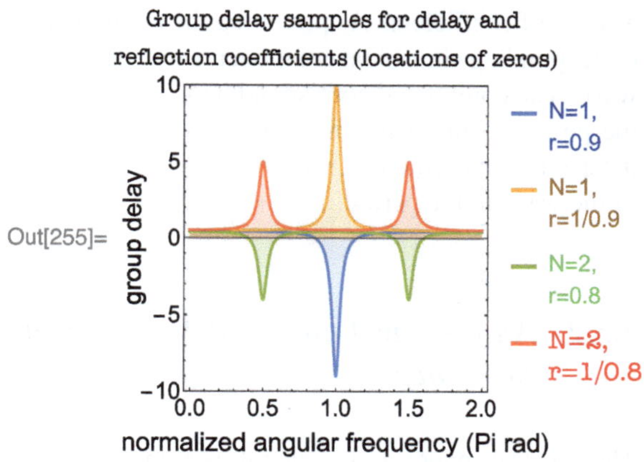

Fig. 3.15 Group delay (normalized) samples due to zeros (reflection coefficients and sample of delay)

These changes in sign can be interpreted as the locations of zeros. When $|z_0| < 1(> 1)$, the phase jump is positive (negative) because of the zero. The envelope of the modulated waveform is delayed (advanced) by a positive group delay (positive phase jump). Figure 3.15 illustrates several group delay functions.

3.5 Fourier Series Expansion of a Periodic Function

3.5.1 *Fourier Series*

The Fourier series expansion is fundamental in the analysis of sound and perception. A periodic wave in the time domain can be expressed as

$$x(t) = \sum_{k=-\infty}^{\infty} A_k e^{ik\frac{2\pi}{T}t} \tag{3.140}$$

where $x(t)$ is a periodic function with period T. The expression above indicates that a periodic function can be expressed as a harmonic superposition of sinusoidal waves with fundamental frequency $1/T(\mathrm{H_z})$. The coefficients A_k are called the spectral sequence in the frequency domain and are determined using

$$A_k = \frac{1}{T}\int_0^T x(t)e^{-ik\frac{2\pi}{T}t}dt = |A_k|e^{i\theta_k} \tag{3.141}$$

given that

$$\frac{1}{T}\int_0^T e^{i(k'-k)\frac{2\pi}{T}t}dt = \delta(k'-k). \tag{3.142}$$

The set of coefficients A_k providing the magnitudes $|A_k|$ and initial phases θ_k to the corresponding sinusoidal components is called the spectral sequence of the periodic wave. The component corresponding to $k = \pm 1$ determines the fundamental giving the period of the wave in the time domain. The higher components with $|k| > 1$ determine the harmonics, whereas the component with $k = 0$ gives the time average of the wave in the period.

3.5.2 Auto-Convolution and Auto-Correlation of Spectral Sequence

The spectral sequence, in general, is composed of complex numbers. Suppose that a periodic function is a real function in the time domain. Taking only the power spectral sequence $|A_k|^2$, the auto-correlation function of $x(t)$ is derived yielding

$$\begin{aligned} r(t) &= \sum_{k=-\infty}^{\infty} |A_k|^2 \mathrm{e}^{\mathrm{i}k\frac{2\pi}{T}t} \\ &= \frac{1}{T}\int_0^T \frac{1}{T}\int_0^T dt_1 dt_2 x(t_1)x(t_2) \sum_{k=-\infty}^{\infty} \mathrm{e}^{\mathrm{i}\frac{2\pi}{T}k(t+t_2-t_1)} \\ &= \frac{1}{T}\int_0^T x(t_1)x(t_1-t)dt_1 \end{aligned} \tag{3.143}$$

where

$$\sum_{k=-\infty}^{\infty} \mathrm{e}^{\mathrm{i}k\frac{2\pi}{T}(t+t_2-t_1)} = \delta(t_2-(t_1-t)). \tag{3.144}$$

The auto-correlation function is given by the power spectral sequence independent of the phase information.

In contrast, taking the square of the function $x(t)$, then

$$\begin{aligned} &\frac{1}{T}\int_0^T x^2(t)\mathrm{e}^{-\mathrm{i}k\frac{2\pi}{T}t}dt \\ &= \sum_{l=-\infty}^{\infty} A_l \sum_{m=-\infty}^{\infty} A_m \frac{1}{T}\int_0^T \mathrm{e}^{\mathrm{i}(l+m-k)\frac{2\pi}{T}t}dt \\ &= \sum_l A_l A_{k-l} = A * A(k) \end{aligned} \tag{3.145}$$

is obtained, and using Eq. 3.142,

$$\frac{1}{T}\int_0^T e^{i(l+m-k)\frac{2\pi}{T}t}dt = \begin{cases} 1 & l+m-k=0 \\ 0 & \text{otherwise.} \end{cases} \tag{3.146}$$

The signal dynamics $x^2(t)$ yields the auto-convolution of the spectral sequence in the frequency domain. In other words, the correlation function in the time domain is determined by the power spectral sequence without the phase; however, the signal dynamics representing the envelope in the time domain is determined by the auto-convolution of the spectral sequence including the phase information. This implies that perceiving the phase in the spectral domain is possibly because of the signal dynamics in the time domain.

3.5.3 Dirichlet Function

A periodic function can be expanded as a Fourier series in which the coefficients for the sinusoidal components are called the Fourier coefficients or the line spectral components. A signal that produces the sensation of pitch is periodic in general. A sound wave can be represented using the line spectral components when the pitch of the sound is identified.

The signal dynamics in the time domain is then interpreted as a temporal variation in the magnitude and/or period. For the fluctuations in period, the auto-correlation function plays a key role; in addition, the envelope is a significant candidate in governing the variation in magnitude. The Dirichlet function represents the frequency-band-limited response in the time domain for a periodic signal represented by line spectral components, whereas the sinc function is defined for a signal that has a band-limited continuous spectrum.

The sinc function is representative of an auto-correlation function of a sinusoidal signal for which the frequency fluctuates randomly in a narrow frequency band. The Dirichlet function in contrast is representative of fluctuations in the line spectral components. The Dirichlet function is still periodic with a much longer period than that for the central sinusoidal component in the frequency band of interest.

Suppose that a periodic function has a central frequency component of angular frequency $K\omega_0(\text{rad/s})$ in a narrowband of frequencies of interest. The auto-correlation function for the narrowband signal can be expressed as

$$r(t) = \frac{1}{N+1}\left(\sum_{k=-N/2}^{N/2} \cos k\omega_0 t\right)\cos K\omega_0 t, \tag{3.147}$$

assuming that the power spectrum of the line spectral component in the fluctuation is $1/(N+1)$. In this equation, the function

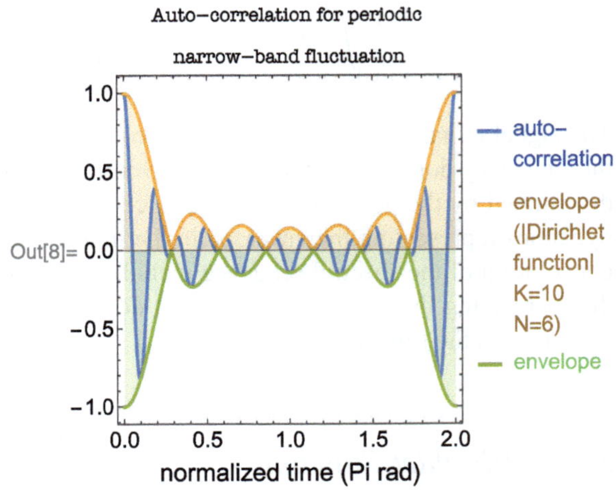

Fig. 3.16 Auto-correlation function for narrowband fluctuations

$$D_N(t) = \frac{1}{N+1}\sum_{k=-N/2}^{N/2} \cos k\omega_0 t \qquad (3.148)$$
$$= \frac{1}{N+1}\frac{\sin \frac{N+1}{2}\omega_0 t}{\sin \frac{\omega_0 t}{2}}$$

is called the Dirichlet function [4]; here N is an even number. Introducing the Dirichlet function into Eq. 3.147, the auto-correlation function can be rewritten as

$$r(t) = \frac{1}{N+1}\frac{\sin \frac{N+1}{2}\omega_0 t}{\sin \frac{\omega_0 t}{2}} \cos K\omega_0 t \qquad (3.149)$$
$$= D_N(t)\cos K\omega_0 t$$
$$= r_e(t)\cdot r_c(t),$$

which is periodic with period $2\pi/\omega_0$. Figure 3.16 illustrates an auto-correlation function for narrowband fluctuations composed of line spectral components with flat magnitude.

The carrier of the auto-correlation, i.e.,

$$r_c(t) = \cos K\omega_0 t, \qquad (3.150)$$

has period $2\pi/(K\omega_0)$ corresponding to the period of the central frequency in the absence of fluctuations. The envelope (Dirichlet function) is periodic with period $2\pi/\omega_0$ as long as the fluctuations are composed of line spectral components. The period of the envelope is much longer than that for the carrier. The fluctuations composed of line spectral components produce very slow fluctuations over a single period for the carrier or central sinusoid in the time domain.

3.6 Exercises

1. Confirm following example below. Consider a pair of sequences

$$a_n = \{1, 2, 3, 4\} \quad (3.151)$$
$$b_n = \{0, 2, 1\}. \quad (3.152)$$

The convolution of a_n and b_n can be written in matrix form,

$$\begin{pmatrix} 1 & 0 & 0 \\ 2 & 1 & 0 \\ 3 & 2 & 1 \\ 4 & 3 & 2 \\ 0 & 4 & 3 \\ 0 & 0 & 4 \end{pmatrix} \begin{pmatrix} 0 \\ 2 \\ 1 \end{pmatrix} = \begin{pmatrix} 0 \\ 2 \\ 5 \\ 8 \\ 11 \\ 4 \end{pmatrix}. \quad (3.153)$$

2. Given a pair of sequences a_n and b_n, derive the generating function for the cross-correlation sequence $r(n) = a \otimes b(n)$. Similarly, derive the generating functions for auto-correlation sequences $a \otimes a(n)$ and $b \otimes b(n)$.
3. Again, given a pair of sequences a_n and b_n, obtain the Fourier transform for sequences $c(n) = a * b(n)$ and $a \otimes b(n)$.
4. Calculate the Fourier transform for each of the following sequences:
(1) $h(0) = 1, \quad h(n) = 0 \quad (n \neq 0)$
(2) $h_1(N) = 1, \quad h_1(n) = 0 \quad (n \neq N)$
(3) $h_2(-N) = 1, \quad h_2(n) = 0 \quad (n \neq -N)$
(4) $h_1(n) + h_2(n)$
(5)

$$\begin{aligned} h_1(N_1) = 1, &\quad h_1(n) = 0 \quad (n \neq N_1) \\ g_1(N_2) = 1, &\quad g_1(n) = 0 \quad (n \neq N_2) \\ h_2(-N_1) = 1, &\quad h_2(n) = 0 \quad (n \neq -N_1) \\ g_2(-N_2) = 1, &\quad g_2(n) = 0 \quad (n \neq -N_2) \\ y(n) = h_1(n) &+ h_2(n) + g_1(n) + g_2(n). \end{aligned}$$

5. Derive the inverse Fourier transform for each of the functions

$$\begin{aligned} &\cos \Omega \quad \sin \Omega \\ &\cos 2\Omega \quad \sin 2\Omega \\ &\cos^2 \Omega \quad \sin^2 \Omega \\ &\cos^4 \Omega \quad \sin^4 \Omega. \end{aligned}$$

6. Confirm that the relationship between the convolution and the Fourier transform for each of the functions

$$\cos 2\Omega$$
$$\cos \Omega \cos 2\Omega$$
$$\sin \Omega \sin 2\Omega$$
$$\cos \Omega \cos \Omega$$
$$\sin \Omega \sin \Omega.$$

7. Obtain the Fourier transform of each of the following sequences,

$$a(-1) = 1, \quad a(0) = 1 \quad a(1) = 1$$
$$a(-1) = 1, \quad a(0) = 2 \quad a(1) = 1$$
$$a(-1) = -1, \quad a(0) = 2 \quad a(1) = -1$$
$$a(-1) = -1, \quad a(0) = 2 \quad a(1) = 1$$
$$a(-1) = -1, \quad a(0) = 2 \quad a(1) = -1.$$

8. Generalizing $\mathrm{e}^{-\mathrm{i}\Omega}$ to an arbitrary complex variable $z = r\mathrm{e}^{\mathrm{i}\Omega}$ in the Fourier transformation is called the $z-$ transformation with

$$H(z^{-1}) = \sum_n h(n) z^{-n}.$$

A z_0 that satisfies

$$H(z_0^{-1}) = 0$$

is called a zero of $H(z^{-1})$. Find the zeros for the following functions

$$H(z^{-1}) = 1 + z^{-1}$$
$$H(z^{-1}) = 1 - z^{-1}$$
$$H(z^{-1}) = 1 + \frac{1}{2} z^{-1}$$
$$H(z^{-1}) = 1 - \frac{1}{2} z^{-1}$$
$$H(z^{-1}) = 1 + 2z^{-1}$$
$$H(z^{-1}) = 1 - 2z^{-1}$$
$$H(z^{-1}) = 1 + z^{-2}$$
$$H(z^{-1}) = 1 - z^{-2}$$
$$H(z^{-1}) = (1 + 2z^{-1})^2$$
$$H(z^{-1}) = (1 - 2z^{-1})^2$$

$$H(z^{-1}) = \left(1 + \frac{1}{2}z^{-1}\right)^2$$

$$H(z^{-1}) = \left(1 - \frac{1}{2}z^{-1}\right)^2.$$

9. The Fourier transform of the auto-correlation sequence is real and nonnegative. Determine whether the following sequences are auto-correlations.

$$a(-1) = 1, \quad a(0) = 2 \quad a(1) = 1$$
$$a(-1) = -1, \quad a(0) = 2 \quad a(1) = 1$$
$$a(-1) = 1, \quad a(0) = 1 \quad a(1) = 1$$
$$a(-1) = 1, \quad a(0) = 1/2 \quad a(1) = 1$$
$$a(-1) = 1, \quad a(0) = 2 \quad a(1) = -1$$
$$a(-1) = 1, \quad a(0) = 3 \quad a(1) = 1.$$

10. What is the difference between the Dirichlet function and sinc function?

References

1. R. Nelson, *Probability, Stochastic Processes, and Queueing Theory* (Springer, Berlin, 1995)
2. P.J. Nahin, *Dr. Euler's Fabulous Formula: Cures Many Mathematical Ills* (Princeton University Press, Princeton, 2006)
3. Y. Ando, *Auditory and Visual Sensation* (Springer, Berlin, 2009)
4. M. Tohyama, *Wavefor m Analysis of Sound* (Springer, 2015)

Chapter 4
Differential Equations for Sinusoidal Waves

Abstract Within the theory of linear systems, the differential equation associated with sinusoidal waves has a central importance in signal analysis. Free and forced oscillations or transient and steady-state responses can be understood as general and particular solutions of linear differential equations. Resonances may be formulated as solutions of the differential equation for a system with a single degree of freedom. Given observation records of a system, estimating source information is a fundamental issue in machine diagnostics. Taking an observation signal as a solution of a differential equation, the waveform of an external source acting on a system of a single degree of freedom can be recovered using the differential equation because in principle the transient response is independent of the source effect. A mixture of transient and steady-state solutions creates beats if the frequency of the source is close to the frequency of the resonance. The differential equation can be extended to the wave equation that governs wave propagation in a medium. A typical phenomenon of wave propagation is the standing wave, which is a superposition of direct and reflection waves, subject to there being no magnitude differences between the two waves. If the coefficient of reflection is given by a complex number, the phase difference moves the nodes; however, the standing waves disappear as the magnitude difference increases.

Keywords Single-degree-of-freedom system · Impulse response · Transient response · Resonance · Source waveform · Estimation of source waveform · Travelling wave · Standing wave

4.1 Differential Equations and Linear Systems

4.1.1 Single-Degree-of-Freedom System

A vibrating system composed of a mass and a spring as shown in Fig. 4.1 is called a single-degree-of-freedom system. The position of the mass at $t = 0$ establishes a condition setting the initial state of the mass's vibration. Suppose $x(t)|_{t=0} = x(0)$ which is determined by stretching or compressing the spring from its equilibrium

M. Tohyama, *Sound in the Time Domain*, Signals and Communication Technology,
https://doi.org/10.1007/978-981-10-5889-9_4

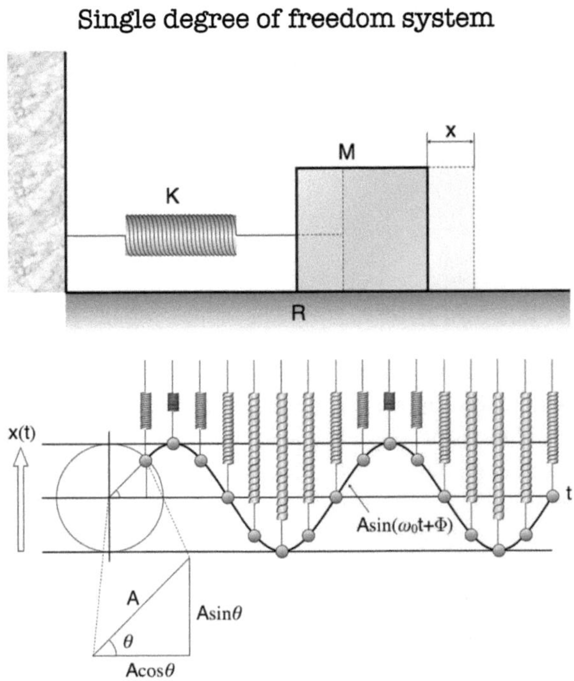

Fig. 4.1 A vibrational system with a single degree of freedom

position. In release, the mass vibrates back and forth governed by its equation of motion (obtained from Newton's Law of Motion)

$$M\frac{d^2x(t)}{dt^2} + Kx(t) = 0 \qquad \text{(N)} \tag{4.1}$$

where M (kg) is the mass and K (N/m) is the spring constant. This type of vibration is called the free vibration or free oscillation.

The equation of motion above can be rewritten in a simplified form

$$D[x] = -bx \tag{4.2}$$

where

$$D = \frac{d^2}{dt^2} \qquad (1/\text{s}^2) \tag{4.3}$$

$$b = \frac{K}{M}. \qquad (1/\text{s}^2) \tag{4.4}$$

Regarding D as a linear operator, the equation can be interpreted as an equation involving an eigenvalue and an eigenvector

$$A\mathbf{x} = \lambda \mathbf{x}, \tag{4.5}$$

where A denotes the linear operator, which acts on an $\mathbf{x}$ denoting an eigenvector and produces λ an eigenvalue. In operating on the vector $\mathbf{x}$ by operator A, the vector $\mathbf{x}$ remains unchanged to within a scaling by λ. The scaling factor, including the sign λ, is called the eigenvalue for operator A and the vector $\mathbf{x}$ is called an eigenvector for the operator corresponding to the eigenvalue.

According to this definition of eigenvalue, the vibrating mass follows a sinusoidal function (eigen-function) of angular frequency (eigen-angular-frequency) such that

$$\omega_{00} = \sqrt{\frac{K}{M}}. \qquad (\text{rad/s}) \tag{4.6}$$

The sinusoidal function can be interpreted as an eigenfunction for the free oscillation of a vibrating system with a single degree of freedom.

4.1.2 Sinusoidal Function and General Solution of Single-Degree-of-Freedom System

A free oscillation, as stated in the previous subsection, does not oscillate forever, but decays as time elapses and will eventually stop. This is because energy, which is being exchanged between potential and kinetic energy, is not retained for long. Dynamical energy is lost, the loss resulting from the interaction between the oscillating system and its surroundings. Gradually, the dynamical energy that an oscillating system possesses converts to thermal energy, such as friction heat. In general, the loss of dynamical energy is proportional to the speed of oscillation. Hence the friction force (N), denoted here by f_R, that causes energy conversion (loss) can be assumed to have the form

$$f_R = Rv, \qquad (\text{N}) \tag{4.7}$$

where v denotes the oscillation velocity (m/s) and R is called the friction constant $(\text{N} \cdot \text{s/m})$.

The Newtonian law that specifies the relationship between acceleration of a mass and force acting on the mass can be expressed by including friction:

$$M\frac{d^2x(t)}{dt^2} + R\frac{dx(t)}{dt} + Kx(t) = F(t) \quad (\text{N}) \tag{4.8}$$

where $F(t)$ denotes an external force. By dividing through by the mass M, the equation above can be rewritten as

$$\frac{d^2x(t)}{dt^2} + a\frac{dx(t)}{dt} + bx(t) = f(t) \qquad (\mathrm{m/s^2}), \tag{4.9}$$

a very general form where

$$a = \frac{R}{M} \qquad (1/\mathrm{s}) \tag{4.10}$$

$$b = \frac{K}{M}. \qquad (1/\mathrm{s}^2) \tag{4.11}$$

Suppose $x(t) = x_f(t)$ is a solution for the equation above, then

$$x(t) = x_f(t) + x_0(t) \tag{4.12}$$

is also one of the solutions, with the condition that

$$\frac{d^2x_0(t)}{dt^2} + a\frac{dx_0(t)}{dt} + bx_0(t) = 0. \qquad (\mathrm{m/s^2}) \tag{4.13}$$

The solution $x_f(t)$ is called the particular solution that satisfies Eq. 4.9 and $x_0(t)$ is called a general solution for Eq. 4.13.

The general solution $x_0(t)$ can be expressed by introducing the Fourier transform into the equation. Setting

$$x_0(t) = \frac{1}{2\pi}\int_{-\infty}^{\infty} X_0(\omega)\mathrm{e}^{\mathrm{i}\omega t}d\omega \tag{4.14}$$

$$\frac{d}{dt}x_0(t) = \mathrm{i}\omega\frac{1}{2\pi}\int_{-\infty}^{\infty} X_0(\omega)\mathrm{e}^{\mathrm{i}\omega t}d\omega \tag{4.15}$$

$$\frac{d^2}{dt^2}x_0(t) = -\omega^2\frac{1}{2\pi}\int_{-\infty}^{\infty} X_0(\omega)\mathrm{e}^{\mathrm{i}\omega t}d\omega \tag{4.16}$$

and introducing them into the differential equation, Eq. 4.13,

$$\omega^2 - a\mathrm{i}\omega - b = 0 \tag{4.17}$$

can be derived. Being a quadratic equation, Eq. 4.17 yields $\omega = \omega_{p_1}$ and ω_{p_2} such that

$$\omega_{p_1} = \omega_0 + \mathrm{i}\delta_0 \qquad (1/\mathrm{s}) \tag{4.18}$$

$$\omega_{p_2} = -\omega_0 + \mathrm{i}\delta_0 \qquad (1/\mathrm{s}) \tag{4.19}$$

where

$$\omega_0 = \sqrt{b - a^2/4} \tag{4.20}$$

$$\delta_0 = a/2. \tag{4.21}$$

The complex angular frequency for the free oscillation is confined to ω_{p_1} and ω_{p_2}. Consequently, the general solution to Eq. 4.13 can be written as

$$\begin{aligned} x_0(t) &= A\mathrm{e}^{\mathrm{i}\omega_{p_1}t} + B\mathrm{e}^{\mathrm{i}\omega_{p_2}t} \\ &= (A\mathrm{e}^{\mathrm{i}\omega_0 t} + B\mathrm{e}^{-\mathrm{i}\omega_0 t})\mathrm{e}^{-\delta_0 t} \end{aligned} \tag{4.22}$$

where A and B are complex numbers that are determined in accordance with the initial conditions. For example, setting the initial conditions such that

$$x_0(0) = u_0 \qquad \text{(m)} \tag{4.23}$$

$$\frac{d}{dt}x_0(0) = v_0 \qquad \text{(m/s)} \tag{4.24}$$

then

$$A + B = u_0 \qquad \text{(m)} \tag{4.25}$$

$$(\mathrm{i}\omega_0 - \delta_0)A + (-\mathrm{i}\omega_0 - \delta_0)B = v_0 \qquad \text{(m/s)} \tag{4.26}$$

must hold for A and B. Assuming that $\omega_0 \gg \delta_0$, then A and B can be approximately determined following

$$A + B = u_0 \qquad \text{(m)} \tag{4.27}$$

$$A - B = \frac{v_0}{\mathrm{i}\omega_0} \qquad \text{(m)} \tag{4.28}$$

such that

$$A = \frac{1}{2}\left(u_0 + \frac{v_0}{\mathrm{i}\omega_0}\right) \qquad \text{(m)} \tag{4.29}$$

$$B = \frac{1}{2}\left(u_0 - \frac{v_0}{\mathrm{i}\omega_0}\right). \qquad \text{(m)} \tag{4.30}$$

By simply setting $v_0 = 0$, the coefficients A and B then are

$$A = \frac{1}{2}u_0 = B. \tag{4.31}$$

The general solution is rewritten in the form

$$x_0(t) = u_0 e^{-\delta_0 t} \cos \omega_0 t \qquad \text{(m)} \tag{4.32}$$

where

$$\omega_0 = \sqrt{b - a^2/4} \cong \sqrt{b} = \omega_{00} \qquad \text{(rad/s)} \tag{4.33}$$

under the proviso that

$$b \gg a^2/4 \tag{4.34}$$

and ω_{00} is the eigen-angular-frequency.

4.1.3 Impulse Response

The free oscillation subject to the initial condition can be formulated as an impulse response. Assuming the external force $f(t)$ takes the form of an impulse

$$f(t) = G_0\delta(t) \qquad (\text{m/s}^2), \tag{4.35}$$

where G_0 is a constant, the differential equation Eq. 4.9 can be rewritten as

$$\frac{d^2h(t)}{dt^2} + a\frac{dh(t)}{dt} + bh(t) = G_0\delta(t). \qquad (\text{m/s}^2) \tag{4.36}$$

Introducing the inverse Fourier transforms of $h(t)$ and $f(t)$

$$h(t) = \frac{1}{2\pi}\int_{-\infty}^{\infty} H(\omega)e^{i\omega t}d\omega \qquad \text{(m)} \tag{4.37}$$

$$G_0\delta(t) = G_0\frac{1}{2\pi}\int_{-\infty}^{\infty} e^{i\omega t}d\omega \qquad (\text{m/s}^2) \tag{4.38}$$

into the differential equation, then

$$H(\omega) = \frac{-G_0}{(\omega - \omega_{p_1})(\omega - \omega_{p_2})} \qquad (\text{m} \cdot \text{s}) \tag{4.39}$$

is obtained. The impulse response can be determined by calculating explicitly the inverse Fourier transform of $H(\omega)$. That is,

$$\begin{aligned}
h(t) &= \frac{1}{2\pi}\int_{-\infty}^{\infty} H(\omega)\mathrm{e}^{\mathrm{i}\omega t}d\omega \\
&= \frac{1}{2\pi}\int_{-\infty}^{\infty} \frac{-G_0}{(\omega-\omega_{p_1})(\omega-\omega_{p_2})}\mathrm{e}^{\mathrm{i}\omega t}d\omega \\
&= G_0\frac{\sin\omega_0 t}{\omega_0}\mathrm{e}^{-\delta_0 t} \\
&= G_0\frac{\sin\omega_0 t}{\omega_0 t}\cdot t\mathrm{e}^{-\delta_0 t} \\
&= G_0 h_0(t) \qquad (t \geq 0) \quad (\mathrm{m})
\end{aligned} \tag{4.40}$$

where $h_0(t)$ is called the impulse response and ω_0 is given by Eq. 4.33 that corresponds to the angular frequency of the free oscillation.

4.1.4 Response to an External Force

The impulse response yields the response to the external force via convolution. Introducing the Fourier transform of the external force

$$f(t) = \frac{1}{2\pi}\int_{-\infty}^{\infty} F(\omega)\mathrm{e}^{\mathrm{i}\omega t}d\omega \qquad (\mathrm{m/s^2}), \tag{4.41}$$

the frequency characteristic function of the response $x(t)$ can be written as

$$\begin{aligned}
X(\omega) &= \frac{-F(\omega)}{(\omega-\omega_{p_1})(\omega-\omega_{p_2})} \\
&= F(\omega)H_0(\omega) \qquad (\mathrm{m\cdot s})
\end{aligned} \tag{4.42}$$

where

$$H_0(\omega) = \frac{-1}{(\omega-\omega_{p_1})(\omega-\omega_{p_2})} \qquad (\mathrm{s^2}) \tag{4.43}$$

that corresponds to Eq. 4.39 by setting $G_0 = 1$. Then the response $x(t)$ can be written as

$$\begin{aligned}
x(t) &= \frac{1}{2\pi}\int_{-\infty}^{\infty} X(\omega)\mathrm{e}^{\mathrm{i}\omega t}d\omega \\
&= \int_{0}^{\infty} h_0(\tau)f(t-\tau)d\tau \qquad (\mathrm{m})
\end{aligned} \tag{4.44}$$

where $h_0(t)$ denotes the impulse response defined by

$$h_0(t) = \frac{1}{2\pi}\int_{-\infty}^{\infty} H_0(\omega)\mathrm{e}^{\mathrm{i}\omega t}d\omega. \qquad \text{(s)} \tag{4.45}$$

Suppose that the response of a system to two inputs x_1 and x_2 is given by y_1 and y_2, respectively. The system is called linear when the response y to the sum of the inputs $x_1 + x_2$ can be written as the sum $y = y_1 + y_2$. The result expressed by Eq. 4.44 indicates that the response to the external force can be written as the convolution for the impulse response and the external force. This is the fundamental property of any linear system.

4.1.5 Transient and Resonant Responses to a Sinusoidal Function

The convolution formula gives a solution that includes both the particular and transient responses. Suppose a sinusoidal function $A\mathrm{e}^{\mathrm{i}\omega t}$ is an external force (m/s^2) such that

$$f(t) = \begin{cases} A\mathrm{e}^{\mathrm{i}\omega t} & t \geq 0 \\ 0 & t < 0. \end{cases} \tag{4.46}$$

Recall that with the impulse response $h_0(t)$

$$h_0(t) = \frac{\sin\omega_0 t}{\omega_0}\mathrm{e}^{-\delta_0 t}, \qquad \text{(s)} \tag{4.47}$$

the response can be written as

$$\begin{aligned} x(t) &= \int_0^t h_0(\tau)f(t-\tau)d\tau \\ &= A\mathrm{e}^{\mathrm{i}\omega t}\int_0^t h_0(\tau)\mathrm{e}^{-\mathrm{i}\omega\tau}d\tau \\ &= AH(\omega, t)\mathrm{e}^{\mathrm{i}\omega t} \qquad \text{(m)} \end{aligned} \tag{4.48}$$

where

$$H(\omega, t) = \int_0^t h_0(\tau)\mathrm{e}^{-\mathrm{i}\omega\tau}d\tau \qquad (\mathrm{s}^2) \tag{4.49}$$

and

$$f(t) = 0, \quad h_0(t) = 0 \tag{4.50}$$

for $t < 0$. Assuming $t \rightarrow +\infty$ then

$$x(t) \rightarrow AH_0(\omega)\mathrm{e}^{\mathrm{i}\omega t} \qquad \text{(m)} \tag{4.51}$$

where

$$\begin{aligned} H_0(\omega) &= \int_0^{\infty} h_0(\tau)\mathrm{e}^{-\mathrm{i}\omega\tau} d\tau \\ &= |H_0(\omega)|\mathrm{e}^{\mathrm{i}\theta(\omega)}. \qquad (\mathrm{s}^2) \end{aligned} \tag{4.52}$$

The frequency characteristic (or response) function which is defined as the Fourier transform of the impulse response, gives the complex magnitude of the response to a sinusoidal function with a unit magnitude and zero initial phase. The magnitude $|H_0(\omega)|$ is called the magnitude frequency characteristic (or response) function, and $\theta(\omega)$ is called the phase frequency characteristic (or response) function. The complex magnitude function corresponds to the particular solution for the sinusoidal input in the limit that time t approaches infinity.

The transient response, corresponding to the general solution, can be obtained by introducing Eq. 4.47 into Eq. 4.49 yielding

$$H(\omega, t) = H_0(\omega) + H_1(\omega, t) \quad (\mathrm{s}^2) \tag{4.53}$$

$$H_1(\omega, t) = \hat{H}_1(\omega, t)\mathrm{e}^{-\mathrm{i}\omega t}\mathrm{e}^{-\delta_0 t} \tag{4.54}$$

where

$$\hat{H}_1(\omega, t) = \frac{1}{2\omega_0 \mathrm{i}}\left[\frac{\mathrm{e}^{\mathrm{i}\omega_0 t}}{\mathrm{i}(\omega_0 - \omega) - \delta_0} + \frac{\mathrm{e}^{-\mathrm{i}\omega_0 t}}{\mathrm{i}(\omega_0 + \omega) + \delta_0}\right]. \tag{4.55}$$

Consequently,

$$\begin{aligned} x(t) &= AH(\omega, t)\mathrm{e}^{\mathrm{i}\omega t} \\ &= A[H_0(\omega)\mathrm{e}^{\mathrm{i}\omega t} + \hat{H}_1(\omega, t)\mathrm{e}^{-\delta_0 t}]. \qquad \text{(m)} \end{aligned} \tag{4.56}$$

The transient response represented by $H_1(\omega, t)$ decays following $\mathrm{e}^{-\delta_0 t}$ as time elapses. The response approaches the particular solution given by $H_0(\omega)$ after the transient response decays sufficiently in time. The particular solution is often called the steady-state response. Figure 4.2 provides samples of waveforms for decaying (or damped) free oscillations corresponding to the impulse response.

Figure 4.3 presents the magnitude frequency characteristic function for the impulse response. In contrast, Fig. 4.4 illustrates the time waveform from the transient to the steady state (corresponding to the particular solution) after a long elapse time. Note the logarithmic scale for time. The waveform approaches a sinusoidal function with the frequency of the external sinusoidal wave. Similarly, Fig. 4.5 shows examples of the magnitude frequency characteristic function converge to the particular solution at the steady state.

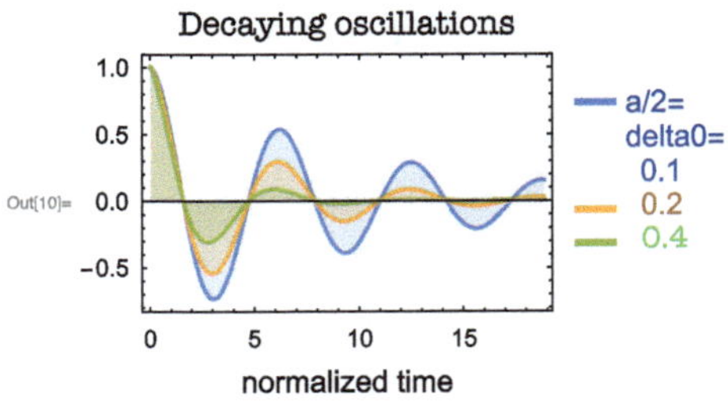

Fig. 4.2 Samples of damped oscillation

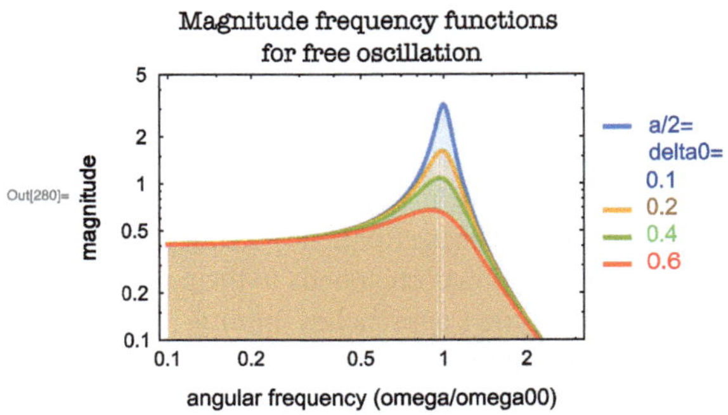

Fig. 4.3 Samples of the magnitude frequency characteristic function for a free oscillation

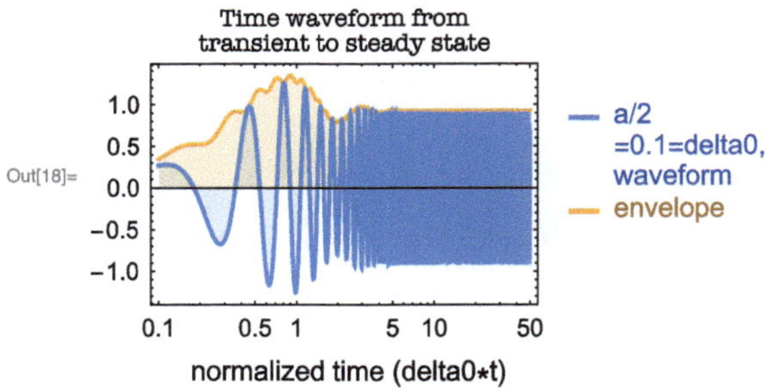

Fig. 4.4 Sample of a time waveform for the steady state over an extended elapse time (log scale)

The magnitude of the sinusoidal wave depends on the frequency of the external one. The frequency-dependent magnitude follows the magnitude frequency characteristic function for $H_0(\omega)$. The magnitude approaches a maximum around the eigenfrequency or natural frequency for the free oscillation. The oscillating system is said to be in resonance.

As for the magnitude, a frequency-dependent response may be observed for the phase and hence

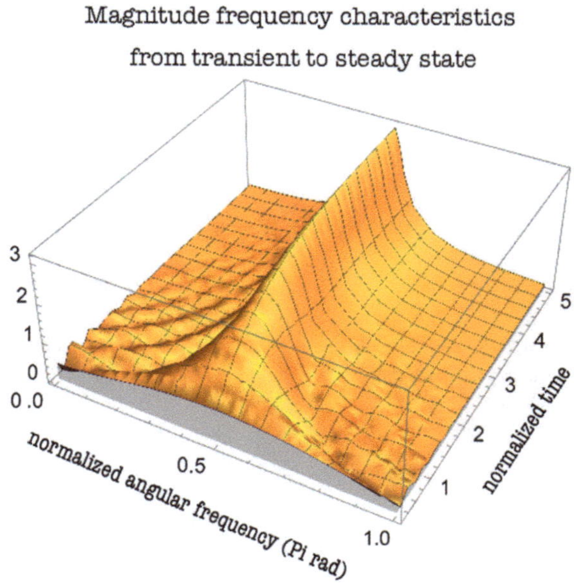

Fig. 4.5 Sample of the magnitude frequency characteristic function converging to the steady-state particular solution

$$H_0(\omega) = \int_0^\infty h_0(\tau)\mathrm{e}^{-\mathrm{i}\omega\tau}d\tau = \frac{1}{\omega_{00}^2 + 2\delta_0\mathrm{i}\omega - \omega^2} \qquad (\mathrm{s}^2) \tag{4.57}$$

$$|H_0(\omega)|^2 = \frac{1}{(\omega_{00}^2 - \omega^2)^2 + 4\delta_0^2\omega^2} \tag{4.58}$$

$$\tan\theta(\omega) = \frac{-2\delta_0\omega}{\omega_{00}^2 - \omega^2} \qquad (\mathrm{rad}) \tag{4.59}$$

where

$$\omega_{00} = \sqrt{b} = \sqrt{\frac{K}{M}} \qquad (\mathrm{rad/s}) \tag{4.60}$$

$$\omega_p = \pm\omega_0 + \mathrm{i}\delta_0 \tag{4.61}$$

$$\omega_0 = \sqrt{\omega_{00}^2 - \delta_0^2} \tag{4.62}$$

$$\delta_0 = \frac{a}{2} = \frac{R}{2M} \tag{4.63}$$

and ω_{00} is called the eigen-angular-frequency. Figures 4.6 and 4.7 show the magnitude and phase characteristics for the particular solution. The magnitude and phase responses are functions of the decay constant $\delta_0(1/s)$. The magnitude response has a maximum at $\omega = \omega_R$ for which

$$\omega_R^2 = b - \frac{a^2}{2} = \omega_{00}^2 - 2\delta_0^2 = \omega_0^2 - \delta_0^2 \qquad (1/\mathrm{s}^2) \tag{4.64}$$

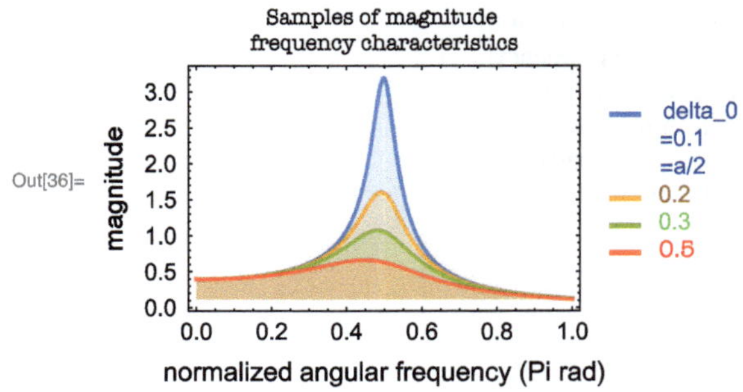

Fig. 4.6 Sample of magnitude frequency characteristic function

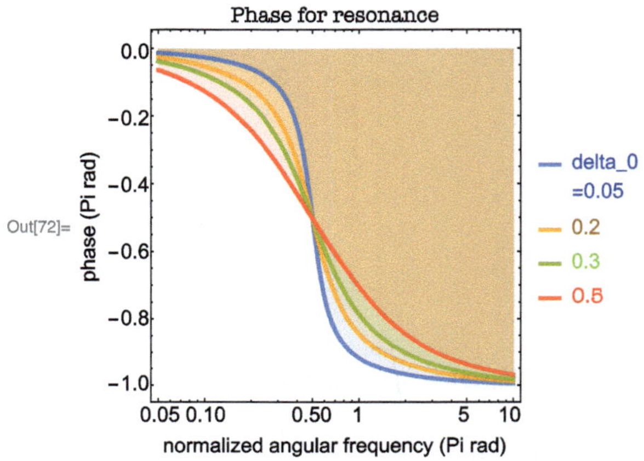

Fig. 4.7 Sample of a phase frequency characteristic function

and therefore depends on the decay constant. Figure 4.6 features resonance in the magnitude response of the displacement. Note that as the damping constant increases, the resonance frequency becomes lower. The dependence of the resonance frequency on the damping constant is interpreted as resonance in the displacement. As seen in reference [1], the frequency dependence is different in regard to resonance in velocity or acceleration. The phase is shifted by $-\pi$ (rad) after the frequency passes the resonance frequency. The phase response also depends on the decay constant; however, it crosses $-\pi/2$ phase-shift point when the frequency passes the eigenfrequency.

The transient response before approaching the steady state is frequency dependent. In particular, when the frequency is close to an eigenfrequency, the time waveform has the form of periodic beats. Intuitively, one understands that the response is a sum of the particular and the transient solutions. The particular solution is the sinusoidal wave of frequency equal to that of the external sinusoidal function; in contrast, the transient response is seen as representing the impulse response, which is the sinusoidal function with a frequency of the free oscillation close to the eigenfrequency. Assuming the frequencies for both the external and the impulse response are very close, the sum of the two responses produce beats. However, with non-vanishing damping constant, the beats weakens as the impulse response decays as shown in Fig. 4.8.

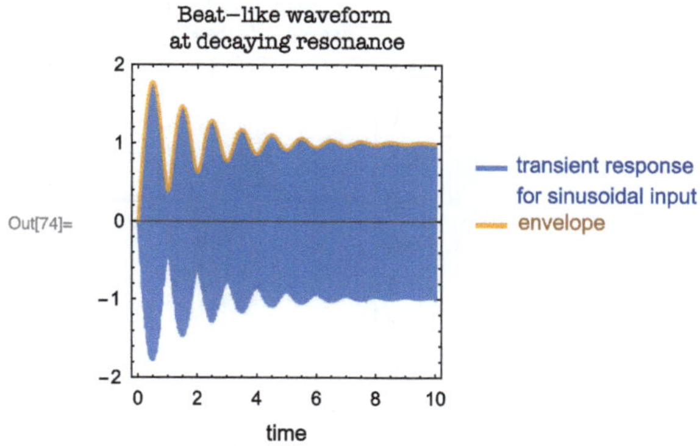

Fig. 4.8 Transient response to a sinusoidal input with frequency close to the resonance frequency

4.2 Source Waveform Estimation Through Differential Equation

4.2.1 *Observed Response from a Single-Degree-of-Freedom System*

Consider once more a vibrating system composed of a spring and a mass. The equation of motion for the mass subject to an external force $F(t)$(N) is written as

$$M\frac{d^2x}{dt^2} + R\frac{dx}{dt} + Kx = F \qquad \text{(N)} \tag{4.65}$$

where $x(t)$(m) denotes the displacement, M(kg) the mass, $R(\text{N}\cdot\text{s/m})$ the friction constant, and K(N/m) the spring constant. The external force elicits from the system three types of forces, an inertial force associated with the acceleration of the mass, a frictional force associated with damping, and an elastic force due to the spring. From the summation of these forces, the external force can be estimated, in principle, from observational measurements of the displacement of the mass.

Suppose that the displacement is given by

$$x_f = A\alpha \mathrm{e}^{-(\alpha t)^2} \qquad \text{(m)} \tag{4.66}$$

where $A(\text{m}\cdot\text{s})$ and $\alpha(1/\text{s})$ are constant scaling factors. Substituting the displacement function for x into Eq. 4.65 and performing the differentiations, the external force is then expressed as

$$\begin{aligned} F &= M\frac{d^2x_f}{dt^2} + R\frac{dx_f}{dt} + Kx_f \\ &= \mathrm{e}^{-(\alpha t)^2}\left(-M\cdot 2A\alpha^3(1-2\alpha^2t^2) - R\cdot 2A\alpha^3 t + K\cdot\alpha A\right). \qquad \text{(N)} \end{aligned} \tag{4.67}$$

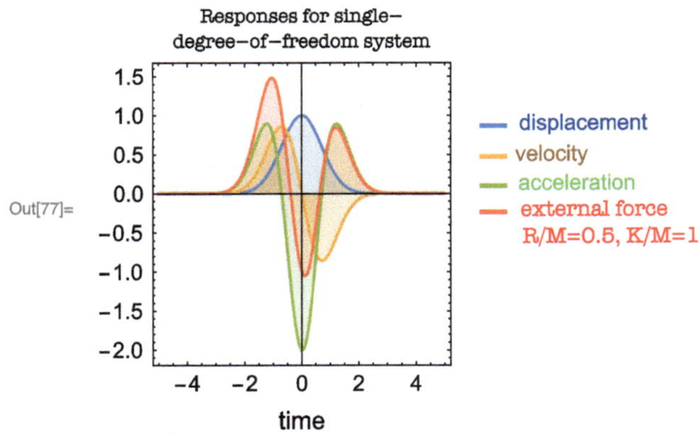

Fig. 4.9 External force, displacement, velocity, and acceleration

Figure 4.9 plots the external force, displacement, velocity, and acceleration.

Inversely, if the external force is given by $F(t)$, then the displacement $x_f(t)$ must be a solution of the differential equation

$$M\frac{d^2x_f}{dt^2} + R\frac{dx_f}{dt} + Kx_f = F. \qquad \text{(N)} \tag{4.68}$$

However, the complete solution is not only $x_f(t)$ given by Eq. 4.66. This is because the solutions of the homogeneous equation

$$M\frac{d^2x}{dt^2} + R\frac{dx}{dt} + Kx = 0 \qquad \text{(N)} \tag{4.69}$$

can be included in the complete solution,

$$\begin{aligned} x &= x_f + x_0 \\ &= x_f + \mathrm{e}^{-\delta_0 t}\sin(\omega_0 t + \phi) \qquad \text{(m)} \end{aligned} \tag{4.70}$$

where $x_0(t)$ is one of the solutions of Eq. 4.69. Introducing $x_0(t)$ into Eq. 4.69 confirms that the equation holds for $x_0(t)$. The observed displacement could take the form given in Eq. 4.70; however, the external force can be estimated as $F(t)$ from Eq. 4.67.

Take another example of $x(t)$; instead of $x_f(t)$ of Eq. 4.66 consider

$$x_{ff} = A\alpha \cdot \mathrm{Erf}(\alpha t) \qquad \text{(m)} \tag{4.71}$$

where

$$\mathrm{Erf}(\xi) = \frac{2}{\sqrt{\pi}}\int_0^{\xi} \mathrm{e}^{-y^2}dy. \tag{4.72}$$

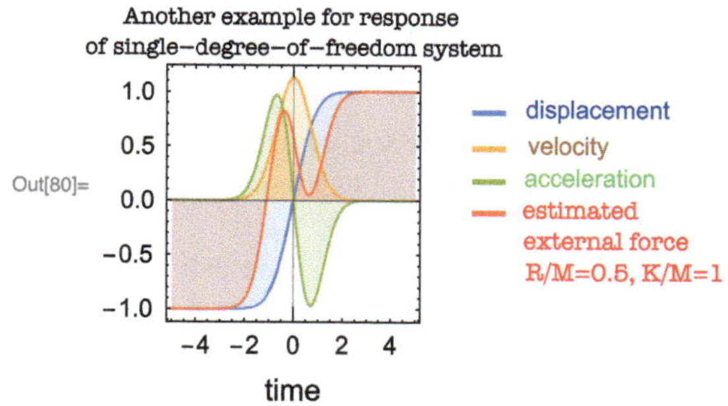

Fig. 4.10 Another example of three functions and estimated external force

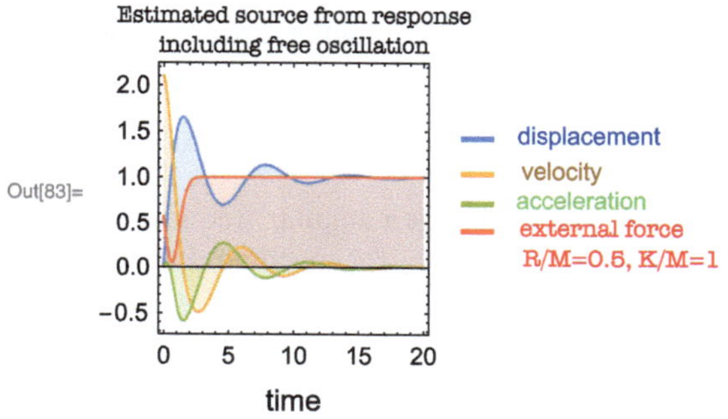

Fig. 4.11 Three functions including eigen-oscillation and estimated source function

By differentiation, the velocity $v_{ff}(t)$ and the acceleration $a_{c_{ff}}(t)$ are

$$v_{ff} = \alpha^2 A \frac{2}{\sqrt{\pi}} \mathrm{e}^{-(\alpha t)^2} \qquad (\mathrm{m/s}) \tag{4.73}$$

$$a_{c_{ff}} = -\alpha^4 A \frac{4t}{\sqrt{\pi}} \mathrm{e}^{-(\alpha t)^2}. \qquad (\mathrm{m/s^2}) \tag{4.74}$$

Figure 4.10 presents plots of the three functions and estimates the external force. Figure 4.11 illustrates the case including the free oscillation similar to Fig. 4.10. The free oscillation is independent of the external force.

4.2.2 *Source Waveform Recovery for a Single-Degree-of-Freedom System*

Source waveform recovery from observational data is an attractive and interesting procedure used in acoustic and vibration signal processing for noise control from machinery and/or diagnostics [2]. Suppose that a signal is received as an output response $x(t)$ from a single-degree-of-freedom system that is governed by an equation of motion as in Eq. 4.68,

$$\left(\frac{d^2}{dt^2} + 2\delta_0 \frac{d}{dt} + \omega_0\right) x = f \qquad (\mathrm{m/s^2}) \tag{4.75}$$

where $f(t)$ denotes the source waveform of interest. The discussion developed in the previous subsection implies the source waveform might be reconstructed provided δ_0 and ω_0 are known a priori.

Indeed, both δ_0 and ω_0 can be estimated if the free vibration of the system is observed. The free vibration is written as

$$x_0 = A\mathrm{e}^{-\delta_0 t} \cos(\omega_d t - \phi) \qquad (\mathrm{m}) \tag{4.76}$$

where ω_0, denoting the angular frequency of the free oscillation, satisfies

$$\omega_0 = \sqrt{\omega_{00}^2 - \delta_0^2} \qquad (\mathrm{rad/s}) \tag{4.77}$$

and ω_{00} gives the eigen-angular frequency. The source waveform recovery would be interesting not only for machine diagnostics [2] but also for design of musical instruments.

4.3 Wave Equations for Sinusoidal Traveling Waves

4.3.1 *Equation of Waves*

Waves are represented by variables that are functions of the spatial and temporal variables. Denoted by ξ or η, these variables take forms like

$$\xi = x \pm ct \qquad (\mathrm{m}) \tag{4.78}$$

$$\eta = ct \pm x \tag{4.79}$$

where c denotes the speed of sound (m/s), x(m) corresponds to the spatial position or propagating distance in the space, and t denotes time (s). Consider now some arbitrary function of ξ, $f(\xi)$. The partial derivatives of the function with respect to t become

$$\frac{\partial f}{\partial t} = \frac{\partial f}{\partial \xi}\frac{\partial \xi}{\partial t} = (\pm c)\frac{\partial f}{\partial \xi} \qquad (\mathrm{m/s}) \tag{4.80}$$

$$\frac{\partial^2 f}{\partial t^2} = \frac{\partial}{\partial t}\frac{\partial f}{\partial t} = c^2 \frac{\partial^2 f}{\partial \xi^2}. \qquad (\mathrm{m/s^2}) \tag{4.81}$$

Similarly, taking the partial derivatives of f with respect to x, then

$$\frac{\partial f}{\partial x} = \frac{\partial f}{\partial \xi}\frac{\partial \xi}{\partial x} = \frac{\partial f}{\partial \xi} \tag{4.82}$$

$$\frac{\partial^2 f}{\partial x^2} = \frac{\partial}{\partial x}\frac{\partial f}{\partial x} = \frac{\partial^2 f}{\partial \xi^2}. \quad (1/\mathrm{m}) \tag{4.83}$$

Consequently, one immediately sees that

$$\frac{\partial^2 f}{\partial t^2} = c^2\frac{\partial^2 f}{\partial \xi^2} = c^2\frac{\partial^2 f}{\partial x^2}. \quad (\mathrm{m/s^2}) \tag{4.84}$$

The partial differential equation

$$\frac{\partial^2 f}{\partial t^2} = c^2\frac{\partial^2 f}{\partial x^2} \qquad (\mathrm{m/s^2}) \tag{4.85}$$

is called the one-dimensional wave equation, which governs the waves propagating in one-dimensionally extended medium like a string or an acoustic tube.

4.3.2 Propagation of Sinusoidal Waves

The wave variable $\eta = ct - x$ indicates that the wave is propagating in the positive x direction. For example, a sinusoidal wave written as

$$\begin{aligned} y^+(t,x) &= A\sin k(ct - x) \\ &= A\sin ck\left(t - \frac{x}{c}\right) \\ &= A\sin\omega\left(t - \frac{x}{c}\right) \qquad (\mathrm{m}) \end{aligned} \tag{4.86}$$

corresponds to a wave propagating in the positive x direction with angular frequency $ck = \omega$ and constant wave number k (1/m). For propagation in the negative x direction

$$y^-(t,x) = A\sin\omega\left(t + \frac{x}{c}\right). \qquad (\mathrm{m}) \tag{4.87}$$

An interesting phenomenon occurs on taking the superposition of the two waves,

$$\begin{aligned} y &= y^+ + y^- \\ &= A\left(\sin\omega\left(t - \frac{x}{c}\right) + \sin\omega\left(t + \frac{x}{c}\right)\right) \\ &= 2A\sin\omega t\cos kx \\ &= 2Af(t)g(x) \qquad (\mathrm{m}) \end{aligned} \tag{4.88}$$

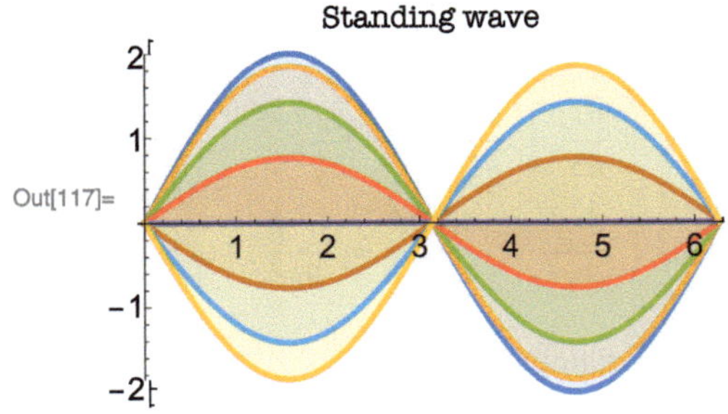

Fig. 4.12 Example of standing waves

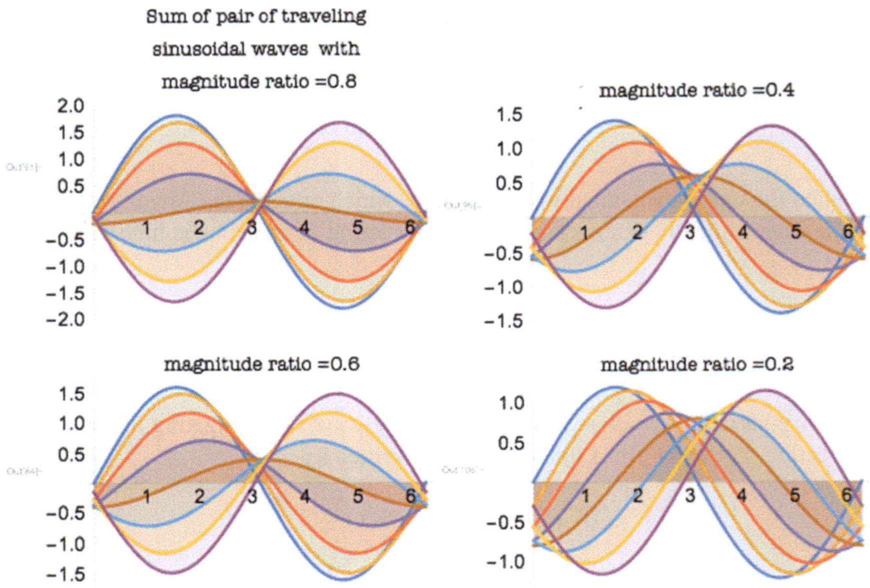

Fig. 4.13 Sum of a pair of progressive waves with the same frequency but different magnitudes, propagating in opposite directions

where well-known trigonometric formulae is used. Note that the superposition now becomes a product of a spatial and a temporal function. The superposed wave looks like an oscillation at fixed positions in space rather than a propagating or progressive wave. Indeed, from its very nature, it is called a standing wave in space. This type of wave is constructed approximately in a one-dimensional space where a direct wave and its reflected wave are superposed. A standing wave is characterized by peaks (loops) and troughs (nodes) observed along the spatial direction as illustrated in Fig. 4.12.

A standing wave is constructed by the sum of a pair of waves with the same frequency and magnitude propagating in opposite directions. If the magnitudes are different, then a standing wave does not form, as depicted in Fig. 4.13.

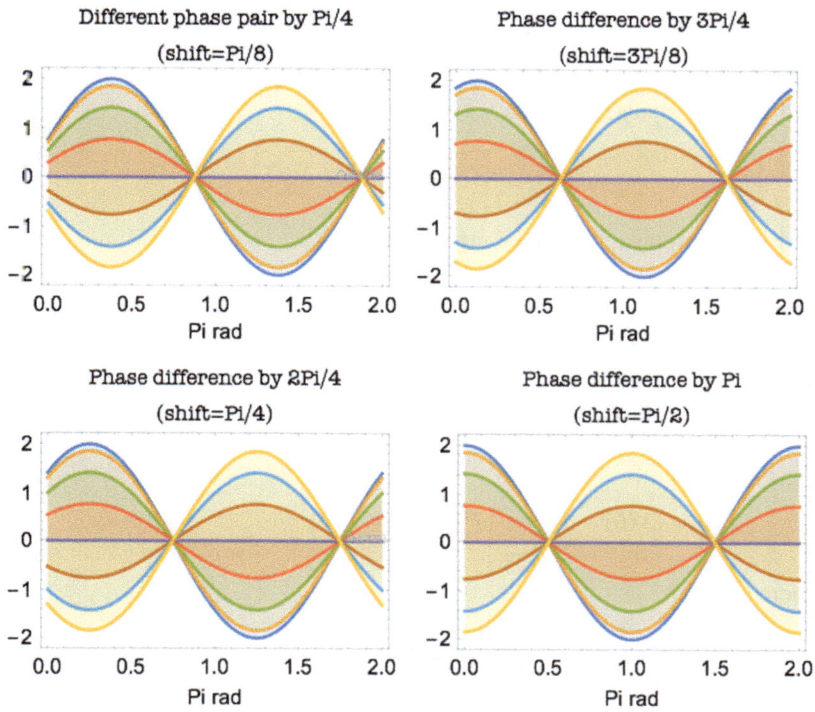

Fig. 4.14 Sum of a pair of progressive waves with the same magnitude but different phase, propagating in opposite directions

The trough becomes shallow as the difference in magnitude becomes large. Therefore the nodes, that is, points where vibrations do not appear, can no longer form. The envelopes and nodes that represent the standing wave no longer are established if the magnitude is different between the pair of progressive waves, even if the frequency is the same. The standing wave disappears as the magnitude difference becomes larger, and the wave begins to look like a progressive wave. The effect of the magnitude and phase difference represents the effect of complex coefficients of reflection on wave propagation. A pair of different phases moves the positions of the nodes keeping the standing wave as shown in Fig. 4.14.

4.4 Exercises

1. Obtain the eigenfrequency, the frequency of free oscillation, and resonance frequency of the following system of a single degree of freedom,

$$M\frac{d^2}{dt^2}x + R\frac{d}{dt}x + Kx = 0.$$

2. Given the impulse response

$$h(t) = A\sin(\omega_0 t + \theta).$$

Derive the response $y(t)$ to the input signals $x(t)$

$$\begin{aligned}
&(1) \quad B\cos\omega t \quad (t \geq 0) \\
&(2) \quad B\cos\omega_0 t \quad (t \geq 0) \\
&(3) \quad \delta(t) + a\delta(t-T) \quad (t \geq 0) \\
&(4) \quad \delta(t) + a\delta(t-T) + b\delta(t-2T) \quad (t \geq 0) \\
&(5) \quad \mathrm{e}^{-\alpha t} \quad (t \geq 0) \\
&(6) \quad \frac{1}{T} \quad (-T/2 \leq t \leq T/2)
\end{aligned} \tag{4.89}$$

where

$$\int f(t)\delta(t-T)dt = f(t).$$

3. What are the responses in magnitude and phase frequency of a system of a single degree of freedom in a steady state to a sinusoidal input where

$$M\frac{d^2}{dt^2}x + R\frac{d}{dt}x + Kx = F.$$

4. Provide the responses obtain above for the following conditions,

$$\begin{aligned}
&(1) \quad \omega << \omega_R \\
&(2) \quad \omega \cong \omega_R \\
&(3) \quad \omega >> \omega_R
\end{aligned} \tag{4.90}$$

where ω_R denotes the resonance frequency assuming that $\omega_R \cong \omega_{00}$.
5. Confirm Eq. 4.40.
6. Confirm Eq. 4.56.
7. Obtain the external force $F(t)(t \geq 0)$ given in the single-degree-of-freedom system above when the observed responses $x(t)$ are

$$\begin{aligned}
&(1) \quad e^{-\alpha t} \quad (t \geq 0) \\
&(2) \quad \alpha t \cdot e^{-\alpha t} \quad (t \geq 0) \\
&(3) \quad e^{-(\alpha t)^2} \quad (t \geq 0) \\
&(4) \quad (\alpha t)^2 e^{-(\alpha t)^2} \quad (t \geq 0).
\end{aligned} \tag{4.91}$$

6. Consider a compound wave composed of two waves such that

$$y(x, t) = u(x + ct) + u(x - ct).$$

Obtain the wave $y(t)$ for each of the following components u,

$$\begin{aligned}
&(1) \quad \cos x \\
&(2) \quad \sin x \\
&(3) \quad \cos 2x \\
&(4) \quad \sin 2x \\
&(5) \quad \sin^2 x \\
&(6) \quad \cos^2 x \\
&(7) \quad e^{-(\alpha x)^2} \\
&(8) \quad \alpha x \cdot e^{-(\alpha x)^2} \\
&(9) \quad (\alpha x)^2 e^{-(\alpha x)^2}.
\end{aligned} \tag{4.92}$$

References

1. J. Blauert, N. Xiang, *Acoustics for Engineers* (Springer, Berlin, 2008)
2. R.H. Lyon, *Machinery Noise and Diagnostics* (Butterworth, Boston, 2000)

Chapter 5
Discrete Signals and Linear Systems

Abstract Signal analysis of discrete sequences is formulated using linear equations. This chapter summarizes the least-squares error solutions and minimum-norm solutions of linear equations. Linear and orthogonal regression analyses are in particular significant topics in discrete signal analysis in the time domain. Signal detection or source signal separation provides a good example that gives an intuitive understanding of regression analysis. Orthogonality or decorrelation is the basic notion underscoring the theoretical tools of discrete signal analysis. The Fourier transform of a sequence provides the spectral representation of the sequence based on the orthogonality of sinusoidal functions; however, the discrete Fourier transform yields the discrete spectral sequence that represents a periodic discrete sequence based on the orthogonality of vectors. The discrete Fourier transform is a practical way to discretize the signal representation. In addition to this discretization, it aids understanding the sampling theorem in a more intuitive manner.

Keywords Linear equations · Linear combination of vectors · Minimum-norm solution · Least-squares error solution · Linear regression · Orthogonal regression · Quadratic form · Diagonal matrix · Eigenvectors · Orthogonality · Principal component analysis · Decorrelation · Source separation · Discrete fourier transform · Sampling theorem

5.1 Linear Equations

5.1.1 Linear Combination of Vectors

Suppose $\mathbf{x}_1$ and $\mathbf{x}_2$ are two vectors and a_1 and a_2 are two scalars. Scalar multiplication yields two further vectors $a_1\mathbf{x}_1$ and $a_2\mathbf{x}_2$. The sum of these vectors produces yet another vector,

$$\mathbf{y} = a_1\mathbf{x}_1 + a_2\mathbf{x}_2, \tag{5.1}$$

called a linear combination of vectors. If there are no scalars a_1 and a_2 other than $a_1 = a_2 = 0$ that satisfy

M. Tohyama, *Sound in the Time Domain*, Signals and Communication Technology,
https://doi.org/10.1007/978-981-10-5889-9_5

Out[336]=

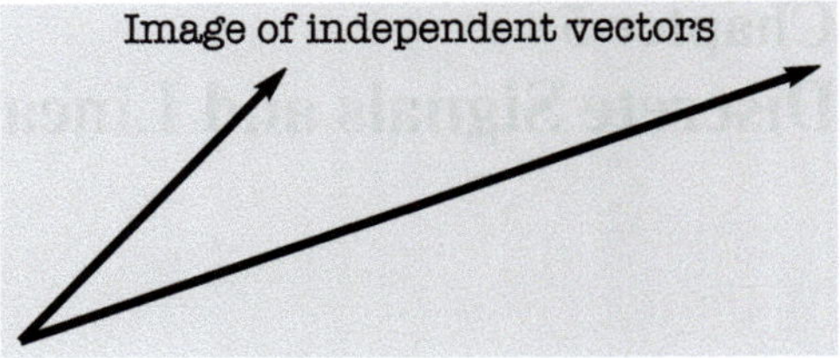

Fig. 5.1 Diagrammatic representation of linearly independent vectors

Out[338]=

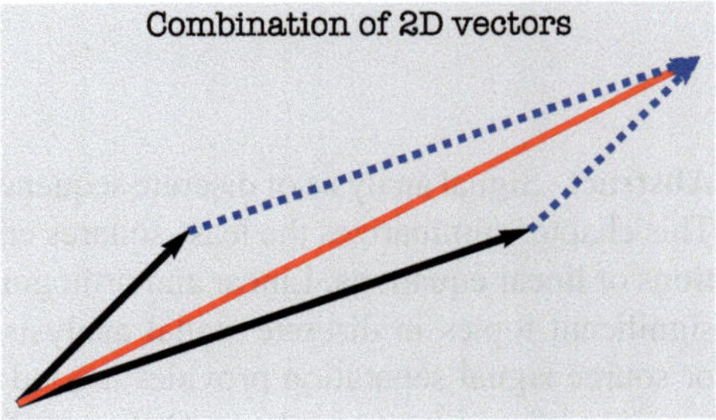

Fig. 5.2 Diagram representing a linear combination of a pair of two-dimensional vectors

$$\mathbf{y} = a_1\mathbf{x}_1 + a_2\mathbf{x}_2 = \mathbf{0}, \tag{5.2}$$

the vectors are said to be linearly independent. Figure 5.1 presents a representation of a pair of linearly independent vectors; note that orthogonality of the pair is not necessary.

Setting vector $\mathbf{y}$ to a given $\mathbf{b}$,

$$a_1\mathbf{x}_1 + a_2\mathbf{x}_2 = \mathbf{b} \tag{5.3}$$

establishes a linear equation for which vectors $\mathbf{x}_1$ and $\mathbf{x}_2$ are the two known vectors. Solving this linear vector equation implies finding the unknown pair of scalars for vector $(a_1\ a_2)^{\mathrm{T}}$. Assuming $\mathbf{x}_1$ and $\mathbf{x}_2$ are N-dimensional vectors, their linear combination also determines another N-dimensional vector. Figure 5.2 shows how to combine a pair of two-dimensional vectors.

5.1.2 Linear Equation and Linear Combination

Take a set of N-dimensional vectors $\mathbf{v}_1 \cdots \mathbf{v}_M$. A general linear combination of the vectors is written as [1]

$$\sum_{i=1}^{M} x_i\mathbf{v}_i = \mathbf{b}. \tag{5.4}$$

If vector $\mathbf{b}$ is given, then the linear combination above is called the linear equation for the unknown vector $\mathbf{x}$ defined by

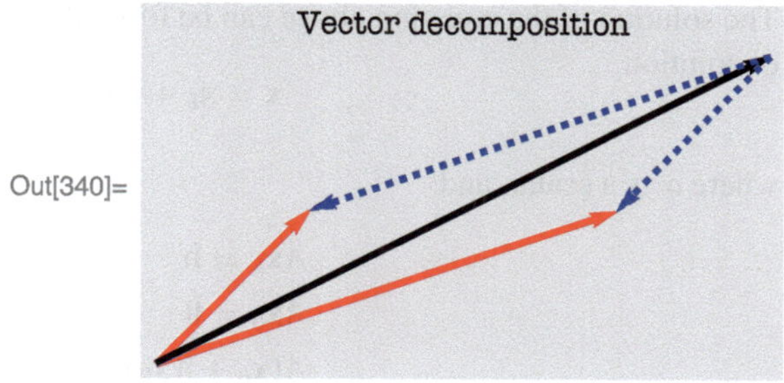

Fig. 5.3 Depiction of vector decomposition

$$\mathbf{x} = (x_1 \quad x_2 \quad \cdots \quad x_M)^{\mathrm{T}}. \tag{5.5}$$

Equation 5.4 can be interpreted as the decomposition of a vector $\mathbf{b}$ into a sum of vectors $\mathbf{v}_i$. If x_i can be determined, then the vector $\mathbf{b}$ can be decomposed into the vectors $\mathbf{v}_i$.

The decomposition of a vector into other known vectors has already been developed in regard to the Fourier series expansion in Subsection 3.5.1. In the Fourier series expansion, the vectors are mutually orthogonal. Therefore, the unknown vector $\mathbf{x}$ can be determined uniquely from the orthogonality for any pair of vectors $\mathbf{v}_i$ and $\mathbf{v}_j$ such that

$$\mathbf{v}_i \cdot \mathbf{v}_j = v_0 \delta_{ij} = \begin{cases} v_0 & i = j \\ 0 & i \neq 0. \end{cases} \tag{5.6}$$

For linear equations in general, the vectors may not all be orthogonal. Figure 5.3 depicts the decomposition of a vector into a pair of vectors in a two-dimensional space.

This difference between orthogonal and linearly independent pairs of vectors can be understood intuitively. If a pair of the vectors cannot be constructed from linearly independent vectors, then the given vector $\mathbf{b}$ cannot be decomposed into a sum of vectors in general, with the exception of $\mathbf{b} = x\mathbf{v}_i$. Therefore, sets of linearly independent vectors are key in the theory of linear vector equations rather than vector orthogonality.

Take the example of two-dimensional vectors. A general linear equation can be written as

$$A\mathbf{x} = \mathbf{b} \tag{5.7}$$

where

$$A = (\mathbf{v}_1 \quad \mathbf{v}_2 \quad \cdots \quad \mathbf{v}_M)^{\mathrm{T}} \tag{5.8}$$

$$\mathbf{x} = (x_1 \quad x_2 \quad \cdots \quad x_M)^{\mathrm{T}} \tag{5.9}$$

$$\mathbf{b} = (b_1 \quad b_2)^{\mathrm{T}} \tag{5.10}$$

$$\mathbf{v}_i = (v_{i_1} \quad v_{i_2})^{\mathrm{T}}. \tag{5.11}$$

The solution of the equation above can be formally expressed as a sum of two types of solution,

$$\mathbf{x} = \mathbf{x}_b + \alpha\mathbf{x}_0, \tag{5.12}$$

where α is a scalar, and

$$A\mathbf{x}_b = \mathbf{b} \tag{5.13}$$

$$A\mathbf{x}_0 = \mathbf{0} \tag{5.14}$$

$$A(\mathbf{x}_b + \alpha\mathbf{x}_0) = \mathbf{b}. \tag{5.15}$$

The two solutions $\mathbf{x}_b$ and $\mathbf{x}_0$ correspond to the particular and general solutions for a differential equation described in Chap. 4. If the solution $\mathbf{x}_0$ exists, then other conditions (such as the initial conditions for a differential equation) are necessary so that the solution for the linear vector equation might be uniquely determined.

The linear equation 5.7 is interpreted as a linear combination of M 2-dimensional vectors assuming $M \geq 2$. The vector $\mathbf{x}_0$ is the solution in an M-dimensional space for which

$$A\mathbf{x}_0 = \begin{pmatrix} \mathbf{u}_1^{\mathrm{T}} \\ \mathbf{u}_2^{\mathrm{T}} \end{pmatrix} \mathbf{x}_0 = \mathbf{0}, \tag{5.16}$$

where

$$\mathbf{u}_1^{\mathrm{T}} = (v_{11} \quad v_{21} \quad \cdots \quad v_{M1}) \tag{5.17}$$

$$\mathbf{u}_2^{\mathrm{T}} = (v_{12} \quad v_{22} \quad \cdots \quad v_{M2}). \tag{5.18}$$

The expression above indicates that the vector $\mathbf{x}_0$ is orthogonal in the M-dimensional space to the pair $\mathbf{u}_1$ and $\mathbf{u}_2$. A vector orthogonal to a pair of vectors may be found in the M-dimensional space when $M > 2$. Figure 5.4 shows an image of a three-dimensional vector that is orthogonal to the two-dimensional plane spanned by the pair of three-dimensional vectors. The three-dimensional vector orthogonal to the plane in the figure is an example of a solution vector $\mathbf{x}_0$ for $M = 3$.

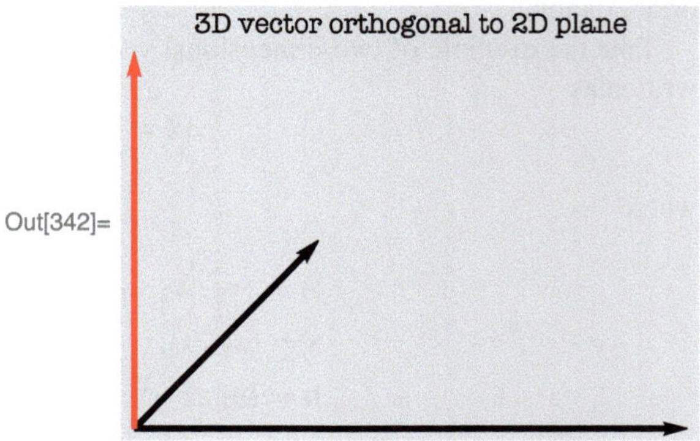

Fig. 5.4 Depiction of a three-dimensional vector orthogonal to a two-dimensional plane

Take the following example:

$$A\mathbf{x} = \begin{pmatrix} 1 & -2 & -1 \\ -3 & 1 & 9 \end{pmatrix} \begin{pmatrix} x_1 \\ x_2 \\ x_3 \end{pmatrix} = \begin{pmatrix} -8 \\ 77 \end{pmatrix} = \mathbf{b}. \tag{5.19}$$

Its solution is formally written as

$$\begin{aligned} \mathbf{x} = \begin{pmatrix} x_1 \\ x_2 \\ x_3 \end{pmatrix} &= \frac{1}{5} \begin{pmatrix} 17x_3 - 146 \\ 6x_3 - 53 \\ 5x_3 \end{pmatrix} \\ &= \frac{1}{5} x_3 \begin{pmatrix} 17 \\ 6 \\ 5 \end{pmatrix} + \frac{1}{5} \begin{pmatrix} -146 \\ -53 \\ 0 \end{pmatrix} \\ &= x_3 \frac{1}{5} \mathbf{x}_0 + \mathbf{x}_b, \end{aligned} \tag{5.20}$$

where

$$A\mathbf{x}_b = \mathbf{b} \tag{5.21}$$

$$A\mathbf{x}_0 = \mathbf{0} \tag{5.22}$$

$$A(\mathbf{x}_b + \mathbf{x}_0) = \mathbf{b}. \tag{5.23}$$

However, the decomposition of the solution $\mathbf{x}$ into $\mathbf{x}_0$ and $\mathbf{x}_b$ is not unique, even if $A(\mathbf{x}_b - \mathbf{x}_0) = \mathbf{b}$.

On the contrary, suppose that one of the vectors $\mathbf{x}$ is observed. The vector $\mathbf{b}$ corresponding to an external input may be estimated from the linear equation $A\mathbf{x} = \mathbf{b}$ as developed in Chap. 4, provided that A is known, even if the particular solution itself has not been extracted. In this example, the vector $\mathbf{b}$ can be estimated by one of the solutions

$$\mathbf{x}_p = \frac{1}{5} \begin{pmatrix} -129 \\ -47 \\ 5 \end{pmatrix} \tag{5.24}$$

such that

$$\begin{pmatrix} 1 & -2 & -1 \\ -3 & 1 & 9 \end{pmatrix} \begin{pmatrix} -129 \\ -47 \\ 5 \end{pmatrix} \frac{1}{5} = \begin{pmatrix} -8 \\ 77 \end{pmatrix} = \mathbf{b}. \tag{5.25}$$

5.1.3 Minimum-Norm Solution

Initial conditions are necessary to determine the general solution to a differential equation uniquely. A candidate for the unique solution to a linear equation that has

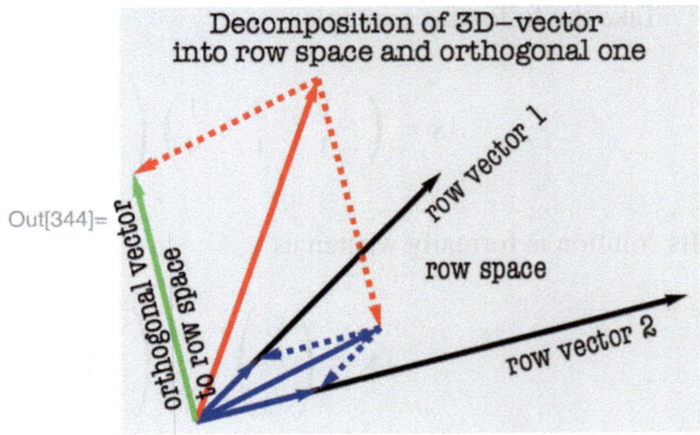

Fig. 5.5 Decomposition of the three-dimensional space into row and complementary spaces

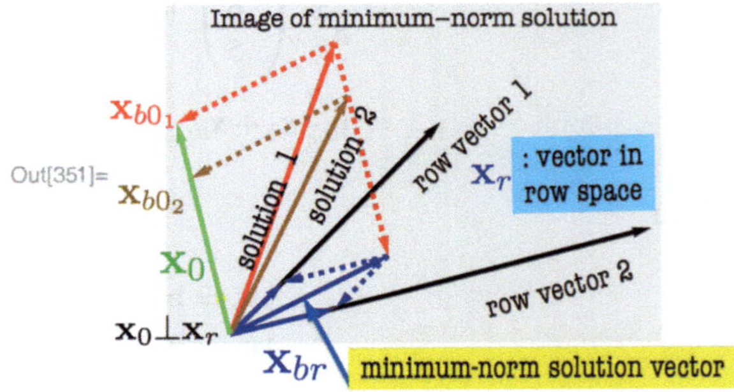

Fig. 5.6 Decomposition of the solution vector into row and its orthogonal space vectors

general solutions is the minimum-norm solution [2–4]. Any M-dimensional space in which the linear equation is defined can be decomposed into a row space spanned by N-dimensional row vectors and another space orthogonal to the row space assuming $M > N$. The solution vector $\mathbf{x}_0$ exists in the space orthogonal to the row space. Figure 5.5 illustrates the two spaces when $M = 3 > N$ as exemplified in the previous subsection. A particular solution $\mathbf{x}_b$ is decomposed into $\mathbf{x}_{br}$ and $\mathbf{x}_{b0}$ that are located in the row space and complementary space orthogonal to the row space. Figure 5.6 shows an image of the minimum-norm solution.

In the decomposition, $\mathbf{x}_{br}$ is always identically independent of $\mathbf{x}_{b0}$ where

$$A\mathbf{x}_{br} = \mathbf{b} \tag{5.26}$$

$$A\mathbf{x}_{b0} = \mathbf{0} \tag{5.27}$$

hold. The solution $\mathbf{x}_{br}$ is said to have the minimum length of all solution vectors $\mathbf{x}$ if

$$|\mathbf{x}|^2 = |\mathbf{x}_{br}|^2 + |\mathbf{x}_{b0}|^2. \tag{5.28}$$

Therefore, the solution $\mathbf{x}_{br}$ that is determined uniquely is called the minimum-norm solution.

Take the example of the previous subsection once more. The vector $\mathbf{x}_b$ is decomposed in vectors $\mathbf{x}_{b0}$ and $\mathbf{x}_{br}$. From the orthogonality between these two vectors, the component vector $\mathbf{x}_{b0}$ along $\mathbf{x}_0$ is given by

$$\begin{aligned}\mathbf{x}_{b0} &= \frac{\mathbf{x}_b \cdot \mathbf{x}_0}{|\mathbf{x}_0|^2}\mathbf{x}_0 \\ &= \frac{\begin{pmatrix}-146/5 & -53/5 & 0\end{pmatrix}}{17^2+6^2+5^2}\begin{pmatrix}17\\6\\5\end{pmatrix}\mathbf{x}_0 \\ &= \frac{-8}{5}\mathbf{x}_0 = \frac{-8}{5}\begin{pmatrix}17\\6\\5\end{pmatrix}.\end{aligned} \tag{5.29}$$

Hence, the component vector $\mathbf{x}_{br}$ is given by

$$\begin{aligned}\mathbf{x}_{br} &= \mathbf{x}_b - \mathbf{x}_{b0} \\ &= \frac{1}{5}\begin{pmatrix}-146\\-53\\0\end{pmatrix} - \frac{-8}{5}\begin{pmatrix}17\\6\\5\end{pmatrix} \\ &= \begin{pmatrix}-2\\-1\\8\end{pmatrix}.\end{aligned} \tag{5.30}$$

The vector $\mathbf{x}_{br}$ is located in the row space and satisfies $A\mathbf{x}_{br} = \mathbf{b}$. The solution vector is called the minimum-norm solution having the minimum length of all solution vectors.

5.1.4 Minimum-Norm Solution by Least-Squares Error Criterion

Take the linear equation

$$\sum_{i=1}^{M} x_i \mathbf{v}_i = \mathbf{b}, \tag{5.31}$$

where the vectors $\mathbf{v}_i$ are not linearly independent. The solution for the linear equation above does not exist for a given vector $\mathbf{b}$. Hence, the minimum-norm solution cannot be determined in general. However, a vector that approximately satisfies the linear equation is possible. A candidate for an approximate solution is the least-squares error solution.

Take the linear system of equation of the numerical example above:

$$A\mathbf{x} = \begin{pmatrix} 1 & -2 & -1 \\ -3 & 6 & 3 \end{pmatrix} \begin{pmatrix} x_1 \\ x_2 \\ x_3 \end{pmatrix} = \begin{pmatrix} b_1 \\ b_2 \end{pmatrix} = \mathbf{b}. \quad (5.32)$$

It can be rewritten in the form

$$\begin{aligned} & x_1 \begin{pmatrix} 1 \\ -3 \end{pmatrix} + x_2 \begin{pmatrix} -2 \\ 6 \end{pmatrix} + x_3 \begin{pmatrix} -1 \\ 3 \end{pmatrix} \\ & = (-x_1 + 2x_2 + x_3) \begin{pmatrix} -1 \\ 3 \end{pmatrix} = \mathbf{b}. \end{aligned} \quad (5.33)$$

This linear equation only has solutions when

$$\mathbf{b} = \alpha \begin{pmatrix} -1 \\ 3 \end{pmatrix}, \quad (5.34)$$

where α is a scalar.

However, it is possible to get an approximate solution $\hat{\mathbf{x}}$ that satisfies

$$A\hat{\mathbf{x}} = \hat{\mathbf{b}}, \quad (5.35)$$

where

$$\mathbf{e} = \Delta\mathbf{b} = \hat{\mathbf{b}} - \mathbf{b} \quad (5.36)$$

is called the error vector.

A candidate for the approximate solution is the least-squares error solution that minimizes the length of the error vector. When the linear equation is written as

$$\sum_{i=1}^{M} x_i \mathbf{v}_i = \mathbf{b}, \quad (5.37)$$

the vectors $\mathbf{v}_i$ are called the column vectors of the matrix A, and the space constructed from the linear combination of column vectors is called the column space. If the vector **b** lies in the column space, the linear equation has a solution; otherwise, there is no solution. The error vector $\Delta\mathbf{b}$ illustrated in Fig. 5.7 is a diagrammatic representation of Eq. 5.36.

To minimize the length of the error vector, $\hat{\mathbf{b}}$ may be taken as the orthogonal projection vector onto the column space. An example for **b** in Eq. 5.32 is

$$\mathbf{b} = \begin{pmatrix} -6 \\ 10 \end{pmatrix}. \quad (5.38)$$

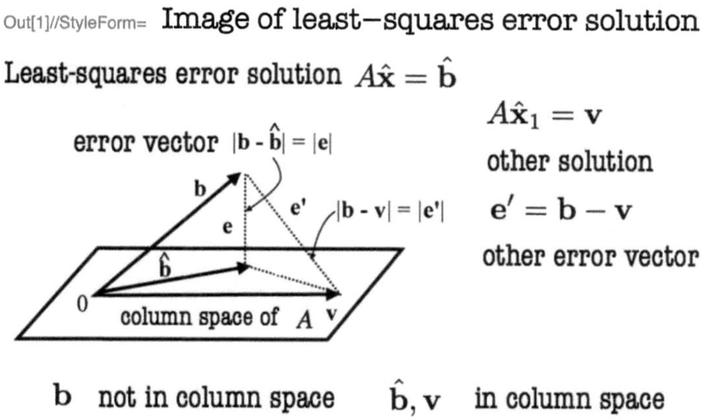

Fig. 5.7 Depiction of the error vector for Eq. 5.36

The linear equation

$$A\mathbf{x} = \begin{pmatrix} 3 & -6 & -3 \\ -3 & 6 & 3 \end{pmatrix} \begin{pmatrix} x_1 \\ x_2 \\ x_3 \end{pmatrix} = \begin{pmatrix} -6 \\ 10 \end{pmatrix} = \mathbf{b} \tag{5.39}$$

has no solution. Defining $\hat{\mathbf{b}}$ as the projection vector of $\mathbf{b}$ into the column space, the error vector in Eq. 5.36, $\Delta\mathbf{b}$, satisfies

$$\begin{pmatrix} 3 & -6 & -3 \\ -3 & 6 & 3 \end{pmatrix}^{\mathrm{T}} \Delta\mathbf{b} = A^{\mathrm{T}} \Delta\mathbf{b} = \mathbf{0} \tag{5.40}$$

because of the orthogonality of the error vector to the column space. Introducing the linear equation for the approximate solution,

$$A\hat{\mathbf{x}} = \hat{\mathbf{b}}, \tag{5.41}$$

the conditions imposed for the error vector can be rewritten as

$$A^{\mathrm{T}}(A\hat{\mathbf{x}} - \mathbf{b}) = \mathbf{0}. \tag{5.42}$$

Consequently,

$$A^{\mathrm{T}} A\hat{\mathbf{x}} = A^{\mathrm{T}}\mathbf{b} \tag{5.43}$$

which provides the formal solution

$$\hat{\mathbf{x}} = (A^{\mathrm{T}} A)^{-1} A^{\mathrm{T}}\mathbf{b}. \tag{5.44}$$

For the example above,

$$\begin{pmatrix} 3 & -6 & -3 \\ -3 & 6 & 3 \end{pmatrix}^{\mathrm{T}} \begin{pmatrix} 3 & -6 & -3 \\ -3 & 6 & 3 \end{pmatrix} \hat{\mathbf{x}} = \begin{pmatrix} 3 & -6 & -3 \\ -3 & 6 & 3 \end{pmatrix}^{\mathrm{T}} \begin{pmatrix} -6 \\ 10 \end{pmatrix} \tag{5.45}$$

is the linear equation for the least-squares error solution.

The linear equation above expands to give

$$-3\hat{x}_1 + 6\hat{x}_2 + 3\hat{x}_3 = 8. \tag{5.46}$$

The solution for this equation is not uniquely determined; however, the minimum-norm solution is determined. The solution takes the explicit form

$$\hat{\mathbf{x}} = \begin{pmatrix} -8/3 + 2\hat{x}_2 + \hat{x}_3 \\ \hat{x}_2 \\ \hat{x}_3 \end{pmatrix}, \tag{5.47}$$

where

$$\hat{\mathbf{x}}_0 = \begin{pmatrix} 2\hat{x}_2 + \hat{x}_3 \\ \hat{x}_2 \\ \hat{x}_3 \end{pmatrix} \tag{5.48}$$

is a solution for

$$-3\hat{x}_1 + 6\hat{x}_2 + 3\hat{x}_3 = 0. \tag{5.49}$$

Setting

$$\hat{\mathbf{x}}_{\hat{b}} = \begin{pmatrix} -8/3 + 2 + 1 \\ 1 \\ 1 \end{pmatrix} \tag{5.50}$$

then the minimum-norm solution can be obtained using the orthogonal projection to the row vector

$$\mathbf{u}_1 = \begin{pmatrix} -3 & 6 & 3 \end{pmatrix}^{\mathrm{T}}. \tag{5.51}$$

The component vector $\hat{\mathbf{x}}_{\hat{b}r}$ becomes

$$\begin{aligned} \hat{\mathbf{x}}_{\hat{b}r} &= \\ &\frac{1}{(-3)^2 + 6^2 + 3^2} \begin{pmatrix} -8/3 + 3 \\ 1 \\ 1 \end{pmatrix}^{\mathrm{T}} \begin{pmatrix} -3 \\ 6 \\ 3 \end{pmatrix} \begin{pmatrix} -3 \\ 6 \\ 3 \end{pmatrix} \\ &= \frac{8}{54} \begin{pmatrix} -3 \\ 6 \\ 3 \end{pmatrix} \end{aligned} \tag{5.52}$$

$$= \frac{-4}{9}\begin{pmatrix} 1 \\ -2 \\ -1 \end{pmatrix}.$$

The vector $\hat{\mathbf{x}}_{\hat{b}r}$ gives the minimum-norm solution for

$$-3\hat{x}_{br_1} + 6\hat{x}_{br_2} + 3\hat{x}_{br_3} = 8. \tag{5.53}$$

5.2 Linear Regression

5.2.1 Least-Squares Error Solution

The least-squares error solution gives the linear regression line between two variables x and y. The cross-correlation coefficients indicate how closely a pair of random variables X and Y are linearly related, that is,

$$\mathrm{Y} = a\mathrm{X} + b. \tag{5.54}$$

As the absolute of the cross-correlation coefficient approaches unity, the linearity between the two variables holds better.

Suppose that this linear relationship is expected in pairs of random variables within some error under the least-squares error criterion. A linear equation

$$A\mathbf{x} = \mathbf{b} \tag{5.55}$$

produces the solution provided the vector **b** lies in the column space spanned by the column vectors of the matrix A. Alternatively, the least-squares error solution is obtained, if the vector **b** is not located in the column space. As an example consider the matrix

$$A = \begin{pmatrix} \mathbf{v}_1 & \mathbf{v}_2 \end{pmatrix}. \tag{5.56}$$

As described in the previous subsection, the approximate solution satisfies the linear equation

$$A\hat{\mathbf{x}} = \hat{\mathbf{b}} \tag{5.57}$$

instead of

$$A\mathbf{x} = \mathbf{b}. \tag{5.58}$$

The error vector

$$\Delta\mathbf{b} = \hat{\mathbf{b}} - \mathbf{b} \tag{5.59}$$

satisfies the relation

$$\begin{pmatrix} \mathbf{v}_1^{\mathrm{T}} \\ \mathbf{v}_2^{\mathrm{T}} \end{pmatrix} \Delta\mathbf{b} = \begin{pmatrix} \mathbf{v}_1^{\mathrm{T}} \\ \mathbf{v}_2^{\mathrm{T}} \end{pmatrix} (\hat{\mathbf{b}} - \mathbf{b}) = \mathbf{0}. \tag{5.60}$$

This equation can be rearranged to give

$$\begin{pmatrix} \mathbf{v}_1^{\mathrm{T}} \\ \mathbf{v}_2^{\mathrm{T}} \end{pmatrix} \hat{\mathbf{b}} = \begin{pmatrix} \mathbf{v}_1^{\mathrm{T}} \\ \mathbf{v}_2^{\mathrm{T}} \end{pmatrix} \mathbf{b} \tag{5.61}$$

$$A^{\mathrm{T}}\hat{\mathbf{b}} = A^{\mathrm{T}}\mathbf{b}, \tag{5.62}$$

and consequently

$$A^{\mathrm{T}}A\hat{\mathbf{x}} = A^{\mathrm{T}}\mathbf{b} \tag{5.63}$$

$$\hat{\mathbf{x}} = (A^{\mathrm{T}}A)^{-1}A^{\mathrm{T}}\mathbf{b} \tag{5.64}$$

the last identity yielding the least-squares error solution.

As a concrete example of a least-squares error solution of $A\mathbf{x} = \mathbf{b}$, let

$$A = \begin{pmatrix} 1 & -2 \\ -3 & 1 \\ -1 & 7 \end{pmatrix} \tag{5.65}$$

$$\mathbf{b} = \begin{pmatrix} -1 \\ -7 \\ 13 \end{pmatrix} \tag{5.66}$$

$$A^{\mathrm{T}}A = \begin{pmatrix} 1 & -2 \\ -3 & 1 \\ -1 & 7 \end{pmatrix}^{\mathrm{T}} \begin{pmatrix} 1 & -2 \\ -3 & 1 \\ -1 & 7 \end{pmatrix} = \begin{pmatrix} 11 & -12 \\ -12 & 54 \end{pmatrix} \tag{5.67}$$

$$A^{\mathrm{T}}\mathbf{b} = \begin{pmatrix} 1 & -3 & -1 \\ -2 & 1 & 7 \end{pmatrix} \begin{pmatrix} -1 \\ -7 \\ 13 \end{pmatrix} = \begin{pmatrix} 7 \\ 86 \end{pmatrix}. \tag{5.68}$$

The least-squares error solution $\hat{\mathbf{x}}$ for

$$A\mathbf{x} = \mathbf{b} \tag{5.69}$$

satisfies the equations

$$A\hat{\mathbf{x}} = \hat{\mathbf{b}} \tag{5.70}$$

$$A^{\mathrm{T}}A\hat{\mathbf{x}} = A^{\mathrm{T}}\mathbf{b}. \tag{5.71}$$

The solution is

$$\hat{\mathbf{x}} = \frac{1}{45}\begin{pmatrix} 141 \\ 103 \end{pmatrix} \tag{5.72}$$

where

$$A\hat{\mathbf{x}} = \hat{\mathbf{b}} = \frac{1}{9}\begin{pmatrix} -13 \\ -64 \\ 116 \end{pmatrix} \tag{5.73}$$

$$A^{\mathrm{T}}\hat{\mathbf{b}} = A^{\mathrm{T}}\mathbf{b} = \begin{pmatrix} 7 \\ 86 \end{pmatrix}. \tag{5.74}$$

The least-squares error solution can be applied in a linear regression analysis.

5.2.2 *Example of Linear Regression Analysis*

The regression line

$$y = ax + b \tag{5.75}$$

approximating the observation data as a linear function of x can be determined provided the sum of the squares of the errors are minimized [4, 5]. Figure 5.8 shows a scheme for the determination of this linear regression line [5].

As shown in the figure, the error is defined by the distance between each datum and the expected linear function y. The linear regression line can be formulated as the least-squares error solution, here denoted as $\mathbf{c}$, for the set of linear equations

$$\begin{aligned} a\xi_1 + b &= \eta_1 \\ &\cdots\cdots\cdots\cdots \\ a\xi_N + b &= \eta_N \end{aligned} \tag{5.76}$$

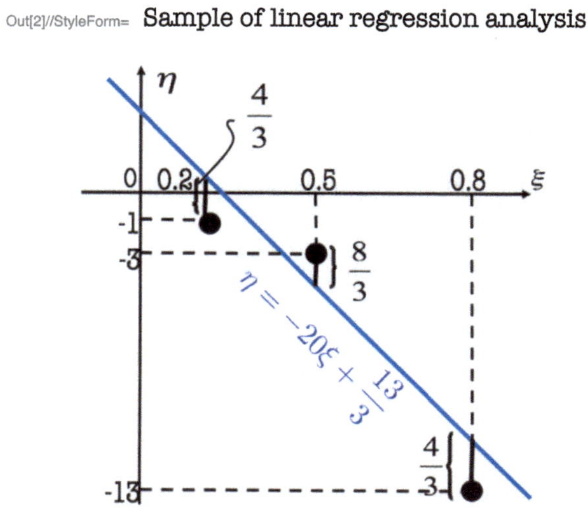

Fig. 5.8 Schematic of linear regression analysis from Fig. 7 in [5]

or in matrix form

$$\begin{pmatrix} \xi_1 & 1 \\ \vdots & \\ \xi_N & 1 \end{pmatrix} \begin{pmatrix} a \\ b \end{pmatrix} = \begin{pmatrix} \eta_1 \\ \vdots \\ \eta_N \end{pmatrix} \tag{5.77}$$

$$A\mathbf{c} = \eta \tag{5.78}$$

for $N > 2$. The solution vector is given by

$$\mathbf{c} = (A^{\mathrm{T}}A)^{-1}A^{\mathrm{T}}\eta. \tag{5.79}$$

5.3 Orthogonal Regression Analysis

5.3.1 Orthogonal Decomposition and Regression

The binaural perception could be interpreted by displaying the pair of binaural signals. When a pair of random signals is orthogonal and independent, the sound image is subjectively diffuse, provided that the signals are not perceived differently by monaural listening [6–9]. The scattering plots provide a way to decompose the sound pair into in-phase and out-of-phase (or orthogonal) pairs. A way of decomposing the pair of signals is orthogonal decomposition. Figure 5.9 shows a schematic of orthogonal decomposition.

Suppose the ith observation data pair is a vector of component length r_i so that [2–4]

$$\begin{aligned} \sum_{i=1}^{N} r_i^2 &= \sum_{i=1}^{N} (l_i^2 + h_i^2) \\ &= L + H = K \end{aligned} \tag{5.80}$$

as depicted in Fig. 5.9. The orthogonal regression line minimizes H or equivalently maximizes L.

In Fig. 5.9, the set of observation data is expressed as a matrix:

$$B = \begin{pmatrix} x_0 & x_1 & x_2 & x_3 \\ y_0 & y_1 & y_2 & y_3 \end{pmatrix}^{\mathrm{T}} = \begin{pmatrix} \mathbf{b}_0 & \mathbf{b}_1 & \mathbf{b}_2 & \mathbf{b}_3 \end{pmatrix}^{\mathrm{T}}. \tag{5.81}$$

Introducing the projection of the unit vector $\mathbf{u}$ by B

$$\mathbf{j} = B\mathbf{u} = \begin{pmatrix} \mathbf{b}_0 & \mathbf{b}_1 & \mathbf{b}_2 & \mathbf{b}_3 \end{pmatrix}^{\mathrm{T}} \mathbf{u}, \tag{5.82}$$

the orthogonal regression line is given by the vector $\mathbf{u}$ that maximizes $|\mathbf{j}|^2$ [2].

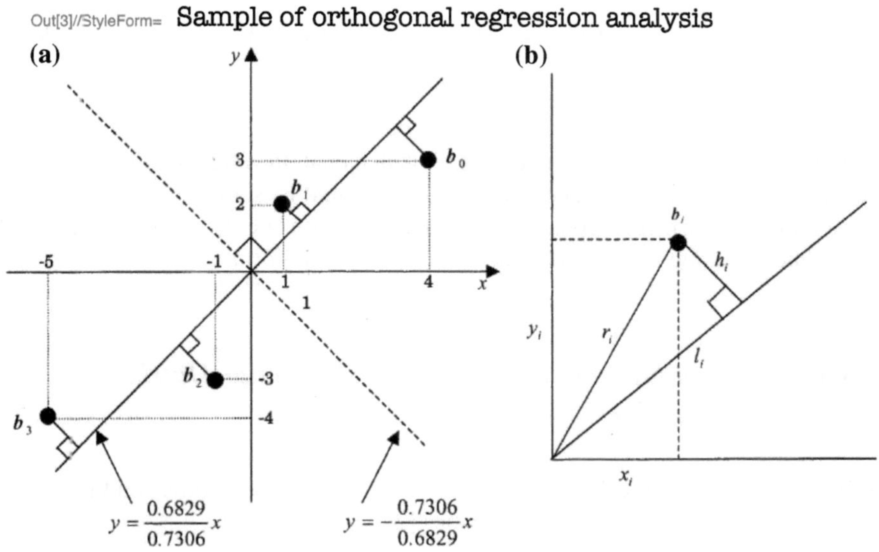

Fig. 5.9 Schematic of orthogonal decomposition from Fig. 11 in [5]

5.3.2 *Quadratic Form*

When a vector **c** is defined by $A\mathbf{x}$, the squared length of vector **c** is given by

$$|\mathbf{c}|^2 = \mathbf{c}^{\mathrm{T}}\mathbf{c} = \mathbf{x}^{\mathrm{T}}(A^{\mathrm{T}}A)\mathbf{x}, \tag{5.83}$$

where matrix $A^{\mathrm{T}}A$ is a symmetric matrix.

The formulation above is called the quadratic form of matrix A and gives a quadratic function. For example,

$$A^{\mathrm{T}}A = \begin{pmatrix} 2 & 1 \\ 1 & 5 \end{pmatrix} \tag{5.84}$$

$$\mathbf{x}^{\mathrm{T}}(A^{\mathrm{T}}A)\mathbf{x} = \begin{pmatrix} x_1 & x_2 \end{pmatrix} \begin{pmatrix} 2 & 1 \\ 1 & 5 \end{pmatrix} \begin{pmatrix} x_1 \\ x_2 \end{pmatrix} \tag{5.85}$$

$$= 2x_1^2 + 2x_1x_2 + 5x_2^2.$$

The quadratic form stated above contains a cross-product term.

As another example, consider

$$A^{\mathrm{T}}A = \begin{pmatrix} 2 & 0 \\ 0 & 5 \end{pmatrix} \tag{5.86}$$

$$\mathbf{x}^{\mathrm{T}}(A^{\mathrm{T}}A)\mathbf{x} = \begin{pmatrix} x_1 & x_2 \end{pmatrix} \begin{pmatrix} 2 & 0 \\ 0 & 5 \end{pmatrix} \begin{pmatrix} x_1 \\ x_2 \end{pmatrix} \tag{5.87}$$
$$= 2x_1^2 + 5x_2^2.$$

Note that when the symmetric matrix $A^{\mathrm{T}}A$ is diagonal, the quadratic from does not contain cross-product terms such as $x_1 x_2$. In general, if by some change of variables the symmetric matrix is diagonal, then the quadratic form does not have cross-product terms; the quadratic form is then said to be in canonical form.

5.3.3 Representation of a Matrix by Its Diagonal Matrix of Eigenvalues

Suppose a square matrix A has eigenvectors $(\mathbf{e_1}, \ldots, \mathbf{e_N})$ with corresponding eigenvalues $(\lambda_\mathbf{1}, \ldots, \lambda_\mathbf{N})$. The matrix A satisfies [2]

$$AE = A\begin{pmatrix} \mathbf{e_1} \cdots \mathbf{e_N} \end{pmatrix} \tag{5.88}$$
$$= \begin{pmatrix} \mathbf{e_1} \cdots \mathbf{e_N} \end{pmatrix} \begin{pmatrix} \lambda_1 & \ldots & 0 \\ & \ddots & \\ 0 & \ldots & \lambda_N \end{pmatrix}$$
$$= E\Lambda.$$

If the set of eigenvectors are linearly independent, then the matrix E is invertible and the equation above can be written as

$$A = E\Lambda E^{-1} \tag{5.89}$$

or equivalently

$$\Lambda = E^{-1}AE. \tag{5.90}$$

5.3.4 Quadratic Form and Orthogonal Eigenvectors

Take the quadratic form such as

$$|\mathbf{c}|^2 = \mathbf{x}^{\mathrm{T}}(A^{\mathrm{T}}A)\mathbf{x} = \mathbf{x}^{\mathrm{T}}C\mathbf{x}. \tag{5.91}$$

Substituting $C = E_c\Lambda_c E_c^{-1}$ for the quadratic form, it can be rewritten as

$$|\mathbf{c}|^2 = \mathbf{x}^{\mathrm{T}}C\mathbf{x} = \mathbf{x}^{\mathrm{T}}E_c\Lambda_c E_c^{-1}\mathbf{x}. \tag{5.92}$$

Recall that matrix C is a symmetric matrix and has orthogonal eigenvectors [2]. Then

$$E_c^{-1} = E_c^{\mathrm{T}}. \tag{5.93}$$

Hence, the quadratic form becomes

$$|\mathbf{c}|^2 = \mathbf{x}^{\mathrm{T}} E_c \Lambda_c E_c^{-1} \mathbf{x} = \mathbf{x}^{\mathrm{T}} E_c \Lambda_c E_c^{\mathrm{T}} \mathbf{x} = \left(E_c^{\mathrm{T}} \mathbf{x}\right)^{\mathrm{T}} \Lambda_c E_c^{\mathrm{T}} \mathbf{x}. \tag{5.94}$$

Introducing new vectors

$$\mathbf{y} = E_c^{\mathrm{T}} \mathbf{x}, \tag{5.95}$$

Equation 5.91 can be written as with a diagonal quadratic form

$$|\mathbf{c}|^2 = \mathbf{y}^{\mathrm{T}} \Lambda_c \mathbf{y} \tag{5.96}$$

as Λ_c is the diagonal matrix. This quadratic form produces no cross-product terms from the components of the new vector $\mathbf{y}$.

As an example of a quadratic form, consider

$$\begin{pmatrix} x_1 & x_2 \end{pmatrix} \begin{pmatrix} 2 & -1 \\ -1 & 2 \end{pmatrix} \begin{pmatrix} x_1 \\ x_2 \end{pmatrix} = 2x_1^2 - 2x_1x_2 + 2x_2^2. \tag{5.97}$$

The eigenvalues of the matrix

$$C = \begin{pmatrix} 2 & -1 \\ -1 & 2 \end{pmatrix} \tag{5.98}$$

are

$$\lambda_1 = 3, \quad \lambda_2 = 1 \tag{5.99}$$

each of which is associated with an eigenvector

$$\mathbf{e}_1 = \frac{1}{\sqrt{2}} \begin{pmatrix} 1 \\ -1 \end{pmatrix} \tag{5.100}$$

$$\mathbf{e}_2 = \frac{1}{\sqrt{2}} \begin{pmatrix} 1 \\ 1 \end{pmatrix} \tag{5.101}$$

$$E = \frac{1}{\sqrt{2}} \begin{pmatrix} 1 & 1 \\ -1 & 1 \end{pmatrix}. \tag{5.102}$$

With a change in variables

$$\mathbf{x} = E\mathbf{y}, \tag{5.103}$$

the quadratic form can be rewritten as

$$\begin{aligned}|\mathbf{c}|^2 &= \mathbf{x}^{\mathrm{T}}A^{\mathrm{T}}A\mathbf{x} = \mathbf{x}^{\mathrm{T}}C\mathbf{x} \\ &= (E\mathbf{y})^{\mathrm{T}}E\Lambda E^{\mathrm{T}}E\mathbf{y} \\ &= \mathbf{y}^{\mathrm{T}}\Lambda\mathbf{y},\end{aligned} \tag{5.104}$$

where

$$\begin{aligned}\mathbf{y} &= E^{\mathrm{T}}\mathbf{x} \\ &= \frac{1}{\sqrt{2}}\begin{pmatrix}1 & -1\\ 1 & 1\end{pmatrix}\begin{pmatrix}x_1\\ x_2\end{pmatrix} \\ &= \frac{1}{\sqrt{2}}\begin{pmatrix}x_1 - x_2\\ x_1 + x_2\end{pmatrix} \\ &= \begin{pmatrix}y_1\\ y_2\end{pmatrix}\end{aligned} \tag{5.105}$$

$$\Lambda = \begin{pmatrix}3 & 0\\ 0 & 1\end{pmatrix}. \tag{5.106}$$

Consequently, the quadratic form can be rewritten as

$$\mathbf{y}^{\mathrm{T}}\Lambda\mathbf{y} = 3y_1^2 + y_2^2 = \lambda_1 y_1^2 + \lambda_2 y_2^2 \tag{5.107}$$

for which no cross-product terms appear. The quadratic form for matrix A becomes maximal for the eigenvector of the symmetric matrix $A^{\mathrm{T}}A$ corresponding to its maximum eigenvalue, provided the squared length of the eigenvector is unity, that is, the eigenvectors are normalized. For example, taking the vector $\mathbf{e}_1$, then

$$\begin{aligned}&\mathbf{e}_1^{\mathrm{T}}(A^{\mathrm{T}}A)\mathbf{e}_1 \\ &= \frac{1}{\sqrt{2}}\begin{pmatrix}1 & -1\end{pmatrix}\begin{pmatrix}2 & -1\\ -1 & 2\end{pmatrix}\frac{1}{\sqrt{2}}\begin{pmatrix}1\\ -1\end{pmatrix} \\ &= \frac{1}{2}\begin{pmatrix}3 & -3\end{pmatrix}\begin{pmatrix}1\\ -1\end{pmatrix} \\ &= 3 = \lambda_1.\end{aligned} \tag{5.108}$$

Similarly

$$\begin{aligned}&\mathbf{e}_2^{\mathrm{T}}(A^{\mathrm{T}}A)\mathbf{e}_2 \\ &= \frac{1}{\sqrt{2}}\begin{pmatrix}1 & 1\end{pmatrix}\begin{pmatrix}2 & -1\\ -1 & 2\end{pmatrix}\frac{1}{\sqrt{2}}\begin{pmatrix}1\\ 1\end{pmatrix} \\ &= \frac{1}{2}\begin{pmatrix}1 & 1\end{pmatrix}\begin{pmatrix}1\\ 1\end{pmatrix} \\ &= 1 = \lambda_2.\end{aligned} \tag{5.109}$$

The eigenvector is called the principal axis. Compare the quadratic form above with that of Eq. 5.97

$$\begin{aligned}\hat{q}(y_1, y_2) &= 3y_1^2 + y_2^2 \\ &= (3/2)(x_1 - x_2)^2 + (1/2)(x_1 + x_2)^2 \\ &= 2x_1^2 - 2x_1x_2 + 2x_2^2 \\ &= q(x_1, x_2).\end{aligned} \tag{5.110}$$

Orthogonal regression analysis is used to find the eigenvector that maximizes the quadratic form. The quadratic form can be maximized by taking the eigenvector as the regression line that corresponds to the maximum eigenvalue of the matrix $B^T B$, where the matrix B in the figure signifies the observation data matrix.

5.3.5 Binaural Merit and Principal Component Analysis

Binaural listening is sensitive to the orthogonal regression line as well as the linear regression. If two binaural signals are in-phase, a clearly sharp sound image is obtained. The sound becomes indistinct if the single orthogonal regression line does not represent the relationship between the pair well. In the limit case, consider the data set composed of four points represented as corners of a square, as shown in Fig. 5.10. The data can be combined to form a matrix:

$$B = \begin{pmatrix} a & -a & -a & a \\ a & a & -a & -a \end{pmatrix}^T. \tag{5.111}$$

The data points of the binaural signals are equally likely to be in-phase or anti-phase. An in-phase pair creates a sharp sound image, whereas an anti-phase pair yields

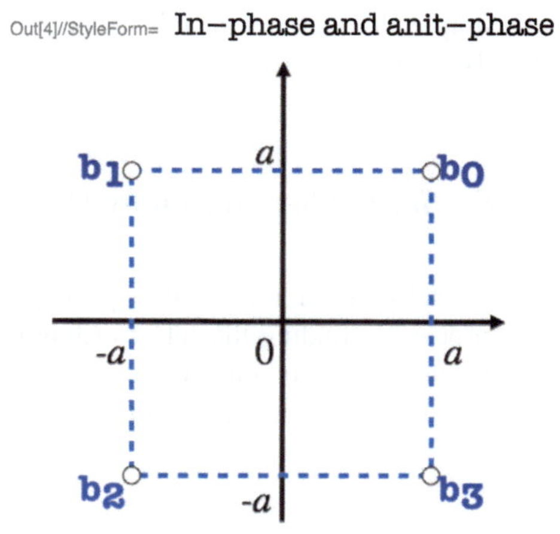

Fig. 5.10 Sample of in-phase and anti-phase data from Fig. 13 in [5]

no compact sound image. Hence, if the pairs assume their phase conditions randomly following Fig. 5.10, then the binaural sound image will not be focussed. Formally, this corresponds to an orthogonal regression line that is not uniquely determined, because eigenvectors are not determined for the matrix $B^{\mathrm{T}}B$; that is

$$B^{\mathrm{T}}B = \begin{pmatrix} a & -a & -a & a \\ a & a & -a & -a \end{pmatrix} \begin{pmatrix} a & a \\ -a & a \\ -a & -a \\ a & -a \end{pmatrix} \tag{5.112}$$
$$= \begin{pmatrix} 4a^2 & 0 \\ 0 & 4a^2 \end{pmatrix}.$$

In contrast to the orthogonal regression line, the other direction orthogonal to the regression line may in certain circumstances prove more informative. Principal component analysis (PCA) is performed for that purpose. The decomposition of mixed sound signals into sound sources is an interesting example of principal component analysis [4, 10, 11]. Consider a pair of mixed signals such that

$$\begin{aligned} x(t) &= s_1(t) + \alpha k s_2(t) \\ y(t) &= \beta s_1(t) + k s_2(t), \end{aligned} \tag{5.113}$$

where α and β denote the mixing ratios and k corresponds to the magnitude ratio of the sound sources $s_1(t)$ and $s_2(t)$. Assuming the pair of mixed signals as a binaural pair, the decomposition of the mixed signal pair into independent source signals would be an example of the binaural merit. The principal component analysis implies in an ideal case the possibility of the decomposition in the time domain, if the time delay is known between the binaural signal pairs.

Figure 5.11 exemplifies the separation of sound sources from a pair of mixed signals in which a speech signal is buried within random noise. The speech sample can be extracted successfully using principal component analysis where $\mathbf{u}_{\mathbf{max}}$ and $\mathbf{u}_{\mathbf{min}}$ denote the eigenvectors corresponding to the larger and smaller eigenvalues. In a simple although not ideal instance from sound recording, the time delay of two signals received by two microphones makes separation complex and difficult.

5.3.6 Source Separation by Decorrelation of Pair of Signals

Orthogonal regression as well as principal component analysis can be interpreted as a process in finding the set of orthogonal eigenvectors and their corresponding eigenvalues. Orthogonality corresponds to two signals being out-of-phase. Consider a mixed signal pair, as shown in Fig. 5.12.

Out[5]//StyleForm= **Separation of signals by PCA**

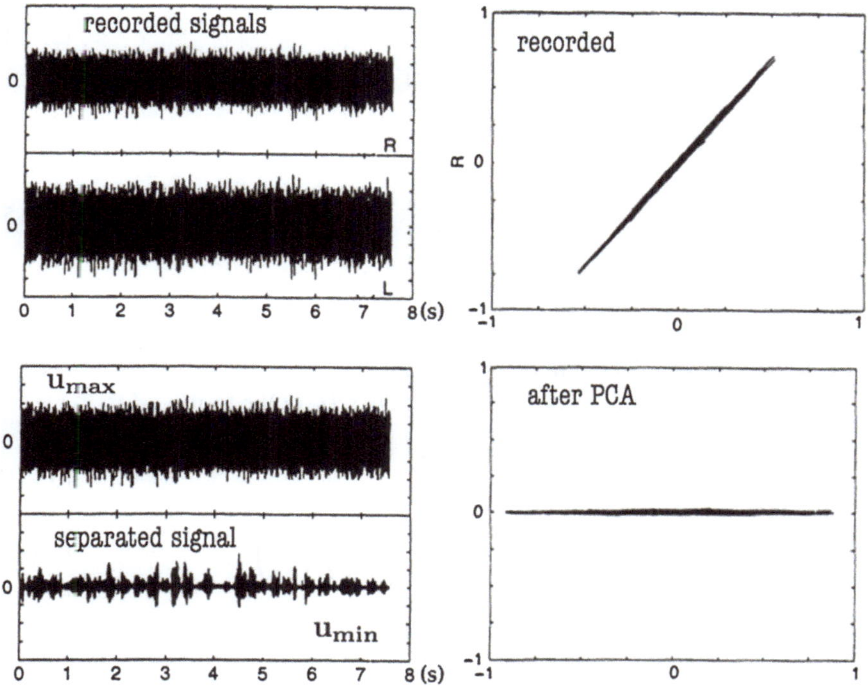

Fig. 5.11 Example of the separation of speech signal buried in random noise with settings $\alpha = 0.96$, $\beta = 0.83$, and $k = 3.6$ from Fig. 10 in [11]

Out[6]//StyleForm= **Recording condition in anechoic room**

Fig. 5.12 Sound reception for a pair of receivers

The pair of received signals can be formally expressed as [12]

$$\begin{aligned} x(t) &= s_1(t) + \alpha s_2(t) \\ y(t) &= \beta s_1(t) + s_2(t) \end{aligned} \tag{5.114}$$

or

$$\begin{aligned} \begin{pmatrix} x \\ y \end{pmatrix} &= \begin{pmatrix} 1 & \alpha \\ \beta & 1 \end{pmatrix} \begin{pmatrix} s_1 \\ s_2 \end{pmatrix} \\ &= \begin{pmatrix} s_1 + \alpha s_2 \\ \beta s_1 + s_2 \end{pmatrix} \end{aligned} \tag{5.115}$$

where a time delay can be introduced to compensate for differences in distance to the two receivers. Consider a decorrelation matrix D_c of the forms

$$D_c = \begin{pmatrix} 1 & a \\ b & 1 \end{pmatrix} \tag{5.116}$$

$$\begin{aligned} D_c \begin{pmatrix} x \\ y \end{pmatrix} &= \begin{pmatrix} 1 & a \\ b & 1 \end{pmatrix} \begin{pmatrix} 1 & \alpha \\ \beta & 1 \end{pmatrix} \begin{pmatrix} s_1 \\ s_2 \end{pmatrix} \\ &= A \begin{pmatrix} s_1 \\ s_2 \end{pmatrix}. \end{aligned} \tag{5.117}$$

The decorrelation matrix D_c is determined by

$$D_c = \begin{pmatrix} 1 & -\alpha \\ -\beta & 1 \end{pmatrix} \tag{5.118}$$

which establishes the property

$$\begin{aligned} D_c \begin{pmatrix} 1 & \alpha \\ \beta & 1 \end{pmatrix} &= \begin{pmatrix} 1 & -\alpha \\ -\beta & 1 \end{pmatrix} \begin{pmatrix} 1 & \alpha \\ \beta & 1 \end{pmatrix} \\ &= \begin{pmatrix} 1 - \alpha\beta & 0 \\ 0 & 1 - \alpha\beta \end{pmatrix} \end{aligned} \tag{5.119}$$

from which the scaling factor A in Eq. 5.116 is found to be $1 - \alpha\beta$. Given the assumption that the source signals s_1 (random variable X) and s_2 (Y) are uncorrelated, the decorrelation matrix can be estimated even under conditions when the mixing ratios α and β are unknown. After estimating the decorrelation matrix, the separated signals $\hat{s}_1$ and $\hat{s}_2$ can be also expected to be uncorrelated. The decorrelation matrix may be determined so that

$$\begin{aligned} &\mathrm{E}[(\mathrm{X} + a\mathrm{Y})(b\mathrm{X} + \mathrm{Y})] \\ &= \mathrm{E}[b\mathrm{X}^2 + (1 + ab)\mathrm{XY} + a\mathrm{Y}^2] \end{aligned} \tag{5.120}$$

$$= b + (1 + ab)r_{XY} + a$$
$$= 0$$

by setting

$$E[X^2] = E[Y^2] = 1 \tag{5.121}$$
$$E[XY] = r_{XY}. \tag{5.122}$$

The equation above can be rewritten in closed form for b such that

$$b = -\frac{a + r_{XY}}{1 + ar_{XY}}. \tag{5.123}$$

Assuming that the signals are recorded frame-by-frame in time and the variance in r_{XY} can be obtained frame-by-frame, then the pair of variables a and b can be determined for every pair of successive frames from the cross-over point of two curves [12, 13],

$$b_i = -\frac{a + r_{XY}(i)}{1 + ar_{XY}(i)} \tag{5.124}$$
$$b_j = -\frac{a + r_{XY}(j)}{1 + ar_{XY}(j)},$$

where $r_{XY}(i)$ indicates the observation record for $r_{XY}(i)$ at the ith time frame. If the mixing ratios are constant, then the estimation of a and b converge

$$a = -\alpha \tag{5.125}$$
$$b = -\beta. \tag{5.126}$$

Figure 5.13 shows the result of separated waveforms.

The separation is possible under ideally simplified conditions for which a compensation has been given for the time delay [12]. If the time delay is left uncompensated, that is, the recorded signals may be already uncorrelated, then b becomes

$$b = -\frac{a + r_{XY}}{1 + ar_{XY}} = -a. \tag{5.127}$$

The source signals are no longer separated because the extracted signals are

$$s_1 = x + ay \tag{5.128}$$
$$s_2 = -ax + y \tag{5.129}$$
$$E[S_1S_2] = -aE[X^2] + aE[Y^2] = 0 \tag{5.130}$$

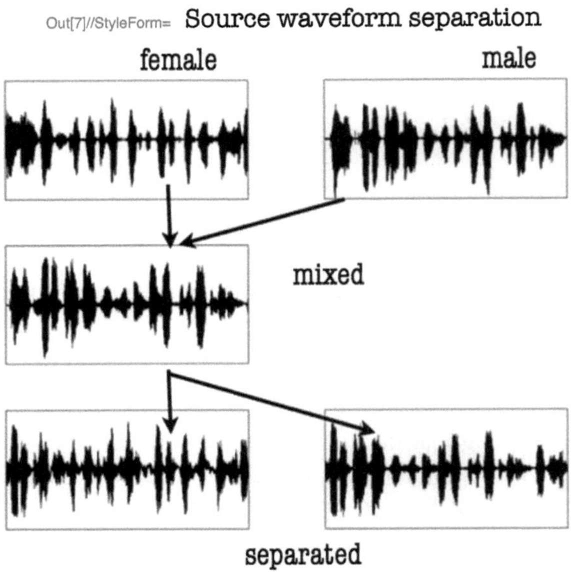

Fig. 5.13 Recorded and separated waveforms from Fig. 4 [12]

providing

$$E[X^2] = E[Y^2]. \tag{5.131}$$

5.3.7 *Decorrelation of Signals and Estimation of Impulse Response*

Consider a two-channel sound reproducing system schematically depicted in Fig. 5.14 where h_{LR} denotes the impulse response from the left loudspeaker to the right receiver. It might be desirable for sound field control to get an impulse response using a pair of normal-stereo-recorded sound samples in the situations. The signal received by the left receiver can be expressed as

$$\begin{aligned} y_L(t) &= x_L * h_{LL}(t) + x_R * h_{RL}(t) \\ &= x_{LL}(t) + x_{RL}(t), \end{aligned} \tag{5.132}$$

where x_L and x_R produce a pair of stereo-signals. Figure 5.15 illustrates an example of a two-channel plot for a stereo-recorded pair of signals.

The signals are highly correlated. If the pair of signals x_{LL} and x_{RL} were uncorrelated, the cross-correlation function is expected to be

$$r_{Y_L X_L} \cong r_{X_{LL} X_L}. \tag{5.133}$$

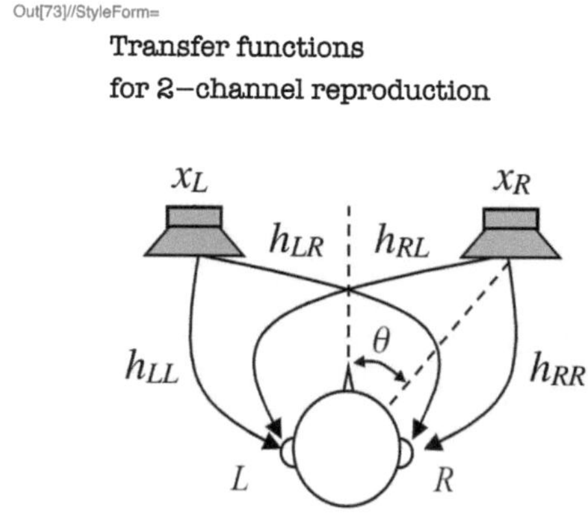

Fig. 5.14 Two-channel stereo reproduction

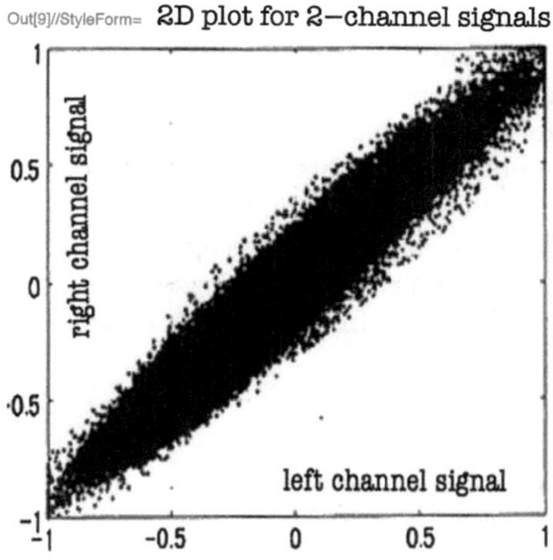

Fig. 5.15 2-dimensional plot for pair of stereo-recorded signals from Fig. 2a [14]

The Fourier transformation defining the cross-spectral density function is

$$P_{Y_L X_L}(\omega) \cong \mathrm{E}\left[\frac{X_{LL}(\omega)X_L^*(\omega)}{T}\right]. \tag{5.134}$$

Introducing the formal relation

$$X_{LL}(\omega) = X_L(\omega)H_{LL}(\omega), \tag{5.135}$$

the cross-spectral density function can be written as

$$P_{Y_L X_L}(\omega) = \mathrm{E}\left[\frac{|X_L(\omega)|^2 H_{LL}(\omega)}{T}\right] = H_{LL}(\omega)\mathrm{E}\left[\frac{|X_L(\omega)|^2}{T}\right]. \quad (5.136)$$

The spectral characteristic function of the impulse response can be estimated from the cross-spectral density function provided that

$$P_{X_L X_L}(\omega) = \mathrm{E}\left[\frac{|X_L(\omega)|^2}{T}\right] \quad (5.137)$$

is known.

Figure 5.16 shows the two-dimensional plot of the two-channel signals after principal component analysis. As shown in Fig. 5.15, the pair of signals is highly correlated, and thus it might be difficult to estimate the impulse responses between the source and receivers. The principal component analysis decomposes the signals into an orthogonalized pair. Figure 5.16 shows components that look like they have been orthogonalized by the principal component analysis. Figure 5.17 presents an example of the estimation of the impulse responses. The responses are well estimated using the orthogonalized signal pairs obtained from principal component analysis.

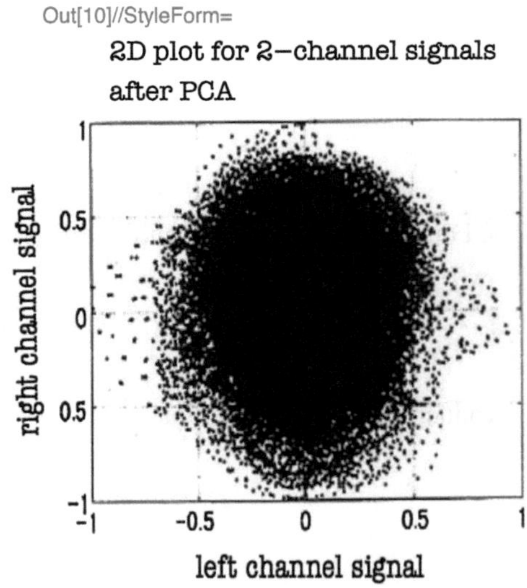

Fig. 5.16 Two-dimensional plot for a pair of stereo-recorded signals after orthogonalization using principal component analysis from Fig. 2b [14]

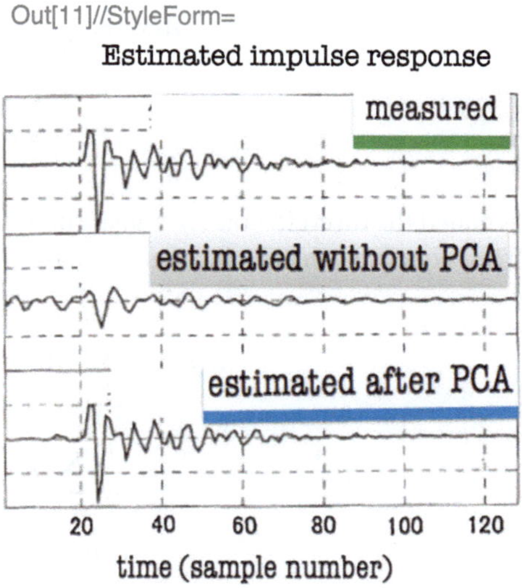

Fig. 5.17 Example of the estimation of impulse response after principal component analysis from Figs. 4 [14]

5.4 Discrete Fourier Transform and Spectral Representation of Sequences

5.4.1 Fourier and Discrete Fourier Transforms of a Sequence

Suppose a discrete sequence $a(n)$ of finite length with $n = 0, 1, \ldots, N-1$. The generating function (introduced in Subsection 3.1.1) of $a(n)$ can be defined as

$$A(X) = \sum_{n=0}^{N-1} a(n)X^n \tag{5.138}$$

using a formal variable X. Substituting

$$X = \mathrm{e}^{-\mathrm{i}\Omega}, \tag{5.139}$$

the Fourier transform of the sequence $a(n)$ is defined as

$$A(\mathrm{e}^{-\mathrm{i}\Omega}) = \sum_{n=0}^{N-1} a(n)\mathrm{e}^{-\mathrm{i}\Omega n}, \tag{5.140}$$

as developed in Subsection 3.1.2.

The Fourier transform of the sequence is a continuous and periodic function of Ω. The sequence $a(n)$ is generated by sampling a continuous function $a(t)$ at $t = nT_s$ where T_s is the sampling interval or period. The sequence $A(k)$ sampled from $A\mathrm{e}^{-\mathrm{i}\Omega}$ such as

$$
\begin{aligned}
A(k) &= \frac{1}{N} A(\mathrm{e}^{-\mathrm{i}\Omega})|_{\Omega=2\pi k/N} \\
&= \frac{1}{N} \sum_{n=0}^{N-1} a(n) \mathrm{e}^{-\mathrm{i}\frac{2\pi}{N}kn}
\end{aligned}
\tag{5.141}
$$

is called the discrete Fourier transform of the sequence $a(n)$ where $0 \le k \le N-1$. The sequence $a(n)$ can be expressed as

$$
\begin{aligned}
\hat{a}(n) &= \sum_{k=0}^{N-1} A(k) \mathrm{e}^{\mathrm{i}\frac{2\pi}{N}kn} \\
&= \frac{1}{N} \sum_{m=0}^{N-1} a(m) \sum_{k=0}^{N-1} \mathrm{e}^{\mathrm{i}\frac{2\pi}{N}k(n-m)} \\
&= a(n)
\end{aligned}
\tag{5.142}
$$

for $0 \le n \le N-1$, where

$$
\begin{aligned}
&\frac{1}{N} \sum_{k=0}^{N-1} \mathrm{e}^{\mathrm{i}\frac{2\pi}{N}k(n-m)} \\
&= \begin{cases} 1 & n = m \\ \frac{1-\mathrm{e}^{\mathrm{i}2\pi(n-m)}}{1-\mathrm{e}^{\mathrm{i}\frac{2\pi}{N}(n-m)}} = 0 & n \ne m \end{cases}
\end{aligned}
\tag{5.143}
$$

that shows the orthogonality of $\mathrm{e}^{\mathrm{i}\frac{2\pi}{N}kn}$. The formulation above for $\hat{a}(n)$ holds even when $n > N-1$; however, $\hat{a}(n)$ is only the periodic sequence of n outside $0 \le n \le N-1$. This is confirmed by noting

$$
\begin{aligned}
\hat{a}(n+pN) &= \sum_{k=0}^{N-1} A(k) \mathrm{e}^{\mathrm{i}\frac{2\pi}{N}k(n+pN)} \\
&= \sum_{k=0}^{N-1} A(k) \mathrm{e}^{\mathrm{i}\frac{2\pi}{N}kn} \mathrm{e}^{\mathrm{i}\frac{2\pi}{N}kpN} \\
&= \sum_{k=0}^{N-1} A(k) \mathrm{e}^{\mathrm{i}\frac{2\pi}{N}kn} \\
&= a(n)
\end{aligned}
\tag{5.144}
$$

where p is an integer. The signal representation in the time domain given by the discrete Fourier transform does not provide estimates of the signal outside the observation interval $0 \le N \le N-1$ unless the signal is periodic with period N. The discrete Fourier transformation provides a representation of the periodic sequence in period N, that is, the length of the interval of the observed sequence. This periodic property is identical to that for a signal representation using a Fourier series expansion.

Indeed, not only $\hat{a}(n)$ but also $A(k)$ is periodic so that

$$\begin{aligned} A(k+pN) &= \frac{1}{N}\sum_{n=0}^{N-1} a(n)\mathrm{e}^{-\mathrm{i}\frac{2\pi}{N}(k+pN)n} \\ &= \frac{1}{N}\sum_{n=0}^{N-1} a(n)\mathrm{e}^{-\mathrm{i}\frac{2\pi}{N}kn}\mathrm{e}^{-\mathrm{i}\frac{2\pi}{N}pNn} \\ &= \frac{1}{N}\sum_{n=0}^{N-1} a(n)\mathrm{e}^{-\mathrm{i}\frac{2\pi}{N}kn} \\ &= A(k) \end{aligned} \tag{5.145}$$

where p is an integer. The Fourier series expansion of a continuous function and the discrete Fourier transform of a sequence generate a periodic signal and sequence, respectively, and yield discrete spectral sequences. Although the discrete Fourier transform preserves periodicity, the period is the length N different from the original one in general.

5.4.2 *Examples of Spectral Sequences by Discrete Fourier Transform*

Figure 5.18 shows examples of the time waveforms of sinusoidal functions and their spectral sequences. The horizontal axis in the upper panel represents time but in the lower panel represents frequency indicated by the number k of the spectral sequences. The inverse Fourier transforms for a continuous function and a discrete sequence are written as

$$\hat{x}(t) = \frac{1}{2\pi}\int_{-\infty}^{\infty} X(\omega)\mathrm{e}^{\mathrm{i}\omega t}d\omega \tag{5.146}$$

$$\hat{x}(n) = \frac{1}{2\pi}\int_{0}^{2\pi} X\left(\mathrm{e}^{-\mathrm{i}\Omega}\right)\mathrm{e}^{\mathrm{i}\Omega n}d\Omega, \tag{5.147}$$

where $\hat{x}(t)$ is a continuous function representing the continuous spectral function $X(\omega)$, whereas $\hat{x}(n)$ is a sequence obtained using a continuous but periodic spectral function, $X(\mathrm{e}^{-\mathrm{i}\Omega})$, of $\Omega = \omega T_s$, with T_s denoting the sampling period. The Fourier transform is based on the complex exponential functions with positive and negative frequencies that combine to give

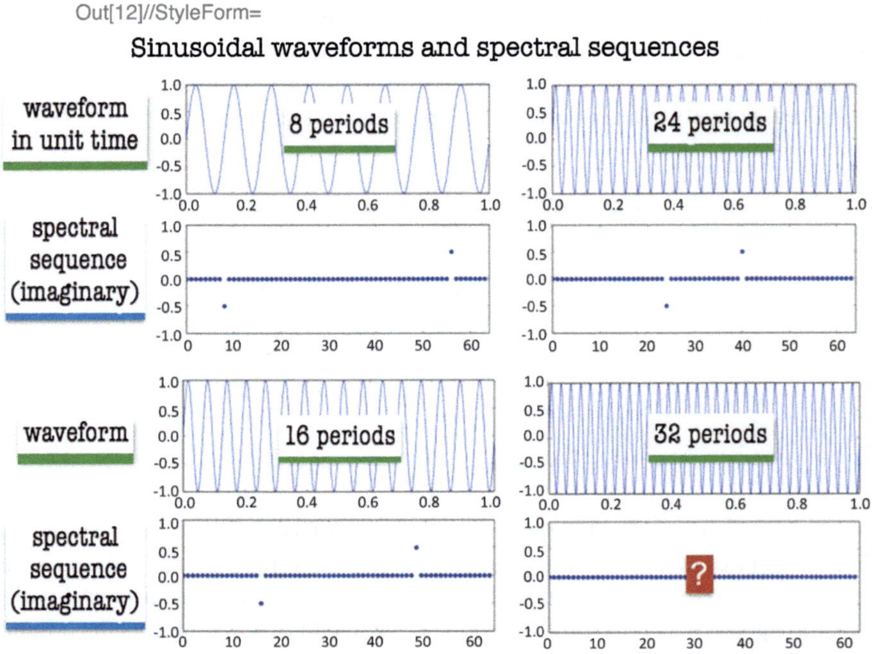

Fig. 5.18 Samples of spectral sequences and sinusoidal waveforms

$$\cos \omega t = \frac{1}{2}\left(\mathrm{e}^{\mathrm{i}\omega t} + \mathrm{e}^{-\mathrm{i}\omega t}\right). \tag{5.148}$$

The negative frequency region, however, might not be explicitly seen in the formulation for $\hat{x}(n)$. This explains the periodicity of $X(\mathrm{e}^{-\mathrm{i}\Omega})$ with period 2π.

A periodic sequence $\hat{x}(n)$ can be also expressed as

$$\hat{x}(n) = \sum_{k=0}^{N-1} X(k)\mathrm{e}^{\mathrm{i}\frac{2\pi}{N}kn}, \tag{5.149}$$

where $X(k)$ denotes the periodic spectral sequence of $x(n)$ of period N. Again no negative frequency numbers are explicitly seen because of this periodicity,

$$X(N-k) = X(-k). \tag{5.150}$$

This line-spectral periodicity is different from the Fourier series expansion of a periodic function. Figure 5.18 shows an instance where the sinusoidal frequency changes $k_0 = 8, 16, 24, 32$. Consider a sinusoidal sequence

$$x(n) = \sin\frac{2\pi}{N}k_0 n = \frac{1}{2\mathrm{i}}\left(\mathrm{e}^{\mathrm{i}\frac{2\pi}{N}k_0 n} - \mathrm{e}^{-\mathrm{i}\frac{2\pi}{N}k_0 n}\right); \tag{5.151}$$

then the spectral sequence $X(k)$ is given by

$$X(k) = \frac{1}{2i}[\delta(k-k_0) - \delta(k-(-k_0))]$$
$$= \frac{1}{2i}[\delta(k-k_0) - \delta(k-(N-k_0))]. \tag{5.152}$$

Taking $k_0 = 8$ with $N = 64$ where N is the number of samples in the observation interval of 1(s), the spectral sequence above becomes

$$X(k) = \frac{1}{2i}[\delta(k-8) - \delta(k-(-8))]$$
$$= \frac{1}{2i}[\delta(k-8) - \delta(k-(64-8))]$$
$$= \frac{1}{2i}[\delta(k-8) - \delta(k-56)], \tag{5.153}$$

which corresponds to Fig. 5.18 by taking the imaginary part.

In contrast, if $k_0 = 32$, Fig. 5.18 shows no responses because

$$X(k) = \frac{1}{2i}[\delta(k-32) - \delta(k-32)] = 0. \tag{5.154}$$

The reason why the spectral sequence for $k_0 = 32$ is lost is the condition $N = 64$. Assuming the observation interval is a single second, N denotes the sampling frequency and determines the sampling interval given by $1/N$. Figure 5.19 shows the same examples as those given in Fig. 5.18 but with a different sampling condition, $N = 128$.

In sampling the waveform at a sampling frequency of $N = 128$, the spectral sequence for $k_0 = 32$ is written as

$$X(k) = \frac{1}{2i}[\delta(k-32) - \delta(k-(-32))]$$
$$= \frac{1}{2i}[\delta(k-32) - \delta(k-(128-32))]$$
$$= \frac{1}{2i}[\delta(k-32) - \delta(k-96)]. \tag{5.155}$$

The result above states that the sampling frequency must be

$$N > 2k_0, \tag{5.156}$$

where k_0 is the frequency bin corresponding to the highest frequency f_0 to be analyzed. Alternatively, given N, the highest frequency to be analyzed must be

$$k_0 < N/2 \tag{5.157}$$

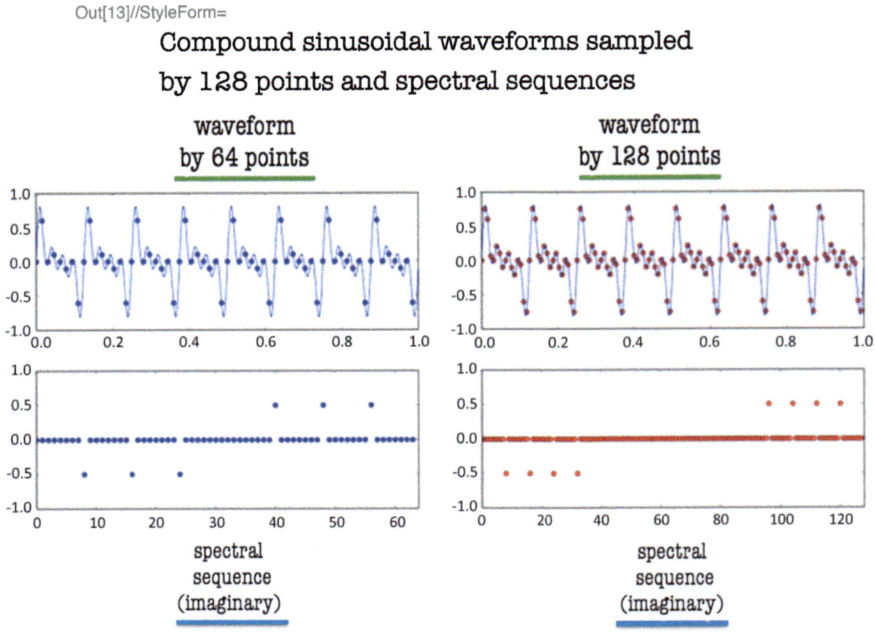

Fig. 5.19 Samples of spectral sequences and compound sinusoidal waveforms obtained by setting $N = 128$

so that both the positive and negative frequency components can be observed separately in the spectral sequence. The condition above can be interpreted as

$$F_s > 2f_0, \tag{5.158}$$

stated in terms of the sampling frequency, because

$$\Omega_k = \frac{2\pi}{N}k = 2\pi \tag{5.159}$$

if $k = N$. Setting

$$\Omega_N = 2\pi = \omega_N T_s \tag{5.160}$$

then

$$\omega_N = \frac{2\pi}{T_s} = 2\pi F_s \tag{5.161}$$

corresponds to the sampling angular frequency. The sampling frequency must satisfy condition Eq. 5.158 where f_0 denotes the highest frequency to be analyzed. Figure 5.19 shows the spectral sequences for $k_0 = 8, 16, 24, 32$ that can be clearly observed when $N = 128$. In this instance, the frequencies to be analyzed must be lower than 64.

5.4.3 *Violation of Sampling Condition*

The violation of the sampling condition makes spectral estimation impossible [4, 15–17]. Set $N = 64$ and $k_0 = 40$ as in the previous sinusoidal sequence. The spectral sequence $X(k)$ becomes

$$X(k) = \frac{1}{2i}\left[\delta(k-40) - \delta(k-(-40))\right] \tag{5.162}$$
$$= \frac{1}{2i}\left[\delta(k+64-40) - \delta(k-(64-40))\right]$$
$$= \frac{1}{2i}\left[\delta(k+24) - \delta(k-24)\right]$$

so that the sinusoidal sequence is

$$-\sin\frac{2\pi}{64}24n = \frac{-1}{2i}\left(e^{i\frac{2\pi}{64}24n} - e^{-i\frac{2\pi}{64}24n}\right) \tag{5.163}$$

with $N = 64$ and $k_0 = 24$ instead of $k_0 = 40$. There is no way to distinguish between

$$x(t) = \sin 2\pi \cdot 40t \tag{5.164}$$
$$y(t) = -\sin 2\pi \cdot 24t \tag{5.165}$$

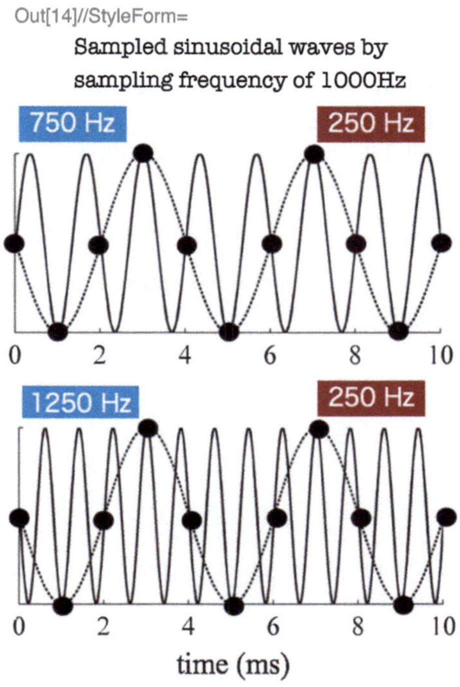

Fig. 5.20 Illustration of the violation of sampling condition from Fig. 21.6 [16]

where $nT_s = n/N = t$, because the sampling condition is violated. Figure 5.20 gives an illustration of the violation of sampling condition for another example [16, 17].

5.5 Exercises

1. Decompose the following vectors into pairs of linearly independent vectors. Also, express the decompositions in matrix form.

$$(1)\ \ (3\ 1)\ \ (2)\ \ (-3\ \ 1)$$
$$(3)\ \ (3\ \ -1)\ \ (4)\ \ (-3\ \ -1).$$

2. Decompose the above vectors into pairs of orthogonal vectors. Express the decompositions in matrix form.
3. Determine whether the following pairs of vectors are orthogonal:

$$\begin{array}{ll} (1) & (3\ 1)\ \ (-3\ \ 1) \\ (2) & (2\ 2)\ \ (-2\ \ 2) \\ (3) & (3\ 1)\ \ (-3\ \ -1) \\ (4) & (1\ 0)\ \ (0\ \ 1) \\ (5) & (-1\ 1)\ \ (-1\ \ -1) \\ (6) & (-1\ 1)\ \ (1\ \ -1). \end{array}$$

4. Confirm the linear regression line shown in Fig. 5.8.
5. Confirm the orthogonal regression line shown in Fig. 5.9.
6. Find the minimum-norm solution for the linear equation

$$3x + 2y + z = 3.$$

7. Obtain the one-point discrete Fourier transform ($N = 1$) for each of the following sequences:

$$\begin{array}{ll} (1) & x(0) = 1 \\ (2) & x(0) = -1 \\ (3) & x(0) = 2 \\ (4) & x(0) = -2. \end{array}$$

8. Obtain the two-point discrete Fourier transform ($N = 2$) for each of the following sequences:

$$
\begin{aligned}
&(1) \quad x(0) = 1, \quad x(1) = 1 \\
&(2) \quad x(0) = -1, \quad x(1) = 1 \\
&(3) \quad x(0) = 2, \quad x(1) = 1 \\
&(4) \quad x(0) = -2, \quad x(1) = 1.
\end{aligned}
$$

9. Obtain the four-point discrete Fourier transform ($N = 4$) for each of the following sequences:

$$
\begin{aligned}
&(1) \quad x(0) = 1, \quad x(1) = 2, \quad x(2) = 1, \quad x(3) = 2 \\
&(2) \quad x(0) = -1, \quad x(1) = -2, \quad x(2) = 1, \quad x(3) = -2.
\end{aligned}
$$

10. Obtain the eight-point discrete Fourier transform ($N = 8$) for each of the following sequences:

$$
\begin{aligned}
(1) \quad & x(0) = 1, \quad x(1) = 3, \quad x(2) = 2, \quad x(3) = 4 \\
& x(4) = 1, \quad x(5) = 4, \quad x(6) = 2, \quad x(7) = 3
\end{aligned}
$$

$$
\begin{aligned}
(2) \quad & x(0) = -1, \quad x(1) = 3, \quad x(2) = -2, \quad x(3) = 4 \\
& x(4) = 1, \quad x(5) = 4, \quad x(6) = -2, \quad x(7) = 3.
\end{aligned}
$$

11. Obtain the eight-point discrete Fourier transforms ($N = 8$) for each of the following sequences:

$$
\begin{aligned}
(1) \quad & x(0) = 1, \quad x(1) = 0, \quad x(2) = 2, \quad x(3) = 0 \\
& x(4) = 1, \quad x(5) = 0, \quad x(6) = 2, \quad x(7) = 0
\end{aligned}
$$

$$
\begin{aligned}
(2) \quad & x(0) = -1, \quad x(1) = 0, \quad x(2) = -2, \quad x(3) = 0 \\
& x(4) = 1, \quad x(5) = 0, \quad x(6) = -2, \quad x(7) = 0
\end{aligned}
$$

$$
\begin{aligned}
(3) \quad & x(0) = 1, \quad x(1) = 2, \quad x(2) = 1, \quad x(3) = 2 \\
& x(4) = 0, \quad x(5) = 0, \quad x(6) = 0, \quad x(7) = 0
\end{aligned}
$$

$$
\begin{aligned}
(4) \quad & x(0) = -1, \quad x(1) = -2, \quad x(2) = 1, \quad x(3) = -2 \\
& x(4) = 0, \quad x(5) = 0, \quad x(6) = 0, \quad x(7) = 0
\end{aligned}
$$

12. Check the orthogonality for the following functions defined over the interval $0 \le x \le 2\pi$.

$$
(1) \quad \sin \frac{2\pi x}{L}, \quad \cos \frac{2\pi x}{L}
$$

$$\begin{array}{ll} (2) & \cos\frac{2\pi x}{L}, \quad \cos\frac{2\pi 2x}{L} \\ (3) & \sin^2 x, \quad \cos^2 x \\ (4) & \sin 2x, \quad \cos 2x \\ (5) & \sin(x+\theta), \quad \cos x \\ (6) & \sin(x+\theta), \quad \cos(x+\phi). \end{array}$$

13. Confirm Eq. 5.143.

References

1. S. Lang, *Linear Algebra* (Springer, Berlin, 1997)
2. D.C. Lay, Linear Algebra and Its Applications (Addison-Wesley Publishing Company, Reading, 1994)
3. G. Strang, *Introduction to Linear Algebra* (Wellesley-Cambridge Press, Wellesley, 1993)
4. M. Tohyama, T. Koike, *Fundamentals of Acoustic Signal Processing* (Academic Press, London, 1998)
5. M. Tohyama, Linear algebraic method for acoustics (in Japanese). J. Acoust. Soc. Jpn. **59**(4), 215–223 (2003)
6. Y. Ando, *Auditory and Visual Sensation* (Springer, Berlin, 2009)
7. P. Damaske, Subjective Untersuchung von Schallfeldern. Acustica **19**, 199–213 (1967/68)
8. M. Tohyama, A. Suzuki, Interaural cross-correlation coefficients in stereo-reproduced sound fields. J. Acoust. Soc. Am. **85**(2), 780–786 (1989)
9. D.H. Cooper, Notions of conventional stereo interaural cross-correlation coefficients in stereo-reproduced sound fields in "Interaural cross-correlation coefficients in stereo-reproduced sound fields [J. Acoust. Soc. Am. 85 780–786 (1989)]". J. Acoust. Soc. Am. **86**(6), 2452–2454 (1989)
10. M. Uchiyama, M. Iwaki, M. Tohyama, A Selective receiving method by using the principal component analysis (in Japanese with English abstract), in Technical report of Institute of Electronics, Information, and Communication Engineers in Japan, EA98-99 (1998–2012)
11. M. Uchiyama, M. Tohyama, Sound image control for internet. Proc. Inst. Acoust. **20**(Part 5) (1998)
12. Y. Takahashi, M. Tohyama, M. Iwaki, Sound source separation by decorrelation of 2-point microphone signals, in *Proceedings of the 17th International Congress on Acoustics, 3D.04.02* (2001)
13. E. Weinstein, M. Feder, A. Oppenheim, Multi-channel signal separation by decorrelation. IEEE Trans. Speech Audio Process. **11**(4), 405–413 (1993)
14. A. Togura, M. Tohyama, T. Hirai, 2-ch impulse response estimation using signal decorrelation (in Japanese). Autumn Meet. Acoust. Soc. Jpn. 3-Q-22 (2001)
15. M. Tohyama, *Waveform Analysis of Sound* (Springer, Berlin, 2015)
16. W.M. Hartmann, *Signals, Sound, and Sensation* (Springer, Berlin, 1997)
17. Y. Hirata, Private communication

Chapter 6
Transfer Functions and Sequences in the Time Domain

Abstract The transfer function of a discrete system is defined by the $z-$transform of the impulse response. This chapter explores the magnitude and phase characteristics by introducing the notion of poles and zeros of a complex function. Phase responses are crucial in establishing the time delay sequences representing the impulse responses. Accumulated phase functions (continuous phase functions of the angular frequency) are defined by integration of the group delay functions. In particular, the linear phase is rendered by a pair of symmetric zeros with respect to the unit circle in the complex frequency plane. Auto-correlation sequences and their Fourier transforms (power spectral functions) are represented by the symmetric pairs of zeros. Time sequences such as impulse responses are identified from the power spectral functions by properly selecting the zeros from the symmetric pairs, subject to the initial value of the time sequence being normalized to unity. Poles, which are identified by the frequencies of the free oscillation or eigenfrequencies of the linear system, determine the transient and decaying time sequences. The cumulative spectral analysis (CSA) is used for estimation of the resonance frequencies from transient responses. CSA is helpful in condition monitoring and/or predicting howling frequencies corresponding to unstable poles. Symmetrically or closely located pairs of poles and zeros are important as these combinations govern the behavior of linear filters such as notch, all-pass, and inverse filters.

Keywords $z-$transform · Pure delay · Accumulate phase · Group delay · Linear phase · Symmetric zeros · Auto-correlation · Resonance · Poles · Cumulative spectral analysis · Howling · Notch filtering · All-pass filtering · Inverse filtering

6.1 z-Transforms

6.1.1 Traveling Waves and the Pure Delay System

A traveling wave (equivalently, a propagating wave) is any disturbance of a medium (wave) for which its peaks (maximum amplitudes) and troughs (minimum amplitudes) travel through a medium. A simple example of a traveling wave is the wave

M. Tohyama, *Sound in the Time Domain*, Signals and Communication Technology,
https://doi.org/10.1007/978-981-10-5889-9_6

pulse traveling along a string that extends infinitely in either direction. A traveling wave is called a plane wave if its frequency and maximum amplitude are constant. Suppose a sound source is located at position A, taken to be the origin of a reference frame. Consider another position B separated at a distance x(m) from the origin. If a sound wave is emitted from the origin at time $t = t_0$, it then reaches position B at elapsed time

$$t = t_0 + \tau = t_0 + x/c \qquad \text{(s)} \tag{6.1}$$

where c(m/s) denotes the speed of sound. If the waveform of the signal from the source is a δ-function

$$p_A(t) = A\delta(t) \tag{6.2}$$

assuming $t_0 = 0$, the waveform

$$\begin{aligned} p_B(t) &= p_A(t-\tau) \\ &= A\delta(t-\tau) = Ah_{AB}(t) \end{aligned} \tag{6.3}$$

is observed at position B. In this example, the expression

$$h_{AB}(t) = \delta(t-\tau) \tag{6.4}$$

describes the impulse response between source at position A and the receiver at position B. The magnitude of the traveling wave is constant and only a time delay τ is observed in the signal between source and receiver. A system for which the impulse response is defined only by its time delay is called a pure delay system.

The system above shows that the response

$$y(t) = x(t-\tau) \tag{6.5}$$

is observed at B from the source signal $x(t)$ at A. When the source signal $x(t)$ is given by

$$x(t) = \mathrm{e}^{\mathrm{i}\omega t} \tag{6.6}$$

the response $y(t)$ is

$$\begin{aligned} y(t) &= x(t-\tau) = \mathrm{e}^{\mathrm{i}\omega(t-\tau)} \\ &= \mathrm{e}^{-\mathrm{i}\omega\tau}\mathrm{e}^{\mathrm{i}\omega t} = H(\omega)\mathrm{e}^{\mathrm{i}\omega t} \end{aligned} \tag{6.7}$$

where

$$\begin{aligned} H(\omega) &= \int_{-\infty}^{\infty} h(t)\mathrm{e}^{-\mathrm{i}\omega t}dt \\ &= \int_{-\infty}^{\infty} \delta(t-\tau)\mathrm{e}^{-\mathrm{i}\omega t}dt \\ &= \mathrm{e}^{-\mathrm{i}\omega\tau} \end{aligned} \tag{6.8}$$

or the Fourier transform of the impulse response, which is called the frequency characteristics (or frequency response) of the system. The frequency characteristics of a pure delay system give simply a complex exponential function of angular frequency ω with magnitude unity. The phase angle is proportional to the angular frequency, i.e., a linear function of ω. The property of the phase as a linear function of ω is called linear phase. A pure delay system is of linear phase.

6.1.2 Discrete System and z-Transform

Suppose that the time delay τ can be written as

$$\tau = NT_s \qquad \text{(s)} \tag{6.9}$$

where T_s denotes the single sample period and N is a positive integer. The delay is called an N-sample delay if the delay time is represented by a discrete number N as in Eq. 6.9. The impulse response corresponding to N-sample delay is given by

$$h(n) = \delta(n - N) \tag{6.10}$$

instead of $h(t)$, where n denotes an integer. The Fourier transform of the impulse response $h(n)$ becomes

$$\begin{aligned} H(\mathrm{e}^{-\mathrm{i}\Omega}) &= \sum_n \delta(n - N)\mathrm{e}^{-\mathrm{i}\Omega n} \\ &= \mathrm{e}^{-\mathrm{i}\Omega N} \end{aligned} \tag{6.11}$$

where $\Omega = \omega T_s$. The complex function $\mathrm{e}^{\mathrm{i}\Omega}$ for $0 \le \Omega < 2\pi$ represents a unit circle of unit radius centered at the origin in the complex plane; Ω denotes the included angle between the real axis and the line from the origin to the point on the unit circle. Scaling the unit circle to include all circles of radii from 0 to ∞ determines the full complex plane, the points for which are denoted $z = r\mathrm{e}^{\mathrm{i}\Omega}$ with $|z| = r$. Then Eq. 6.11 becomes

$$H(z^{-1}) = \sum_n \delta(n - N)z^{-n} = z^{-N}; \tag{6.12}$$

this complex function is called the z-transform of the impulse response for the pure delay system. Its Fourier transform is expressed as

$$H(\mathrm{e}^{-\mathrm{i}\Omega}) = H(z^{-1})\big|_{z=\mathrm{e}^{\mathrm{i}\Omega}}. \tag{6.13}$$

The z-transform $H(z^{-1})$ is called the transfer function for the impulse response, where the Fourier transform $H(e^{-i\Omega})$ is said to be the frequency characteristics. The z-transform $H(z^{-1})$ has a singularity at $z = 0$, called the pole of the transfer function.

In Subsect. 3.1.1, a formal variable X was introduced to define the generating function of a discrete sequence. In this section, X is redefined as a point on the complex plane, such that $X = z^{-1}$. Substituting this complex variable z^{-1} for X, the generating function of a discrete sequence is rewritten as

$$A(z^{-1}) = \sum_n a(n) z^{-n}. \tag{6.14}$$

Here, $A(z^{-1})$ is called the z-transform of the sequence $a(n)$ rather than the generating function.

Consider a system that gives output sequences $y_1(n)$ and $y_2(n)$ corresponding to input sequences $x_1(n)$ and $x_2(n)$, respectively. The system is called a linear system if the output signal $y_1(n) + y_2(n)$ corresponds to the input signal $x_1(n) + x_2(n)$. The impulse response is defined as the response of the linear system to the unit pulse where $u(n) = 0$ for all integers n except zero for which $u(0) = 1$. The transfer function or the impulse response uniquely characterizes the linear system.

A linear system is called a causal system, if the impulse response is a right-hand sequence such that $h(n) = 0$ for $n < 0$. In addition, if the sum of the absolute value squared of $h(n)$, converges, that is, $\sum_{n=0}^{\infty} |h(n)|^2 < +\infty$, the system is called causal and stable. Note that the transfer function $H(z^{-1})$ is not defined on the entire z-plane, even if the system is causal and stable. Points in the z-plane at which the transfer function is undefined are singular points called poles of the transfer function.

6.2 Phase Response and Symmetric Zeros

6.2.1 Accumulation Phase and Group Delay

The phase frequency characteristics are intriguing because, in general, linear phase characteristics are not expected [1, 2]. Consider the example of transfer function

$$H(z^{-1}) = 1 + az^{-N} = z^{-N}(z^N + a). \tag{6.15}$$

Setting $N = 1$ for simplicity and taking the logarithmic function yields

$$\log H = -\log z + \log(z - z_0), \tag{6.16}$$

where

$$z_0 = -a \tag{6.17}$$

denotes the zero, and the magnitude a is formally extended to complex numbers represented by z_0. Substituting

$$z = e^{i\Omega}, \tag{6.18}$$

then

$$\begin{aligned}\log H &= \log |H| e^{i\Phi} \\ &= \log |H_0| - i\Delta\Omega + i\Phi_0\end{aligned} \tag{6.19}$$

obtains, where

$$H_0 = e^{i\Delta\Omega} - |z_0| = |H_0| e^{i\Phi_0} \tag{6.20}$$

$$z_0 = -a = |z_0| e^{i\Omega_0} \tag{6.21}$$

$$\Delta\Omega = \Omega - \Omega_0. \tag{6.22}$$

The imaginary part (the phase) of Eq. 6.19 is

$$\begin{aligned}\Phi &= -\Delta\Omega + \Phi_0 \\ &= \Omega_0 - \Omega + \Phi_0\end{aligned} \tag{6.23}$$

and determines the phase characteristics. The phase increment from that at $\Omega = 0$ is called the accumulation phase, which represents the frequency characteristics of the phase for a linear system. The second term on the right-hand side of the equation above shows the linear phase due to the pole at the origin, whereas Φ_0 indicates the phase characteristics rendered by the zero $z = z_0$.

The phase characteristics of the pole and zero can be interpreted geometrically on the complex plane (Fig. 6.1). Accumulation of the phase angle Φ_0 reaches 2π from the initial phase depending on how angle Ω moves, subject to $|z_0| \leq 1$ where $|z_0| = 1$ is the limit case.

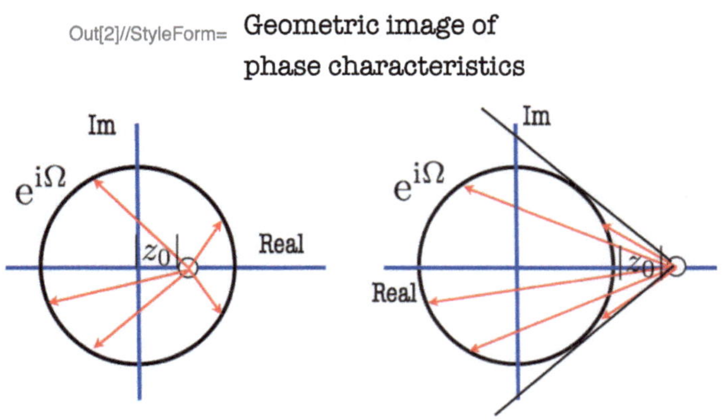

Fig. 6.1 Geometric plot of the phase frequency function due to its zero

In contrast, a phase angle is no longer accumulated if $|z_0| > 1$, even if Ω traces a circle centered on the origin (Fig. 6.1). The phase due to the pole at the origin is the linear phase obtained by taking the phase angle for z^{-1}; that is,

$$z^{-1}\Big|_{z=e^{i\Omega}} = e^{-i\Omega}. \tag{6.24}$$

When the angle Ω moves continuously from 0 to 2π on the unit circle $z = e^{i\Omega}$, the phase angle due to the pole can be written as

$$\Phi_p = -i\Omega, \tag{6.25}$$

giving a linear function of Ω, albeit of opposite sign.

Figure 6.2 illustrates the accumulation phase defined by setting the initial phase for $\Omega = 0$ to be zero. The phase for the pole is a linear function of Ω, but no longer exists for the zero. The deviation from linear phase can be analytically estimated from the derivative of the phase function (or the group delay). The phase angle Φ_0 can be expressed as

$$\Phi_0 = \tan^{-1}\frac{\Im[H_0]}{\Re[H_0]} = \tan^{-1} X \tag{6.26}$$

where

$$\Re[H_0] = \cos\Delta\Omega - |z_0| \tag{6.27}$$
$$\Im[H_0] = \sin\Delta\Omega. \tag{6.28}$$

The first derivative of $\Phi_0(\Omega)$ with respect to Ω is written as

$$\frac{\partial}{\partial\Omega}\Phi_0 = \frac{1}{1+X^2}\frac{\partial X}{\partial\Omega} \tag{6.29}$$

where

$$\frac{\partial X}{\partial\Omega} = \frac{1 - |z_0|\cos\Delta\Omega}{(\cos\Delta\Omega - |z_0|)^2}. \tag{6.30}$$

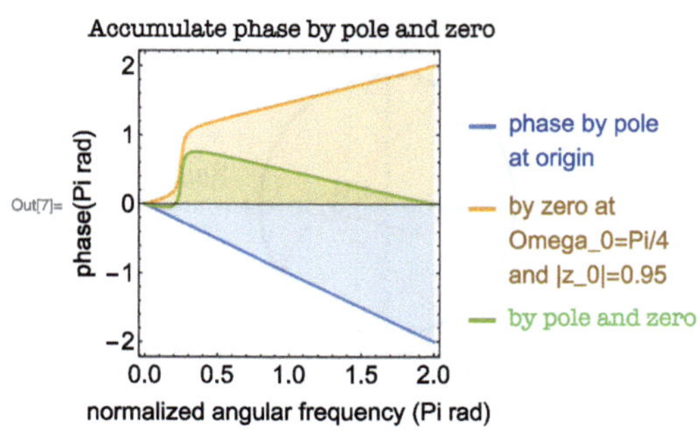

Fig. 6.2 Accumulation phase function resulting from the interference represented by the pole and zero

Substituting $\partial X/\partial \Omega$, then

$$\frac{\partial}{\partial \Omega}\Phi_0 = \frac{1 - |z_0|\cos \Delta\Omega}{1 - 2|z_0|\cos \Delta\Omega + |z_0|^2}. \tag{6.31}$$

is obtained, where the denominator is nonnegative. The expression above generates a positive peak when $\Delta\Omega \to 0$, as illustrated in Fig. 6.3, provided $|z_0| \leq 1$. The accumulation phase can be derived by integrating the derivative of the phase (Fig. 6.2). The phase is not a linear function, and it appears as if it might steeply jump when $\Omega = \Omega_0$. Adding the phase arising from the pole at the origin gives the total phase function plotted in Fig. 6.2. The accumulation phase becomes 0 when $\Omega = 2\pi$, because the phase -2π accumulated by the pole is canceled by that for the zero; a phase jump is nevertheless obtained for the zero when $\Omega = \Omega_0$.

In contrast, the phase associated with the pole is not canceled but accumulates when $|z_0| > 1$ (Fig. 6.4). In addition, the local property of the phase is opposite around the zeros inside and outside the unit circle as drawn in Fig. 6.5.

Following the first derivative of the phase by Eq. 6.31, the sign of the derivative is

$$\frac{\partial}{\partial \Omega}\Phi_0 = \begin{cases} > 0 & |z_0| < 1 \\ < 0 & |z_0| > 1 \end{cases} \tag{6.32}$$

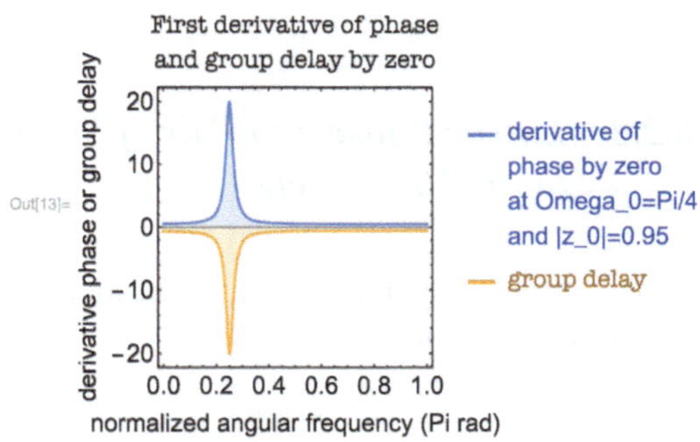

Fig. 6.3 First derivative of the phase function and the group delay for a single zero inside the unit circle

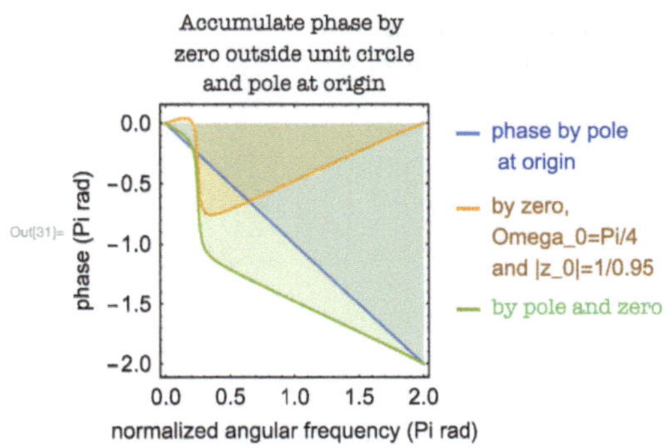

Fig. 6.4 Accumulation phase for a zero outside the unit circle and a pole at the origin

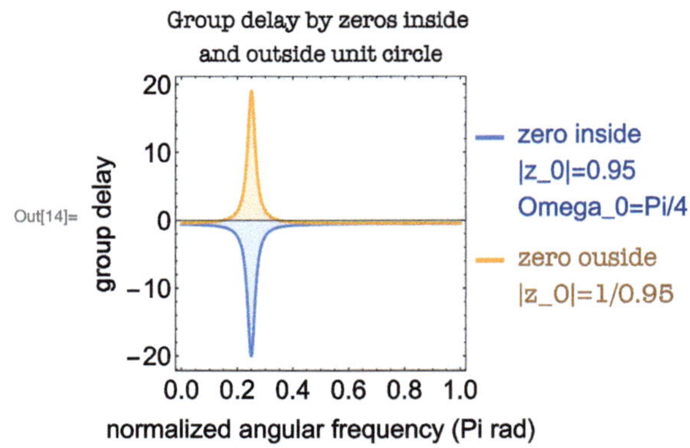

Fig. 6.5 Group delay for zeros inside and outside the unit circle

for $\Delta\Omega = 0$. The jump in phase that is made when Ω passes Ω_0 corresponds to a sign change for zeros inside and outside. A zero located inside the unit circle is called a minimum phase zero that cancels the phase accumulation from the initial phase by the pole at the origin so that

$$\Phi|_{\Omega=2\pi} = \Phi_{\rm p}\big|_{\Omega=2\pi} + \Phi_0|_{\Omega=2\pi} = 0 \tag{6.33}$$

where Φ denotes the accumulation phase (from the initial phase at $\Omega = 0$) for both pole and zero.

6.2.2 *Linear Phase and Pair of Symmetric Zeros with Respect to the Unit Circle*

A pair of zeros symmetrically arranged with respect to the unit circle (Fig. 6.6) provides a linear phase.

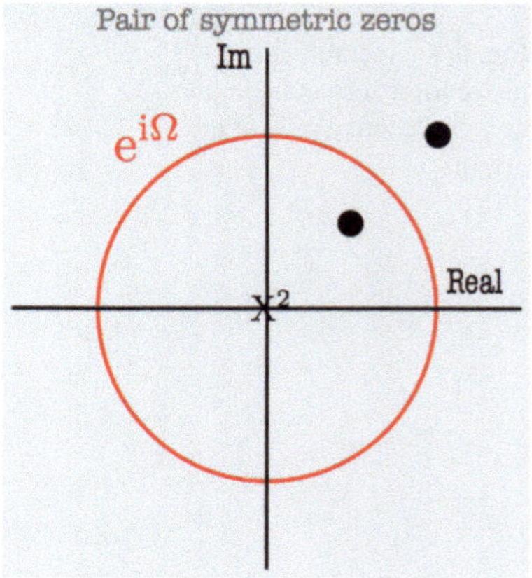

Fig. 6.6 Pair of zeros symmetrically arranged with respect to the unit circle

Symmetric zeros are produced at z_1 and z_2 under conditions

$$z_1 = |z_1|e^{i\Omega_1} \tag{6.34}$$

$$\begin{aligned} z_2 &= |z_2|e^{i\Omega_2} \\ &= |z_1|^{-1}e^{i\Omega_1} = (z_1^{-1})^*. \end{aligned} \tag{6.35}$$

The derivatives of the phase arising from z_1 and z_2 become

$$\frac{\partial}{\partial \Omega}\Phi_1 = \frac{1 - |z_1|\cos\Delta\Omega_1}{1 - 2|z_1|\cos\Delta\Omega_1 + |z_1|^2} \tag{6.36}$$

$$\begin{aligned} \frac{\partial}{\partial \Omega}\Phi_2 &= \frac{1 - |z_1|^{-1}\cos\Delta\Omega_1}{1 - 2|z_1|^{-1}\cos\Delta\Omega_1 + |z_1|^{-2}} \\ &= \frac{|z_1|^2 - |z_1|\cos\Delta\Omega_1}{1 - 2|z_1|\cos\Delta\Omega_1 + |z_1|^2}. \end{aligned} \tag{6.37}$$

The derivative functions of the phase for the pair of zeros sum to unity,

$$\frac{\partial}{\partial \Omega}(\Phi_1 + \Phi_2) = 1, \tag{6.38}$$

and the linear phase generates a single sample of delay due to the poles at the origin. The magnitude frequency characteristics are not independent of Ω for the pair of zeros that are different from the pure delay system. To create an impulse response sequence that is real, one takes a pair of conjugate zeros (see Fig. 6.7).

The transfer function can be written as

$$\begin{aligned} H(z^{-1}) &= (1 - z_0 z^{-1})(1 - z_0^{*^{-1}} z^{-1})(1 - z_0^* z^{-1})(1 - z_0^{-1} z^{-1}) \\ &= z^{-2}(z^2 - az + b - az^{-1} + z^{-2}), \end{aligned} \tag{6.39}$$

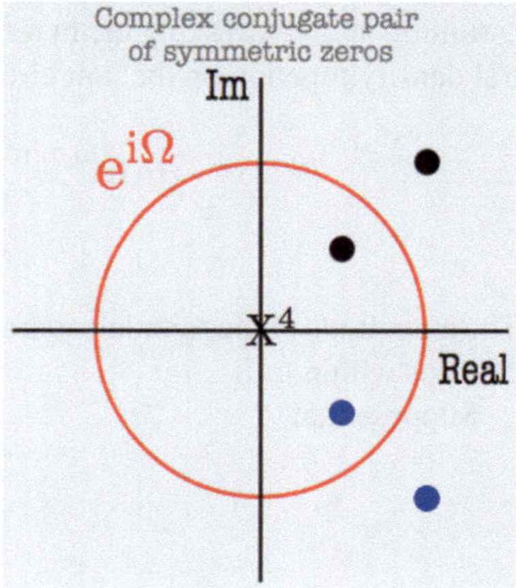

Fig. 6.7 Complex conjugate pairs of zeros symmetrically arranged with respect to the unit circle

where

$$a = z_0 + z_0^* + z_0^{-1} + z_0^{-1^*} \tag{6.40}$$

$$b = 2 + z_0 z_0^{*^{-1}} + z_0^* z_0^{-1}, \tag{6.41}$$

yielding an impulse response that is a real symmetric sequence. Rewriting the transfer function as

$$H(z^{-1}) = |z_0|^{-2} z^{-2} (z - z_0)(z^{-1} - z_0^*)(z - z_0^*)(z^{-1} - z_0) \tag{6.42}$$

then

$$\begin{aligned} H(\mathrm{e}^{-\mathrm{i}\Omega}) &= |z_0|^{-2} \left|(\mathrm{e}^{\mathrm{i}\Omega} - z_0)(\mathrm{e}^{\mathrm{i}\Omega} - z_0^*)\right|^2 \mathrm{e}^{-2\mathrm{i}\Omega} \\ &= R\mathrm{e}^{-2\mathrm{i}\Omega} \end{aligned} \tag{6.43}$$

can be derived where R denotes a real nonnegative function. The phase frequency characteristic is a linear function of Ω for $0 \le \Omega < 2\pi$. A symmetric sequence does not always generate a linear phase; however, the impulse response sequence is symmetric for a linear phase system.

6.2.3 Power Spectral Functions for Pair of Symmetric Zeros

The magnitude frequency characteristics are not independent of Ω for a pair of zeros. Consider the transfer functions

$$H_1(z^{-1}) = 1 - z_0 z^{-1} \tag{6.44}$$

$$H_2(z^{-1}) = 1 - z_0^{*^{-1}} z^{-1} \tag{6.45}$$

assuming $|z_0| < 1$. Taking the magnitude-squared corresponding to the power spectral density function on the unit circle in the z-plane, then

$$\begin{aligned} \left|H_1(\mathrm{e}^{-\mathrm{i}\Omega})\right|^2 &= H_1(\mathrm{e}^{-\mathrm{i}\Omega}) \cdot H_1^*(\mathrm{e}^{-\mathrm{i}\Omega}) \\ &= |z_0|^2 \left|H_2(\mathrm{e}^{-\mathrm{i}\Omega})\right|^2. \end{aligned} \tag{6.46}$$

The magnitude or magnitude-squared frequency characteristics are related to the pair of zeros within $|z_0|$.

Suppose that

$$z_0 = |z_0| \mathrm{e}^{\mathrm{i}\Omega_0}, \tag{6.47}$$

where $|z_0| < 1$, then

$$
\begin{aligned}
\left|H_2(e^{-i\Omega})\right|^2 &= \left|1 - |z_0|^{-1} e^{-i(\Omega - \Omega_0)}\right|^2 \\
&= |z_0|^{-2}(1 - 2|z_0| \cos \Delta\Omega + |z_0|^2) \\
&= |z_0|^{-2} \left|H_1(e^{-i\Omega})\right|^2,
\end{aligned}
\tag{6.48}
$$

which has a minimum at $\Omega = \Omega_0$, where $\Delta\Omega = \Omega - \Omega_0$. One of the power spectral density functions

$$\left|H_1(e^{-i\Omega})\right|^2 = 1 - 2|a| \cos \Delta\Omega + |a|^2 \tag{6.49}$$

$$\left|H_2(e^{-i\Omega})\right|^2 = |a|^{-2} \left|H_1(e^{-i\Omega})\right|^2 \tag{6.50}$$

is given by setting $|z_0| = |a|$. The impulse response or the transfer function is determined by selecting a zero or $|a|$ from the pair of the zeros so that the power spectral density function might be fitted.

The magnitude-squared $|a|^2$ is key in estimating the transfer function or the impulse response from the power spectral density function. Figure 6.8 presents the difference in the magnitude-squared frequency characteristics for the symmetric zeros.

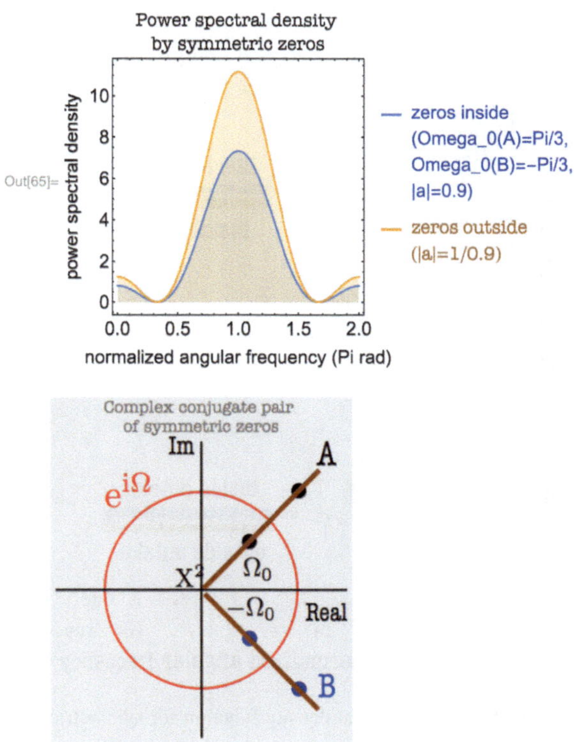

Fig. 6.8 Difference in the magnitude-squared frequency characteristics for the symmetric zeros

The determination of the pair of time pulses that represents the zeros in the transfer function can be extended to factorizing the impulse response into convolutions of pairs of time pulses subject to that $h(0) = 1$. Suppose that the pairs of symmetric zeros are known. Figure 6.9 displays an example of the power spectral density function for

$$
\begin{aligned}
H(z^{-1}) &= H_1(z^{-1}) \cdot H_2(z^{-1}) \\
&= (1 - z_1 z^{-1})(1 - z_1^* z^{-1})(1 - z_2 z^{-1})(1 - z_2^* z^{-1}),
\end{aligned}
\tag{6.51}
$$

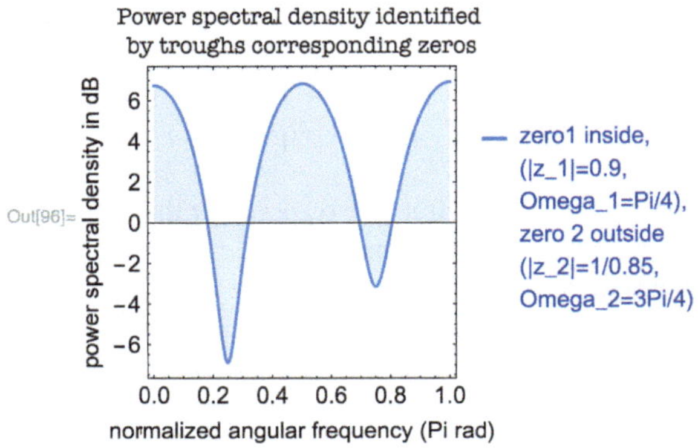

Fig. 6.9 Sample of a power spectral density function where two troughs are identified in the positive frequency domain ($0 < \Omega < \pi$)

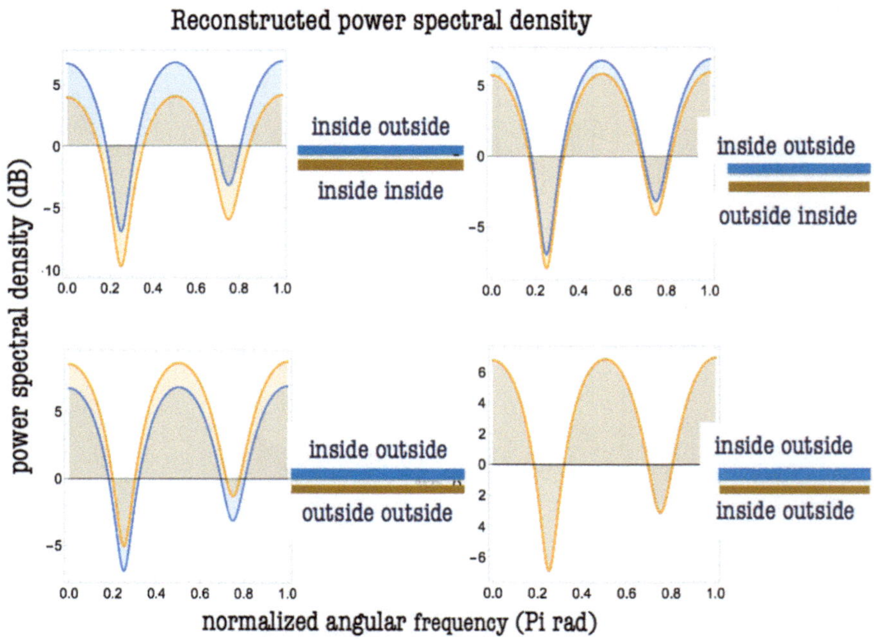

Fig. 6.10 Power spectral density function for symmetric pairs of zeros in the positive frequency domain

where z_1 and z_2 denote the zeros, respectively. Assuming the symmetric pairs of zeros are known, there are four possibilities in selecting the zeros z_1 and z_2 according to whether they are inside or outside the unit circle. Figure 6.10 presents the four possible situations.

The situation in the right panel on the bottom row can be determined by a comparison with the power spectral density function. Consequently, the transfer function or the impulse response could be determined as a convolution $h(n) = h_1 * h_2(n)$ where $h_i(n)$ is composed of two time pulses subject to conditions $h(0) = h_1(0) = h_2(0) = 1$.

6.3 Poles for Resonance Systems and Transient Time Responses

6.3.1 Cumulative Spectral Analysis of the Transient Response for a Resonance System

Steady-state responses of the transfer functions are generally determined by the zeros (or troughs of the magnitude frequency characteristics) as described in the previous subsection. The transient responses, however, are represented by poles determining damped vibrations with frequencies associated with free vibration of the resonance system of interest. The transient responses characterized by such poles are informative in allowing estimates of the resonance frequencies and the decay properties independent of the waveforms executed by the exciting source. This independence offers benefits in health monitoring [3] of structures because the excitation force for the structures quite likely may not be identifiable from a practical point of view.

Damped sinusoidal responses are observable in certain intervals of the response to the excitation source composed of on-sets and off-sets. Damped responses under resonance are usually analyzed using the so-called cumulative spectral analysis (CSA) [4]. Consider a signal $x(n)$ frame by frame. For the on-set and on-going signal, CSA is performed using the Fourier transform of the frame-wise signal so that

$$X_{\text{on}}\left(n, \mathrm{e}^{-\mathrm{i}\Omega}\right) = \sum_{m=0}^{n} u(m)x(m)\mathrm{e}^{-\mathrm{i}\Omega m} \tag{6.52}$$

for $0 \leq n \leq N-1$, where $u(n)$ is the unit step sequence (Heaviside function) defined as

$$u(n) = \begin{cases} 1 & n \geq 0 \\ 0 & n < 0, \end{cases} \tag{6.53}$$

and N denotes the frame length. Instead of $u(n)$, applying the triangular windowing with positive slope, i.e.,

$$w(n) = \begin{cases} n & n \geq 0 \\ 0 & n < 0, \end{cases} \tag{6.54}$$

is also possible [5].

The CSA can be also applied to the signal off-set so that

$$X_{\text{off}}\left(n, \mathrm{e}^{-\mathrm{i}\Omega}\right) = \sum_{m=n}^{N-1} u(m)x(m)\mathrm{e}^{-\mathrm{i}\Omega m}. \tag{6.55}$$

Suppose that a sinusoidal signal in complex form is fed to a resonance system for which the impulse response is $h(n)$, and the sinusoidal signal stops at $n = 0$. The response of the system after the sinusoidal input signal stops can be written as

$$\begin{aligned} y(n) &= \sum_{m=n}^{N-1} \mathrm{e}^{\mathrm{i}\Omega(n-m)} h(m) \\ &= \sum_{m=n}^{N-1} h(m)\mathrm{e}^{-\mathrm{i}\Omega m}\mathrm{e}^{\mathrm{i}\Omega n} \end{aligned} \tag{6.56}$$

where the impulse response has length N. The cumulative analysis of $h(n)$ is similarly defined for the sinusoidal signal off-set yielding

$$H_{\text{off}}\left(n, \mathrm{e}^{-\mathrm{i}\Omega}\right) = \sum_{m=n}^{N-1} h(m)\mathrm{e}^{-\mathrm{i}\Omega m} \tag{6.57}$$

as well as for the signal on-set giving

$$H_{\text{on}}\left(n, \mathrm{e}^{-\mathrm{i}\Omega}\right) = \sum_{m=0}^{n} h(m)\mathrm{e}^{-\mathrm{i}\Omega m}. \tag{6.58}$$

Figure 6.11 illustrates scale models used for cumulative spectral analysis for structural vibration [6, 7]. The models are wooden frames (0.18W, 0.21D, 0.38H in (m)). Each model has four built-up pillars, each pillar composed of four posts of lumber (4cm square). In the experiments, two constructions labeled condition 1, which is the original construction, and condition 4, where a single post is removed from one pillar, are shown in Fig. 6.11.

The middle row presents examples of impulse response recordings under conditions 1 and 4. The impulse responses generated by an impulsive hammer were recorded using a piezoelectric accelerometer. In addition, impulse response recordings were made on the linear interpolation between the responses under conditions

Out[16]//StyleForm=

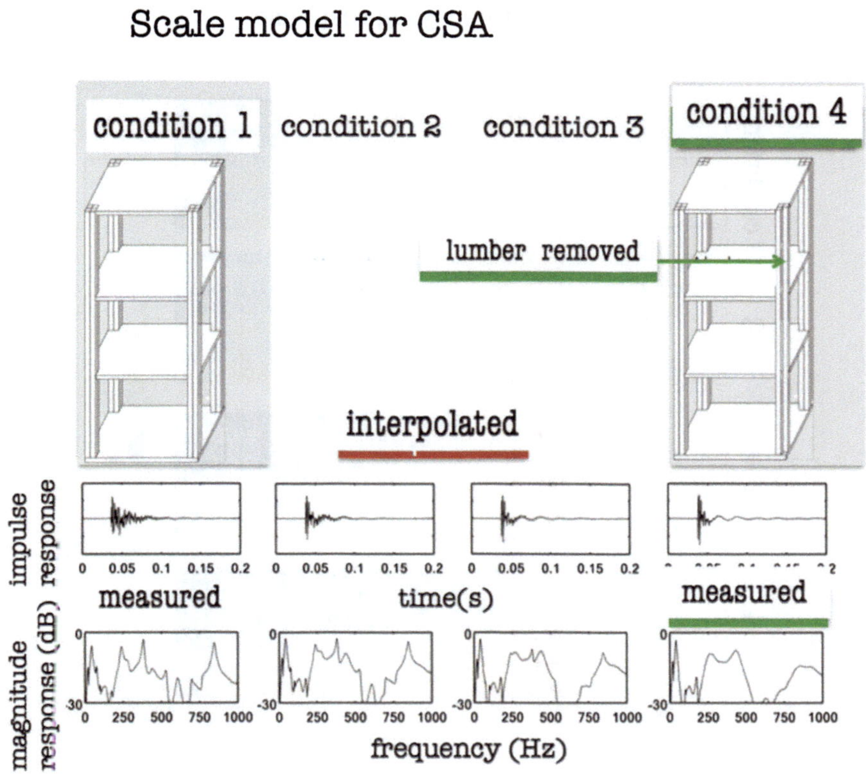

Fig. 6.11 Scale model for experiments on structural vibration from Fig. 1 [7]

1 and 4. Representative samples are shown in the figure as instances of conditions 2 and 3.

The bottom row presents the magnitude frequency characteristics (or magnitude response) for the four conditions. Prominent spectral peaks determined by the free oscillations associated with the resonance frequencies can be seen in the magnitude responses. The spectral peaks are much clearer under the original condition than those in other modified conditions. The damping factor or rate of decay in the resonance response increases as the modification proceeds. The change in damping properties is seen in the waveforms of the impulse response recordings themselves.

Figure 6.12 displays the results of H_{off} under conditions 1–4, where the frame length is set to 60 ms. The left column displays the magnitude frequency characteristics presented in Fig. 6.11 for ease of comparison. Dampening in the time domain is confirmed from the corresponding prominent spectral peaks. The top row clearly shows the slowly decaying dominant resonance responses for condition 1. In contrast, under condition 4, the resonance components are blurred and decay rapidly.

Impulse responses are unlikely to be observed in practical situations. Thus, monitoring the damping effects for the prominent resonance frequencies is difficult in

Out[17]//StyleForm=

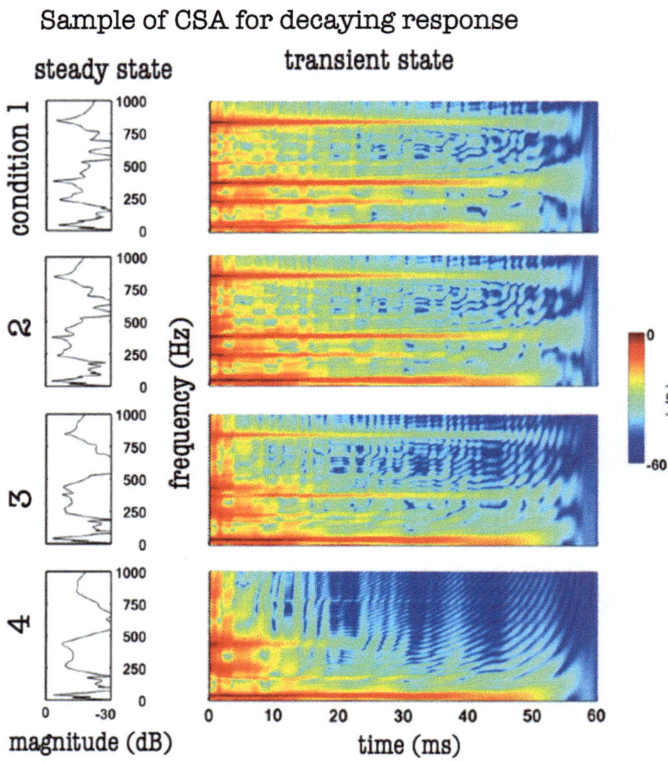

Fig. 6.12 Samples of H_{off} from Fig. 2 [7]

the diagnostics of structural conditions from a practical point of view. Figure 6.13 presents an example for which the resonance response (or hidden resonance [5]) was monitored for the possibility of analyzing resonance responses without impulse responses [6, 7].

This is performed by taking frame-wise signals from the build-up (on-set), steady state (on-going), and damped (off-set) signals in response to a nonstationary signal such as a random burst of white noise of random duration, interval, and magnitude. The top row in the figure is a sample of the waveform obtained by the convolution of the impulse response and a random burst of white noise subject to condition 1. The middle row shows samples of three (off-set, on-set, and on-going) segments taken from the response waveform displayed in the top row. The bottom row illustrates X_{off} obtained from CSA for the corresponding waveforms in the three segments, respectively. Interestingly, the damped segment reveals decay characteristics very similar to ideal values derived using the impulse response given in Fig. 6.12. In contrast,

Out[18]//StyleForm=

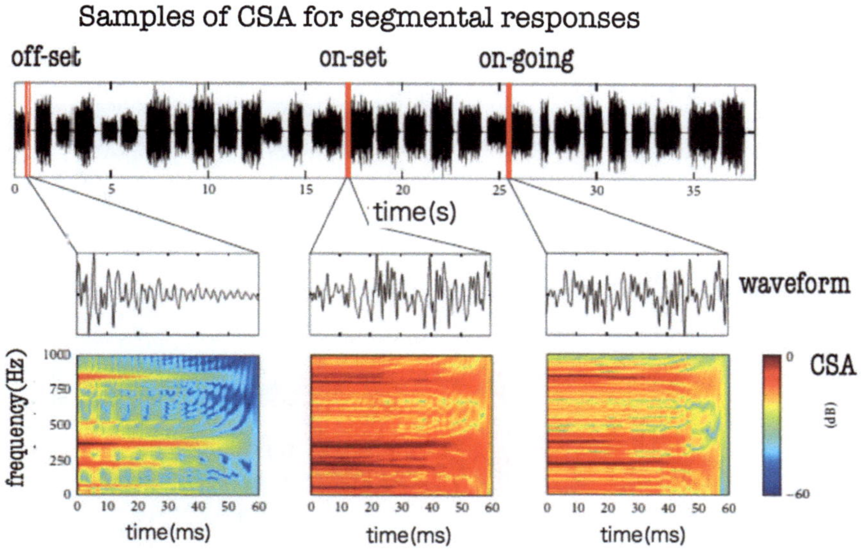

Fig. 6.13 Samples of CSA for dampening (off-set), build-up (on-set), and steady state (on-going) in response to a random burst of white noise from Fig. 3 [7]

Fig. 6.14 Ensemble average of CSA for all damped (off-set) segments in response to a random burst of white noise from Fig. 4 [7]

Out[19]//StyleForm=

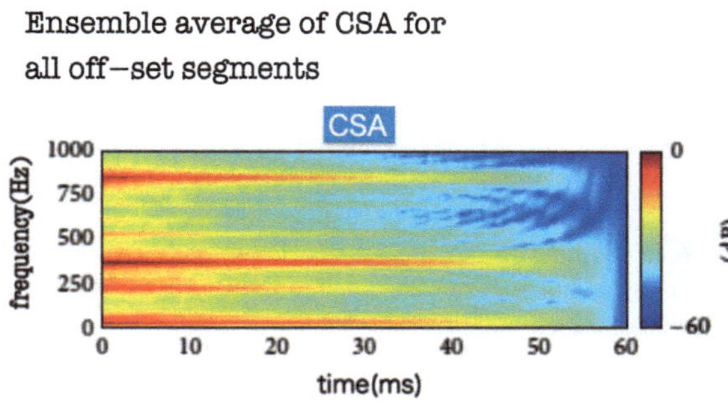

no clear resonance decay characteristics are seen for build-up and steady-state segments. Figure 6.14 shows the ensemble average of X_{off} for all damped segments in response to the random burst of white noise.

Figure 6.15 gives another example under excitation from non-Gaussian noise, such as Cauchy noise given in Fig. 6.16. Cauchy noise can be rendered by making random data Z such as

$$Z = \frac{Y}{X} \tag{6.59}$$

where X and Y are independent Gaussian variables. Similar to a random burst of white noise, all damped segments are taken from the response signal and convolved

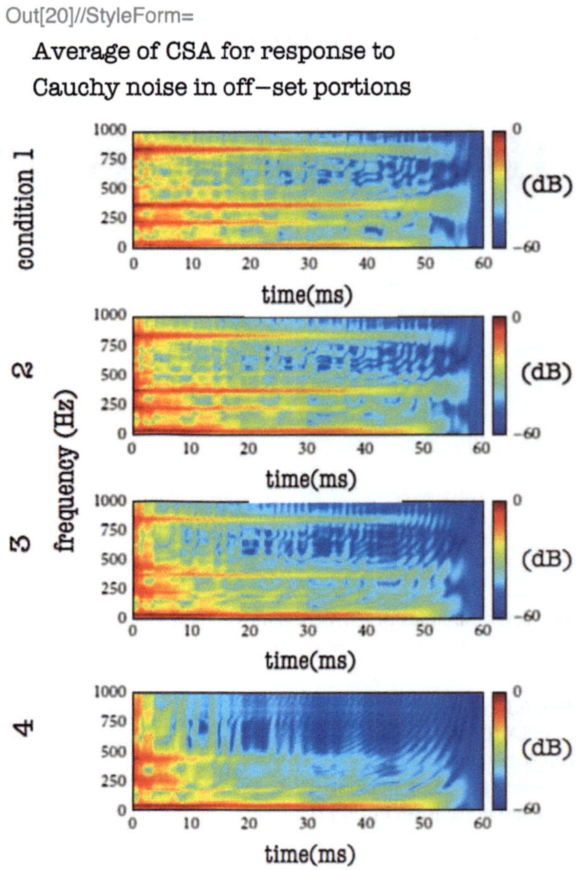

Fig. 6.15 Ensemble average for the damped (off-set) segments in response to non-Gaussian random noise from Fig. 7 [7]

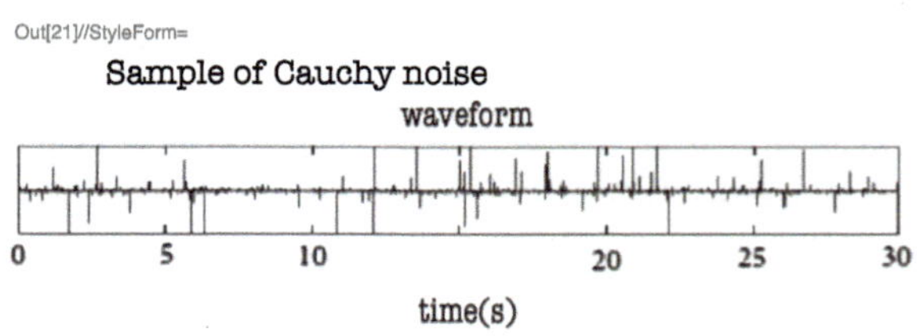

Fig. 6.16 Sample of non-Gaussian noise following a Cauchy distribution from Fig. 5 [7]

with non-Gaussian noise and the impulse responses. The panels in Fig. 6.15 are the ensemble averaged data for X_{off} under the structure modifications specified in conditions 1–4. All of the four displays look quite similar to those in Fig. 6.12 obtained using ideal impulse responses. The results for the damped segments in the responses, even assuming the impulse responses are unknown, would be informative in monitoring the resonance frequencies with decay characteristics under free vibration.

6.4 Unstable Pole for Causal System

For stable causal systems, the poles of the transfer function are distributed over the unit disc on the z-plane. Of interest is knowing in what context unstable poles appear in linear systems. In communication systems, unstable poles appear with howling or singing. Howling sometimes occurs over a public address system because of a sudden change in the sound path between pairs of microphones and loudspeaker systems. Such a change moves the poles from inside to outside the unit circle.

6.4.1 Poles and Howling

Figure 6.17 shows a simplified scheme of a public address system. The system constitutes a closed loop because of a recursive sound path. The transfer function for such a recursive system is written as

$$\begin{aligned} L(z^{-1}) &= H(z^{-1})\frac{1}{1-G(z^{-1})H(z^{-1})} \\ &= H(z^{-1})F(z^{-1}) \\ &= H(z^{-1})\frac{1}{D(z^{-1})} \end{aligned} \tag{6.60}$$

where $D(z^{-1}) = 1 - G(z^{-1})H(z^{-1})$ assuming $|G(z^{-1})H(z^{-1})| < 1$. The system is stable if $\sum_{n=0}^{\infty} |f(n)|^2 < +\infty$. Here, $f(n)$ is the impulse response corresponding to $F(z^{-1})$ for the closed loop shown in Fig. 6.17. System stability is conditioned on whether the poles of the transfer function are located inside the unit disc. In other words, for the stable system, all zeros of $D(z^{-1})$ must be located inside the unit disc.

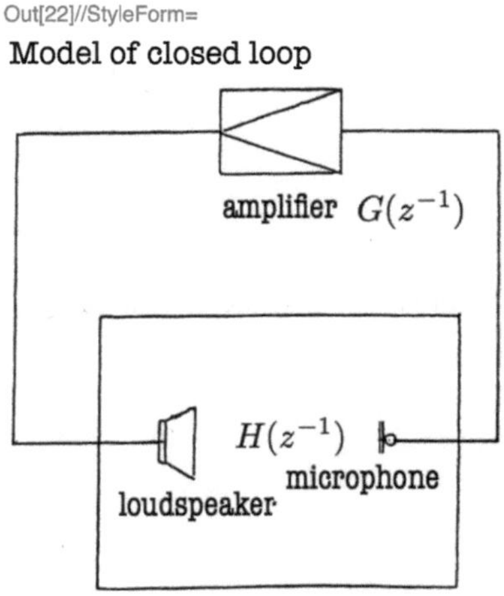

Fig. 6.17 Simplified model of a public address system from Fig. 15.10 [8]

Consider a simplified example of a single-zero system [9]

$$H(z^{-1}) = 1 - az^{-1} \tag{6.61}$$

and

$$G(z^{-1}) = b \tag{6.62}$$

where $0 < b < 1$. The transfer function $L(z^{-1})$ becomes

$$\begin{aligned} L(z^{-1}) &= \frac{1 - az^{-1}}{1 - b(1 - az^{-1})} \\ &= \frac{1 - az^{-1}}{1 - b} \cdot \frac{1}{1 - \alpha z^{-1}} \end{aligned} \tag{6.63}$$

where

$$\alpha = \frac{-ab}{1 - b}. \tag{6.64}$$

The impulse response for the closed loop $f(n)$ can be written as

$$f(n) = \left(\frac{-ab}{1 - b}\right)^n \frac{1}{1 - b} \tag{6.65}$$

where n is a nonnegative integer. The impulse response $f(n)$ is stable provided $|\frac{-ab}{1-b}| < 1$. Note that $\frac{-ab}{1-b}$ determines the pole location of $F(z^{-1})$. The above example therefore illustrates that a system is stable as long as the pole of the transfer function is located inside the unit disc. The pole moves in accordance with the open-loop condition described by $H(z^{-1})$ and/or the loop gain.

Consider a simplified example for the denominator of the transfer function, that is,

$$\begin{aligned} D(z^{-1}) &= 1 - bH(z^{-1}) \\ &= 1 - b(1 + h_1 z^{-1} + h_2 z^{-2}) \\ &= z^{-2}\left((1 - b)z^2 - bh_1 z - bh_2\right) \\ &= z^{-2}(1 - b)(z^2 - \frac{b}{1 - b}h_1 z - \frac{b}{1 - b}h_2). \end{aligned} \tag{6.66}$$

where b is a parameter that denotes the gain. Given the pair of zeros for equation

$$z_p^2 - \frac{b}{1 - b}h_1 z_p - \frac{b}{1 - b}h_2 = 0. \tag{6.67}$$

Equation 6.66 can be rewritten as

$$D(z^{-1}) = z^{-2}(1-b)(z^2 - \frac{b}{1-b}h_1 z - \frac{b}{1-b}h_2) \qquad (6.68)$$
$$= z^{-2}(1-b)(z - z_{p_1})(z - z_{p_2}).$$

When the poles for the closed-loop transfer function are located within the unit circle, that is, $|z_p| < 1$, this system is stable. In general, $D(z^{-1})$ can be formally written as

$$D(z^{-1}) = (1-b)\sum_{n=0}^{N-1} d_n z^{-n}. \qquad (6.69)$$

If $D(z^{-1})$ is of minimum phase, then the system is stable and howling-free.

6.4.2 Detection of Unstable Time Responses

Predicting unstable poles is desirable from a practical point of view, if it is possible when the system of interest is even in a stable condition. The frequency of the pole could be estimated given a stable condition. As described in Sect. 6.3, the resonance response is observed in a damped (off-set) segment of the response to the signal composed of on-set, on-going, and off-set segments. Figure 6.18 presents examples of a CSA using a triangular window with positive slope (CHA) introduced in Subsect. 6.3.1 [5]. The frame (window) length is 20(ms) for both off-set and build-up or on-going segments of random noise samples undergoing a closed loop. For off-set

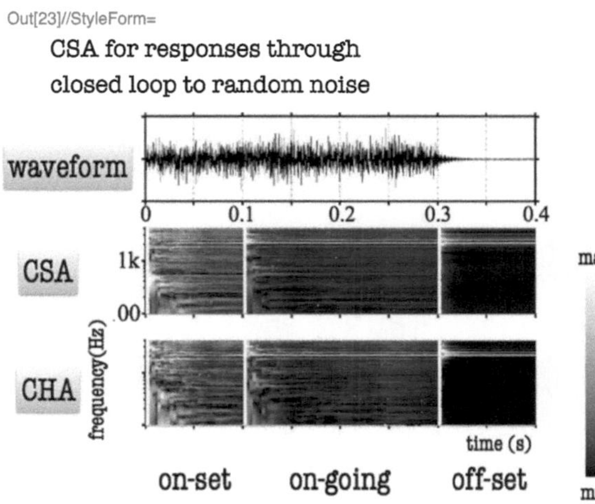

Fig. 6.18 Samples of cumulative spectral analysis using triangular windowing with positive slope (CHA) for in random noise samples undergoing a closed loop from Fig. 15.15 [8]

segments, an isolated spectral peak is observed around 2 kHz that may correspond to a howling frequency [5].

6.5 Poles and Filtering

6.5.1 Notch Filtering

The pole associated with howling yields a spectral distortion in the sound quality even in a stable condition. Such distortion from the original sound can be suppressed by so-called notch filtering. If the frequency of the pole can be estimated, then the distortion associated with the pole can be reduced by arranging a pole and a zero at the pole frequency to be very close to the unit circle, as illustrated in Fig. 6.19.

Consider the transfer function

$$N(z^{-1}) = \frac{1 - z_{00}z^{-1}}{1 - z_{0p}z^{-1}} \tag{6.70}$$

where

$$z_{00} = |z_{00}|e^{i\Omega_{00}} = |z_{00}|e^{i\Omega_p} \tag{6.71}$$

$$z_{0p} = |z_{0p}|e^{i\Omega_{0p}} = |z_{0p}|e^{i\Omega_p} \tag{6.72}$$

$$|z_{00}| \cong |z_{0p}| \tag{6.73}$$

with $1 > |z_{00}| > |z_{0p}|$, as shown in Fig. 6.19, and the poles and zeros are located near the unit circle at the same angular frequency, but at slightly different locations

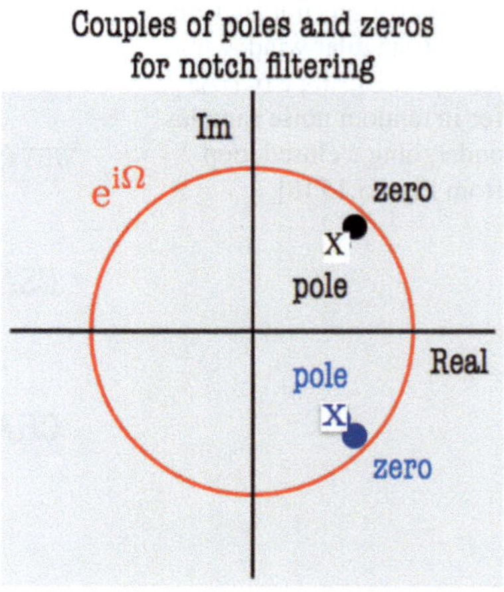

Fig. 6.19 Closely situated pole/zero pairs inside and close to the unit circle

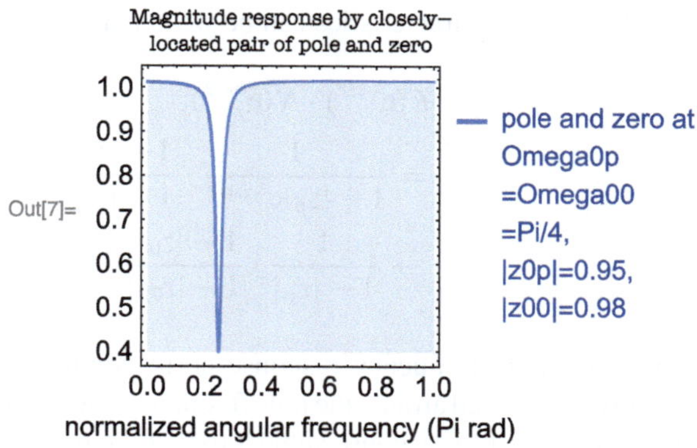

Fig. 6.20 Magnitude frequency characteristics for the pole and zero for notch filtering

between $|z_{00}|$ and $|z_{0p}|$. The frequency characteristics for the transfer function can be written as

$$
\begin{aligned}
N(\mathrm{e}^{-\mathrm{i}\Omega}) &= \frac{1 - z_{00}\mathrm{e}^{-\mathrm{i}\Omega}}{1 - z_{0p}\mathrm{e}^{-\mathrm{i}\Omega}} \\
&= \frac{1 - |z_{00}|\mathrm{e}^{-\mathrm{i}\Delta\Omega}}{1 - |z_{0p}|\mathrm{e}^{-\mathrm{i}\Delta\Omega}} \\
&\cong 0 \qquad (\Delta\Omega \to 0)
\end{aligned}
\tag{6.74}
$$

where

$$\Delta\Omega = \Omega - \Omega_p. \tag{6.75}$$

The magnitude frequency characteristics are illustrated in Fig. 6.20. A very sharp rejection at around $\Omega \cong \Omega_p$ can be seen.

According to the sharp rejection property, the so-called notch filter can be applied to reduce the spectral distortion and hence avoid howling. As an example of the transfer function for a closed loop, consider once more

$$F(z^{-1}) = \frac{1}{1 - z_p z^{-1}}. \tag{6.76}$$

Convolving the transfer function above with that for the notch filter, then

$$F(z^{-1}) \cdot N(z^{-1}) = \frac{1}{1 - z_p z^{-1}} \cdot \frac{1 - z_{00}\mathrm{e}^{-\mathrm{i}\Omega}}{1 - z_{0p}\mathrm{e}^{-\mathrm{i}\Omega}} \tag{6.77}$$

where $= z_p = |z_p| = |z_p|\mathrm{e}^{\mathrm{i}\Omega_p}$.

The frequency characteristics can be written as

$$F(e^{-i\Omega}) \cdot N(e^{-i\Omega}) \tag{6.78}$$
$$= \frac{1}{1-|z_p|e^{-i\Delta\Omega}} \cdot \frac{1-|z_{00}|e^{-i\Delta\Omega}}{1-|z_{0p}|e^{-i\Delta\Omega}}$$
$$\to \frac{1}{1-|z_p|} \cdot \frac{1-|z_{00}|}{1-|z_{0p}|} \cong 0. \quad (\Delta\Omega \to 0)$$

The magnitude frequency response corresponding to the pole of $H(z^{-1})$ can be sharply suppressed around the pole frequency, thereby reducing the spectral distortion due to the impulse response subject to that the pole frequency is estimated.

6.5.2 Inverse and All-Pass Filtering for Predicted Pole

If the pole location could be estimated in addition to the frequency, then inverse filtering is possible in avoiding howling. Under stable conditions, the predicted unstable pole can be canceled by the zero that is located at the same position as that for the pole. Convolving the transfer function given by Eq. 6.76 in the previous subsection with $M(z^{-1})$ of the form

$$M(z^{-1}) = 1 - z_p z^{-1}, \tag{6.79}$$

which has a zero at $z = z_p$, then

$$M(z^{-1}) \cdot F(z^{-1}) = 1. \tag{6.80}$$

A system with transfer function $M(z^{-1})$ is called an inverse system or inverse filter for $F(z^{-1})$ because

$$M(z^{-1}) = 1 - z_p z^{-1} = F^{-1}(z^{-1}). \tag{6.81}$$

The pole has disappeared through inverse filtering.

By setting the zero at the pole location, inverse filtering then works so that the pole response might disappear. Alternatively, it may be also possible to place a symmetric zero at the pole. Arranging the symmetric zero outside the unit circle creates the all-pass filter by pairing a pole and zero, as shown in Fig. 6.21.

The pole/zero combination is made by the convolution of the closed-loop transfer function of Eq. 6.76 with

$$K(z^{-1}) = 1 - z_p^{-1^*} z^{-1} \tag{6.82}$$

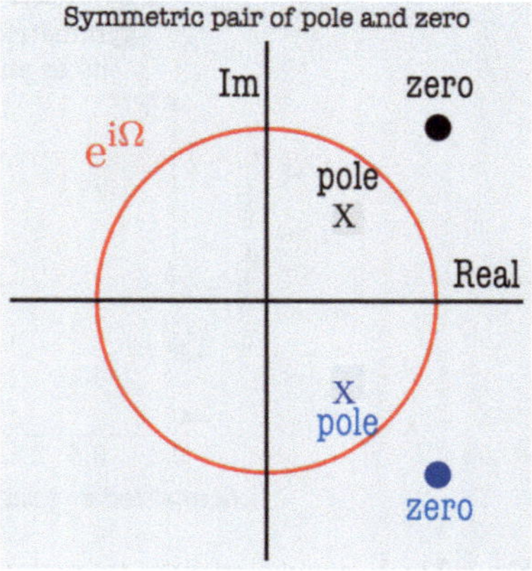

Fig. 6.21 Symmetric zero for a pole inside the unit circle

giving

$$A_p(z^{-1}) = F(z^{-1}) \cdot K(z^{-1}) = \frac{1 - z_p^{-1*} z^{-1}}{1 - z_p z^{-1}}. \tag{6.83}$$

The symmetric zero of the pole is located outside the unit circle.

The magnitude frequency characteristics, which are independent of Ω, can be obtained by the symmetric pair of pole and zero. The transfer function for this symmetric pair can be written as

$$\begin{aligned} A_p(z^{-1}) &= \frac{1 - a^{*^{-1}} z^{-1}}{1 - a z^{-1}} \\ &= -a^{*^{-1}} z^{-1} \frac{1 - a^* z}{1 - a z^{-1}} \end{aligned} \tag{6.84}$$

where

$$z_0 = a^{*^{-1}}, \quad z_p = a \tag{6.85}$$

constitute the zero and pole, respectively. Substituting $z = e^{i\Omega}$ then gives

$$\left| A_p \left(e^{-i\Omega} \right) \right| = |a^{*^{-1}}| \tag{6.86}$$

implying that the magnitude is independent of Ω. The system composed of symmetric pairs of poles and zeros is called an all-pass filter.

The phase frequency characteristic is shown in Fig. 6.22. The phase for the symmetric pair of pole and zero lying inside and outside, respectively, the unit circle is accumulated without cancelation because the zero is outside the circle. The accumulation phase is a function of Ω, although the magnitude characteristics are independent of Ω. This all-pass-type processing works as long as the system is stable. It may not be useful at suppressing howling; however, it helps in reducing the spectral

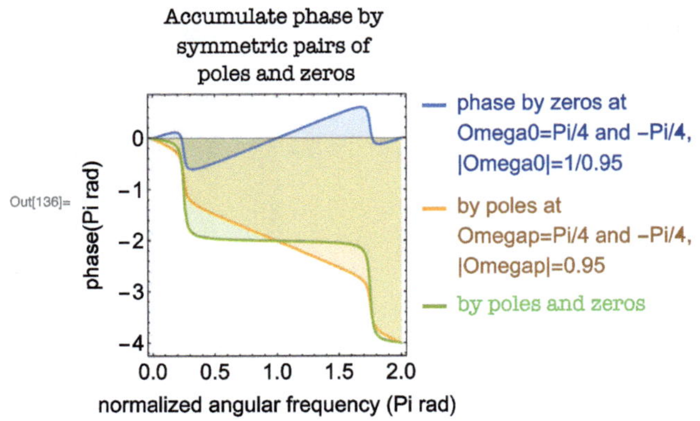

Fig. 6.22 Accumulation phase for symmetric pairs of poles and zeros

distortion because the spectral effect of the pole disappears without losing the time response due to the pole.

6.5.3 Inverse Filter for the Minimum Phase System

A stable and causal system has its poles inside the unit circle; no pole lies outside the unit circle. Consider the transfer function

$$H(z^{-1}) = 1 - z_0 z^{-1} \tag{6.87}$$

once again, where $|z_0| < 1$. The inverse system has transfer function

$$H^{-1}(z^{-1}) = \frac{1}{1 - z_0 z^{-1}} \tag{6.88}$$

with a pole at $z = z_0$ and zero at $z = 0$. The inverse system is stable and causal because there is no pole outside the unit circle. Inverse systems are realizable (or stable and causal) for minimum phase systems in which all zeros are located inside the unit circle. If there is a zero outside the unit circle, that is, the system is non-minimum phase, then its inverse system has a pole outside the unit circle and therefore it is noncausal and not realized. The pole/zero plot of the transfer function can be decomposed into two transfer functions such as the minimum phase and all-pass transfer functions [8, 9].

6.6 Exercises

1. Establish whether the phase characteristic of each of the following impulse responses is linear.

$$\begin{aligned} &(1) \quad h(0) = 1 \quad h(1) = 1/2 \quad h(2) = 1 \\ &(2) \quad h(0) = 1 \quad h(1) = 3/4 \quad h(2) = 1 \\ &(3) \quad h(0) = 1 \quad h(1) = 1 \quad h(2) = 1 \\ &(4) \quad h(0) = 1 \quad h(1) = 2 \quad h(2) = 1. \end{aligned} \tag{6.89}$$

2. Establish whether the phase response of the transfer function

$$H(z^{-1}) = (1 - az^{-1})(1 - a^* z^{-1})(1 - bz^{-1})(1 - b^* z^{-1}).$$

is linear for each of the following instances,

$$\begin{aligned} &(1) \quad b = -a \\ &(2) \quad b = a^{*^{-1}} \\ &(3) \quad b = a^* \\ &(4) \quad b = a^{-1}. \end{aligned} \tag{6.90}$$

3. Derive Eq. 6.31.
4. Confirm Eq. 6.46.
5. Derive Eq. 6.63.
6. Confirm Eq. 6.86.
7. Calculate the power spectral characteristic of the transfer function

$$H(z^{-1}) = (1 - az^{-1})(1 - a^* z^{-1})$$

for each of the following conditions,

$$\begin{aligned} &(1) \quad a = \frac{1}{2} e^{i\frac{\pi}{4}} \\ &(2) \quad a = 2e^{i\frac{\pi}{4}} \\ &(3) \quad a = e^{i\frac{\pi}{4}} \\ &(4) \quad a = 2. \end{aligned} \tag{6.91}$$

8. Show the magnitude frequency characteristic of the transfer function

$$H(z^{-1}) = \frac{1 - bz^{-1}}{1 - az^{-1}} \frac{1 - b^* z^{-1}}{1 - a^* z^{-1}}$$

for the following settings of b,

$$\begin{aligned} &(1) \quad b = a^{-1^*} \\ &(2) \quad b = a^{-1} \\ &(3) \quad b = a^* \\ &(4) \quad b = a. \end{aligned} \tag{6.92}$$

9. Show the phase frequency response geometrically for the zero (Fig. 6.12 in [9]) such as $z_0 = e^{i\frac{\pi}{4}}$.

References

1. R.H. Lyon, Progressive phase trends in multi-degree-of-freedom systems. J. Acoust. Soc. Am. **73**(4), 1223–1228 (1983)
2. R.H. Lyon, Range and frequency dependence of transfer function phase. J. Acoust. Soc. Am. **76**(5), 1435–1437 (1984)
3. Y. Hirata, Estimation of the frequency response of a structure using its non-stationary vibration. J. Sound Vibration **313**, 363–366 (2008)
4. J.M. Berman, L.R. Fincham, The application of digital techniques to the measurement of loudspeakers. J. Audio Eng. Soc. **25**(6), 370–384 (1977)
5. Y. Takahashi, M. Tohyama, Y. Yamasaki, Cumulative spectral analysis for transient decaying signals in a transmission system including a feedback loop. J. Audio Eng. Soc. **54**(7/8), 620–629 (2006)
6. Y. Hirata, M. Tohyama, M. Matsumoto, S. Gotoh, Health monitoring for a structure using its non-stationary vibration. Advances in Acoustics and Vibration, vol. 2010, Article ID 696512 (Hindawi Publishing Corporation, 2010). https://doi.org/10.1155/2010/696512
7. S. Goto, Y. Takahashi, M. Tohyama, Structural vibration monitoring using cumulative spectral analysis. *Advances in Acoustics and Vibration*, vol. 2013, Article ID 158329 (Hindawi Publishing Corporation, 2013). https://doi.org/10.1155/2013/158329
8. M. Tohyama, *Sound and Signals* (Springer, Berlin, 2011)
9. M. Tohyama, *Waveform Analysis of Sound* (Springer, Berlin, 2015)

Chapter 7
Signal Dynamics for Time and Frequency Domains

Abstract This chapter describes the basic nature of signal dynamics (that is, changes in the signal within the time and frequency domains). Typical examples of modifications in the signal dynamics arise from ambient noise and sound path. The random nature of noise is understood as processes in addition and multiplication. Random noise processes are modeled as unpredictable external sources to a linear system. Information concerning the sound path for waves of a linear system can be obtained even when a nondeterministic source excitation is present from the statistical frequencies (histogram) with respect to frame-wise power spectral maxima. The frame-wise maxima interpolated from the frame-wise spectra are also informative in source signature analyses, for example, in the temporally changing fundamental frequency tracking of musical instruments, speech, and singing voices. Power spectral maxima characterize the resonance structures in the source or sound path structures. The sound path is quite likely determining those resonances which are with close frequencies. Overlapping resonance modes require, in general, the decomposition of the modal responses. With inspiration from human hearing, a filter set made of three narrow-band filters is a possible way to decompose the overlapping response. Another way is solving a set of linear equations that represents overlapping modes based on the least-squares criterion—referred here as clustered line spectral modeling (CLSM). The envelope representing the signal dynamics in the time domain can be characterized from CLSM because the envelope is composed of clustered frequency components around the resonance frequency. When extended into the time domain, CLSM is here referred to as CTSM (clustered time sequence modeling). Signature analysis of the piano-string vibration provides a nice example of CLSM and CTSM.

Keywords Enhancement of signal dynamics · Synchronized addition · Synchronized multiplication · Sound path information · Spectral peak selection · Spectral peak estimation · Fundamental frequency tracking · Auto-correlation function · Modal overlap · Envelope · Separation of auto-correlation functions · Spectral representation of time envelope · CLSM · CTSM

M. Tohyama, *Sound in the Time Domain*, Signals and Communication Technology,
https://doi.org/10.1007/978-981-10-5889-9_7

7.1 Enhancement of Signal Dynamics for Noisy Waveforms

7.1.1 Superposition of Noisy Signals

A representative merit of binaural hearing would indicate whether a signal can be heard in random noise. Suppose that two sound signals

$$x(t) = s(t) + n_1(t) \tag{7.1}$$
$$y(t) = s(t) + n_2(t) \tag{7.2}$$

are fed into binaural inputs. Assuming time-synchronized addition,

$$\begin{aligned} z(t) &= x(t) + y(t) \\ &= 2s(t) + n_1(t) + n_2(t) \end{aligned} \tag{7.3}$$

where $s(t)$ denotes the sound of interest whereas $n_1(t)$ and $n_2(t)$ represent independent random noise. The power ratio between sound and noise is given by

$$r_M = \mathrm{E[S^2]/E[N^2]} = s_0^2/n_0^2 \tag{7.4}$$

where

$$\mathrm{E[S^2]} = s_0^2 \tag{7.5}$$
$$\mathrm{E[N^2]} = \mathrm{E[N_1^2]} = \mathrm{E[N_2^2]} = n_0^2. \tag{7.6}$$

Taking the power ratio after the addition yields

$$r_B = \mathrm{E[(2S)^2]/E[(N_1 + N_2)^2]} = 4s_0^2/2n_0^2 \tag{7.7}$$

which is twice that before addition. Figure 7.1 shows an example of the merit by time-synchronized addition where

$$x(t) = \cos\omega_0 t + n_1(t) = s(t) + n_1(t), \tag{7.8}$$

and

$$y(t) = \cos\omega_0 t + n_2(t) = s(t) + n_2(t), \tag{7.9}$$

$n_1(t)$, and $n_2(t)$ are independent random variables $(-1 \leq N \leq 1)$ with $\mathrm{E[N]} = \mu = 0$ and $\mathrm{E[N^2]} = \sigma^2 = 1$. By addition, the change in signal in the time domain (signal dynamics) can be emphasized. Figure 7.2 shows another example of the power ratio between a signal and noise.

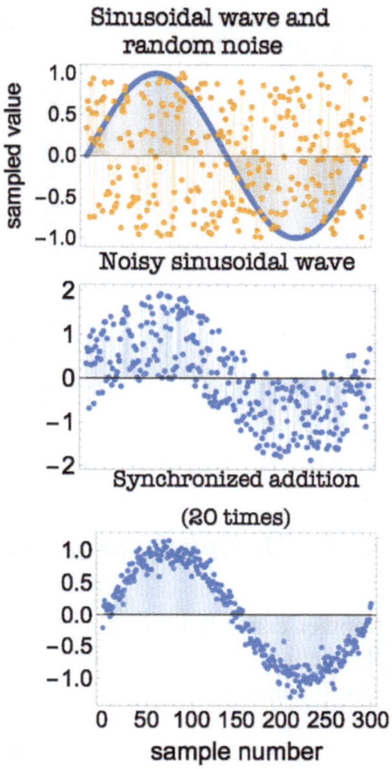

Fig. 7.1 Example of time-synchronized addition

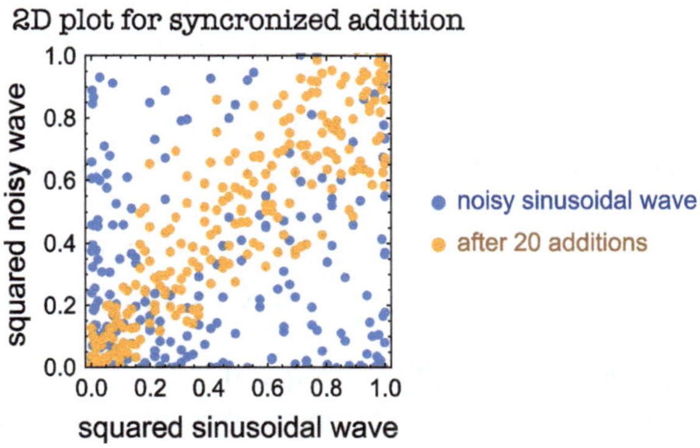

Fig. 7.2 Two-dimensional representation of the power ratio for signal and noise using the sample is given in Fig. 7.1

The horizontal and vertical axes represent $s^2(t)$ and either $x^2(t)$ or $z^2(t)$ with $z(t) = x(t) + y(t)$. The expectation values are

$$\mathrm{E}[\mathrm{Z}^2] = 4s_0^2 + 2n_0^2 \tag{7.10}$$

$$\mathrm{E}[\mathrm{X}^2] = s_0^2 + n_0^2. \tag{7.11}$$

after a single addition.

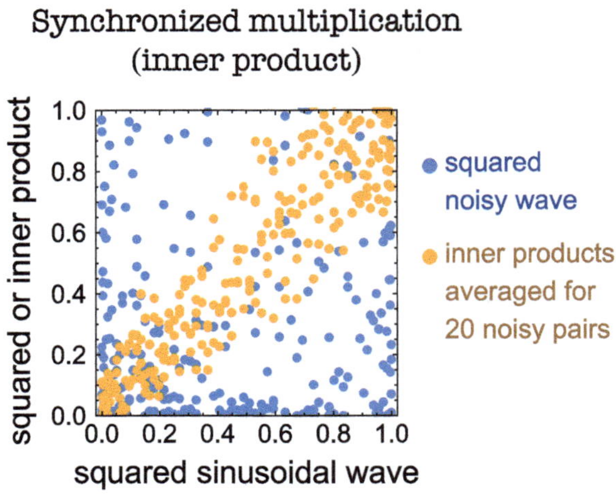

Fig. 7.3 Examples of synchronized multiplication

7.1.2 *Multiplication of Noisy Signals*

Another example illustrating binaural merit relates to synchronized multiplication [1] like the correlation coefficient. Consider once more the pair of variables given in Eqs. 7.8 and 7.9

$$x(t) = \cos \omega_0 t + n_1(t) \quad (7.12)$$
$$= s(t) + n_1(t) \quad (7.13)$$
$$y(t) = s(t) + n_2(t)$$

where $n_1(t)$, and $n_2(t)$ are independent random variables with $\mu = 0, \sigma^2 = 1$. Taking the synchronized multiplication, its expectation then becomes

$$\mathrm{E}[\mathrm{X} \cdot \mathrm{Y}] = \mathrm{E}[\mathrm{S}^2] \quad (7.14)$$
$$\mathrm{E}[\mathrm{X}^2] = \mathrm{E}[\mathrm{S}^2] + \mathrm{E}[\mathrm{N}^2] \quad (7.15)$$

subject to constraint $\mathrm{E}[N_1 \cdot N_2] = 0$.

Figure 7.3 shows an example of synchronized multiplication where the horizontal axis represents $s^2(t)$ and the vertical axis represents either $x^2(t)$ or $z^2(t)$ with $z(t) = x(t)y(t)$. The figure indicates that the signal dynamics in the time domain can be emphasized by taking the expectation after multiplication.

7.2 Estimation of Path Information

In a linear system, the output signal can be written as a convolution between the input signal and the impulse response of the system of interest. Explicitly,

$$y(t) = x * h(t), \tag{7.16}$$

where $x(t)$ and $h(t)$ denote the input signal and the impulse response, respectively. Similarly, the auto-correlation function of the response can be expressed as

$$r_{YY}(t) = r_{XX} * r_{HH}(t). \tag{7.17}$$

That is, the power spectral density function becomes

$$\begin{aligned} \mathrm{E}\left[|Y(\omega)|^2\right] &= \mathrm{E}\left[|X(\omega) \cdot H(\omega)|^2\right] \\ &= \mathrm{E}\left[|X(\omega)|^2\right]|H(\omega)|^2. \end{aligned} \tag{7.18}$$

The power spectral density function or the auto-correlation function depends on both the input signal and the impulse response. The auto-correlation function of the impulse response that represents the path information of the linear system may not be well estimated unless the input signal is known. In addition, if $\mathrm{E}[|X(\omega)|^2]$ is dependent on the observation time interval, as for the signal dynamics of a non-stationary signal, then it would be difficult to estimate the auto-correlation function or the power spectral density function of the impulse response in accordance with relation Eq. 7.18. Daily experiences such as tuning a sound transmission channel in a public address system for an auditorium, however, suggest that path information may be estimated more or less without detailed information of the input signal.

Assume that $|X(\omega)|^2$ in every observation record becomes maximally and equally likely for all angular frequencies ω. If the function $|H(\omega)|^2$ is constant and independent of the angular frequency, then $|Y(\omega)|^2$ might be also maximally and equally likely for all angular frequencies. This suggests that the statistical frequency, as a function of angular frequency, for the maxima in the observation intervals for $|Y(\omega)|^2$ might be expected to be proportional to $|H(\omega)|^2$ [2–4].

Figure 7.4 illustrates examples of estimations of power spectral density functions for an ideal case for which $\mathrm{E}|X(\omega)|^2$ might be expected to be uniform with respect to ω [5]. The impulse responses are sums of a pair of decaying sinusoidal functions such as

$$h_1(t) = A\left(e^{i\omega_1 t} + e^{i\omega_2 t}\right)e^{-\delta t} \tag{7.19}$$

$$h_2(t) = A\left(e^{i\omega_1 t} - e^{i\omega_2 t}\right)e^{-\delta t} \tag{7.20}$$

$$W_L = \frac{W}{L} \tag{7.21}$$

$$L = \frac{2\pi}{\omega_2 - \omega_1} \tag{7.22}$$

where W denotes the length of a single observation record. The statistical frequency for taking the frame-wise maxima of $|Y(\omega)|^2$ as a function of the angular frequency ω gives estimations of the power spectral density functions of $|H_1(\omega)|^2$ and $|H_2(\omega)|^2$, respectively. Estimates of $|H_1(\omega)|^2$ were obtained not only for peaks in the power

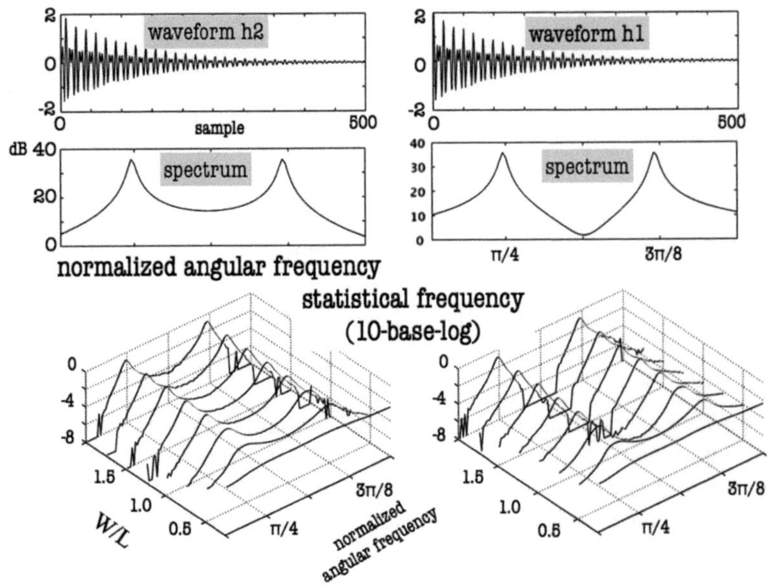

Fig. 7.4 Estimations of several power spectral density functions using statistical frequencies of frame-wise-maxima in the power spectral observation time frames of the output responses [6]

spectral density function but also for troughs. Troughs can be estimated at angular frequencies for which the maxima are most unlikely. The example above is only an ideal case where really nonstationary noise is not taken; however, the power spectral density function can be estimated without requiring estimates of $\mathrm{E}\left[|X(\omega)|^2\right]$.

7.3 Spectral Peak Selection and Representation of Time Waveforms

7.3.1 Estimation of Spectral Sequence of Sinusoidal Function

In Sect. 5.4, examples in which the number of cycles in the observation interval is an integer were only considered. In practice, an observation interval may not necessarily contain an integer number of cycles even for a sinusoidal time waveform. An estimation of the spectral sequence, however, is not a simple matter when $N/P_0 \neq m$ where N denotes the length of the interval and P_0 the period to be estimated for the time waveform. Figure 7.5 gives an example in which $N/P_0 \neq m$ when $N = 64$.

The period of the waveform would appear difficult to estimate using the spectral sequence given by the discrete Fourier transform. No clear spectral peak corresponding to the frequency of the sinusoidal function is observed, although, there may be a possible spectral peak between $k_0 = 4$ and 5.

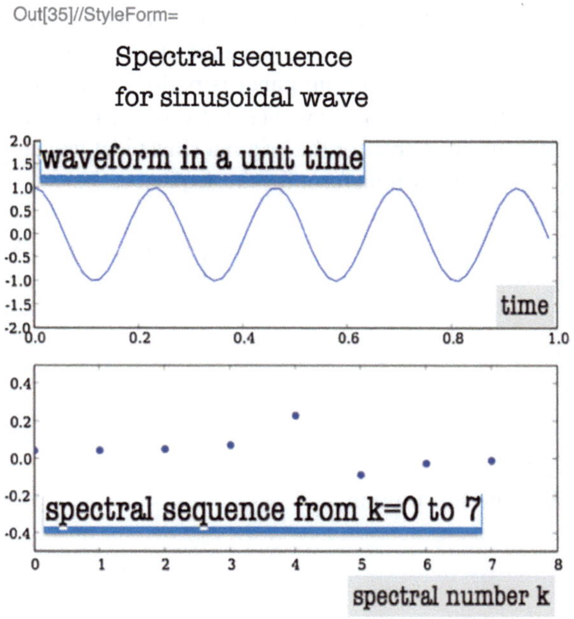

Fig. 7.5 Spectral sequence (real part) for sinusoidal time waveform when cycles in observation interval are not integer

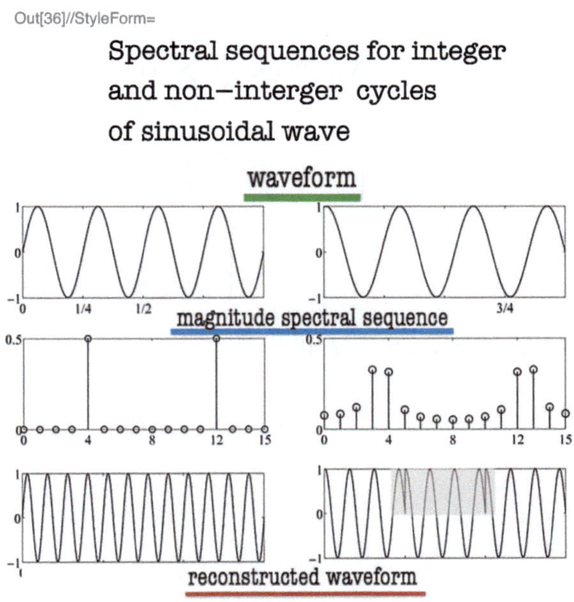

Fig. 7.6 Comparison of instances when N/P_0 is integer and non-integer from Fig. 8.1 in [7]

Figure 7.6 illustrates a formal comparison in the spectral sequences.

The left panels show an instance when N/P_0 is an integer. The spectral sequence presents the peaks corresponding to the period of the time waveform. In contrast, the right panels present an instance when N/P_0 is non-integer, giving the spectral sequence and synthesized waveform in the middle and lower panels. The spectral sequence does not provide the estimation for the period. Note that the synthesized waveform in the bottom panel is periodic with period N but no longer of period P_0. The discontinuous synthesized waveform explains why the spectral sequence is different from that for the original sinusoidal waveform.

The Fourier transform for the sequence,

$$x(n) = \begin{cases} x(n) & 0 \leq n \leq N-1 \\ 0 & \text{otherwise} \end{cases} \tag{7.23}$$

gives

$$X(\mathrm{e}^{-\mathrm{i}\Omega}) = \frac{1}{N}\sum_{n=0}^{N-1} x(n)\mathrm{e}^{-\mathrm{i}\Omega n}. \tag{7.24}$$

With the substitution

$$x(n) = A\mathrm{e}^{\mathrm{i}\Omega_0 n} \tag{7.25}$$

in Eq. 7.24, the Fourier transform can be rewritten as

$$\begin{aligned} X(\mathrm{e}^{-\mathrm{i}\Omega}) &= \frac{1}{N}\sum_{n=0}^{N-1} x(n)\mathrm{e}^{-\mathrm{i}\Omega n} \\ &= \frac{A}{N}\sum_{n=0}^{N-1} \mathrm{e}^{\mathrm{i}\Delta\Omega n} \\ &= \begin{cases} A & \Omega_0 = \Omega(\Delta\Omega = 0) \\ \frac{A}{N}\frac{1-\mathrm{e}^{\mathrm{i}\Delta\Omega N}}{1-\mathrm{e}^{\mathrm{i}\Delta\Omega}} & \Omega_0 \neq \Omega(\Delta\Omega \neq 0) \end{cases} \end{aligned} \tag{7.26}$$

where $\Delta\Omega = \Omega_0 - \Omega$. If

$$N = mP_0 = m\frac{2\pi}{\Omega_0} \tag{7.27}$$

$$\Omega_0 N = 2m\pi \tag{7.28}$$

then

$$\begin{aligned} 1 - \mathrm{e}^{\mathrm{i}\Delta\Omega N} &= 1 - \mathrm{e}^{-\mathrm{i}\Omega N} \\ &= \begin{cases} 0 & \Omega = l\Omega_0 \\ 1 - \mathrm{e}^{-\mathrm{i}\Omega N} & \Omega \neq l\Omega_0 \end{cases} \end{aligned} \tag{7.29}$$

where P_0 denotes the period of the sinusoidal sequence. The result above implies

$$X(\mathrm{e}^{-\mathrm{i}\Omega})|_{\Omega=l\Omega_0} = 0 \tag{7.30}$$

if $l > 1$ is an integer subject to Eq. 7.28. Therefore, the Fourier expansion yields no spectral components except at $\Omega = \Omega_0$, if $\Omega_0 N = 2m\pi$. The spectral function, however, shows the original sinusoidal sequence at $\Omega = \Omega_0$ even in the general situation when $N \neq mP_0$. The angular frequency Ω_0 can be estimated by taking the spectral peak of $|X(\mathrm{e}^{-\mathrm{i}\Omega})|^2$.

The above result is the same as that for its discrete Fourier transform. The sequence under the discrete Fourier transform is the sequence sampled from the Fourier transform that gives the continuous spectral function. The Fourier transform can be estimated by interpolation of the discrete Fourier transform. Suppose a sequence $\hat{x}(n)$ such as

$$\hat{x}(n) = \begin{cases} x(n) & 0 \leq n \leq N-1 \\ 0 & N \leq n \leq M-1. \end{cases} \tag{7.31}$$

The discrete Fourier transform of $\hat{x}(n)$ can be written as

$$\begin{aligned} \hat{X}(k) &= \frac{1}{N}\sum_{n=0}^{N-1} x(n)\mathrm{e}^{-\mathrm{i}\frac{2\pi}{M}kn} \\ &= \frac{1}{N}\sum_{n=0}^{N-1} x(n)\mathrm{e}^{-\mathrm{i}\Omega n}|_{\Omega=\frac{2\pi k}{M}} \end{aligned} \tag{7.32}$$

which corresponds to the sequence obtained by sampling $X(\mathrm{e}^{-\mathrm{i}\Omega})$ at $\Omega = 2\pi k/M$. If $M > N$, the spectral sequence is interpolated from its N-point to M-point discrete Fourier transforms. When $M \to \infty$, the Fourier transform of the sequence exhibits a limit in its discrete Fourier transform. The angular frequency Ω_0 at which the power spectral sequence is maximum and produces a spectral peak can be estimated by interpolation of the spectral sequence given by the discrete Fourier transform. The interpolation can be performed by appending zeros to the original sequence $x(n)$.

Figure 7.7 illustrates the point that a spectral peak can be estimated by interpolation or zero-padding. By finding Ω_0, the spectral sequence that gives the original sinusoidal waveform can then be read from the spectral sequence at $\Omega = \Omega_0$. The estimated original sinusoidal sequence predicts the function outside the observation interval because the original waveform can be estimated independent of the length of the observation interval. Fig. 7.8 illustrates an example of a compound sinusoidal function. The spectral peaks can be estimated and subtracted one-by-one. The sinusoidal component estimated by the spectral peak can be removed by subtracting the component from the time waveform in the observation time interval. The process of estimation and subtraction is repeated until all prominent peaks disappear [8].

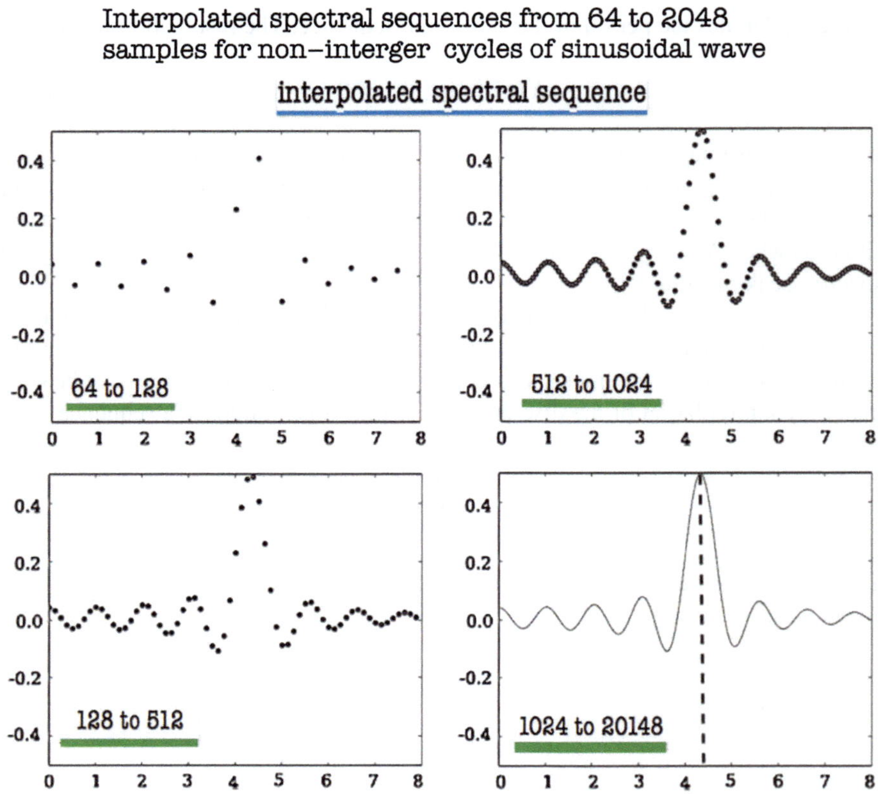

Fig. 7.7 Interpolation of a spectral sequence (real part) for a sinusoidal sequence with $A = 1$ from Fig. 14.13 [5]

However, it might be still difficult to estimate even a single sinusoidal component because a real sinusoidal sequence is a sum of two complex exponential functions. The Fourier transforms of the two complex exponential components overlap when the length of the observation interval or the spacing of the two components is not long or wide enough to discriminate the two responses (Fig. 7.9). Estimating the spectral components adequately may be difficult when the observation interval is too short because of overlapping spectral functions. This issue is developed in next subsection.

7.3.2 Spectral Peak Estimation for Single Sinusoidal Sequence

In the previous subsection, if the observation interval is too short, as illustrated in Fig. 7.9, the spectral estimation of a single real sinusoidal function proves

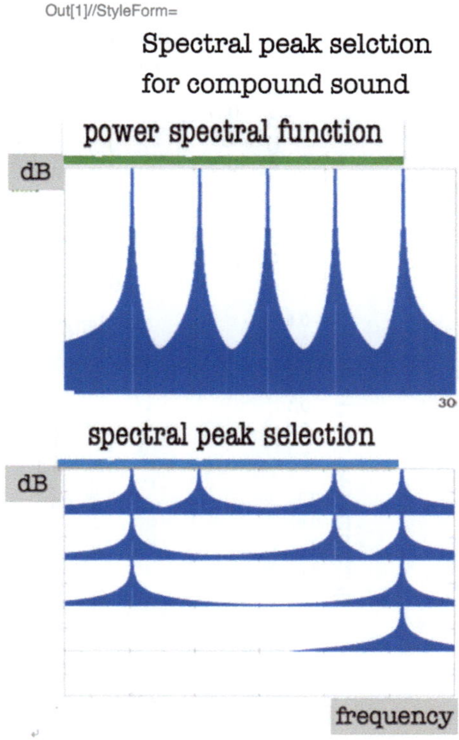

Fig. 7.8 Example of a waveform analysis for a compound sinusoidal function from Fig. 14.14 [5]

problematic. The spectral peak observed at $\Omega = \Omega_0$ for a single real sinusoidal function may be described by

$$X(e^{-i\Omega_0}) = \frac{A}{2}(1 + W(-2\Omega_0)), \tag{7.33}$$

where

$$W(\Omega) = \frac{1}{N}\frac{1 - e^{i\Delta\Omega N}}{1 - e^{i\Delta\Omega}}, \tag{7.34}$$

with $\Delta\Omega = \Omega_0 - \Omega$ and N denoting the observation length. Hence, the true spectral peak $A/2$ can be estimated using

$$\frac{A}{2} = \frac{X(e^{-i\Omega_0})}{1 + W(-2\Omega_0)}. \tag{7.35}$$

Figure 7.10 presents an example illustrating spectral peak estimation following Eq. 7.35. The amplitude of a single real sinusoidal function can be estimated using spectral peak selection even if the observation length is short.

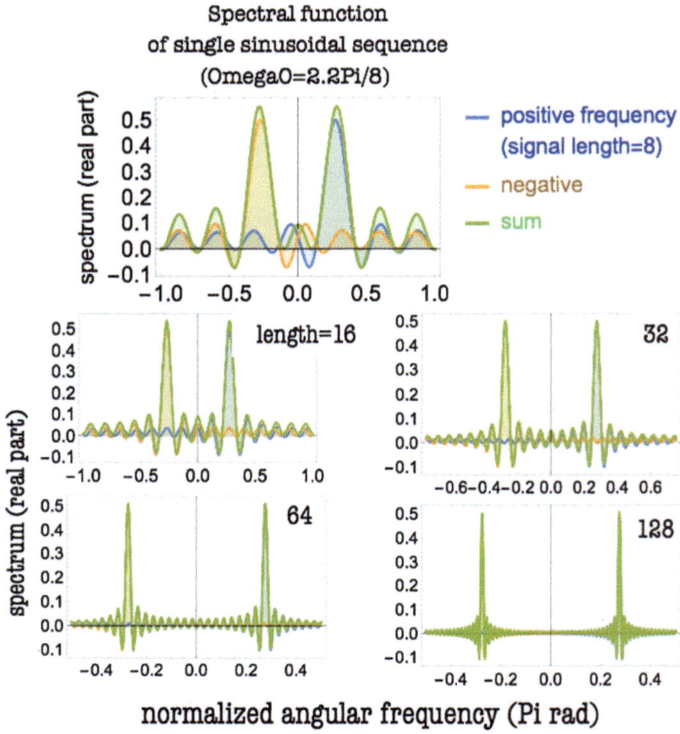

Fig. 7.9 Fourier transform of a single sinusoidal function

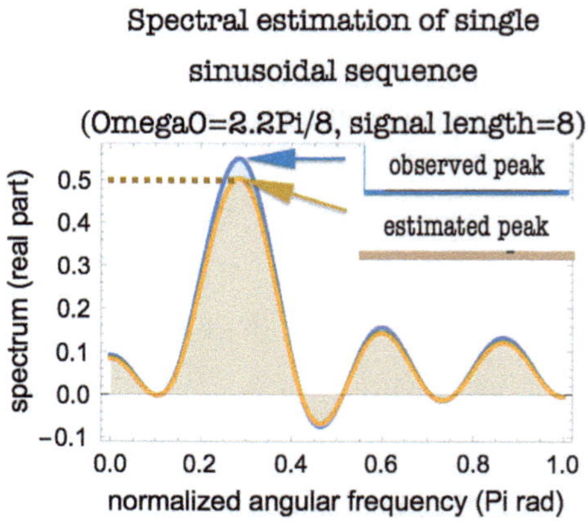

Fig. 7.10 Diagram illustrating a spectral peak estimate obtained from a short observation interval

7.3.3 Time-Variant Spectral Peaks in Time Waveforms

A time waveform is viewed as a series of sinusoidal components that produce prominent spectral peaks in short time frames. This is particularly so for sound from a musical instrument. Figure 7.11 shows an example of sound from a piano [9, 10].

In general, musical instruments generate slowly decaying sounds that do not decrease monotonically but vary in a nonstationary manner. A possible approach to time-dependent signature analysis of signals would be to track the prominent spectral peaks over short time frames. Figure 7.11 exhibits the nonharmonic spectral components over a wide frequency range at the initial onset and shows the time-process in which resonant components build up and slowly decay after onset. Figure 7.12 shows several sinusoidal components clustered around the fundamental component. The clustered components may represent the time envelope of the fundamental component.

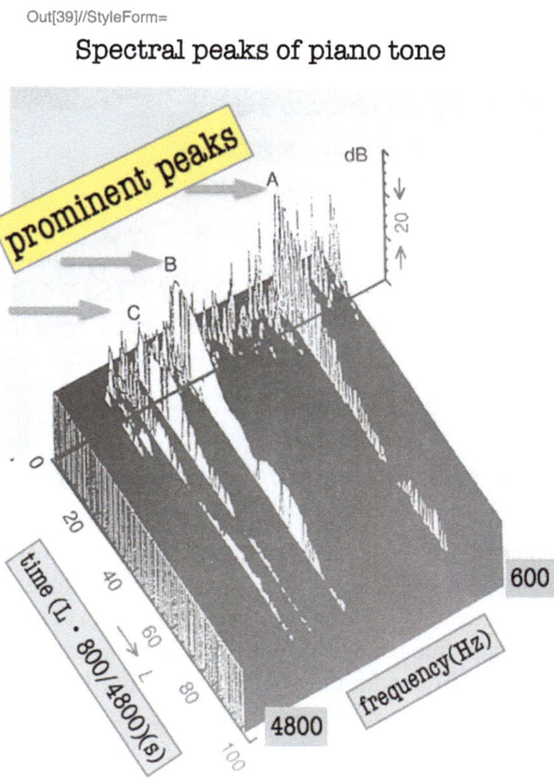

Fig. 7.11 Example of time-dependent spectral peaks in a piano tone from Fig. 2a [9]

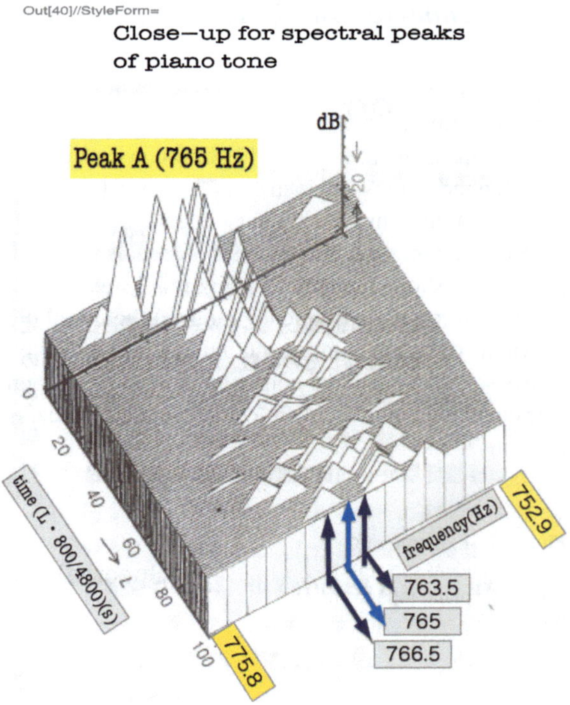

Fig. 7.12 Magnification of the spectral components in Fig. 7.11 from Fig. 2b [9]

The residual component gives the difference between the original and synthesized time waveforms using spectral peak selection in Fig. 7.13. Nonharmonic components distributed over a wide range of frequencies may concentrate around an initial interval.

7.3.4 *Fundamental Frequency Tracking of Time Waveform*

Tracking the fundamental frequencies of a time waveform would be an attractive and significant aspect of waveform analysis in the time and frequency domains. Spectral peak selection would be a relevant approach in tracking. Temporal tracking of the fundamental frequencies including fluctuations can be performed. Figure 7.14 presents time waveforms for the original, synthesized, and error waveforms. The waveforms were mostly represented by repeating the process for spectral peak selection and subtraction.

Figure 7.15 shows another example of the spectral peak selection for speech waveform. The upper panel is the original waveform and the lower panel plots the five dominant frequencies that create the prominent spectral peaks in the corresponding short time frames.

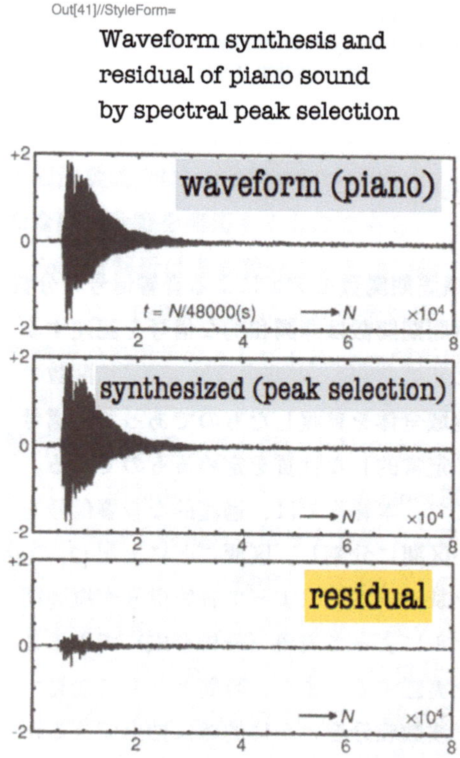

Fig. 7.13 Time waveform of piano sound analyzed in Fig. 7.11 from Fig. 1 [9]

The solid line shows the fundamental frequency estimated from the five sinusoidal components (indicated by dots) using the auto-correlation function. Figure 7.16 displays a sample of the auto-correlation function for the five prominent sinusoidal components, in which the fundamental period and therefore frequency can be read off. The time-dependent fundamental frequencies and their fluctuations would be significant in naturally sounding speech.

The auto-correlation analysis suggests that the best-matched frame length for the segmental time waveform may be the fundamental period. Figure 7.17 compares the power spectral sequences for frame lengths of 512 and 431, where 431 is the length suggested by the auto-correlation function for the five prominent sinusoidal components. The fundamental and harmonics are clearly seen in the spectral sequence using the best-matched window length.

Changing the length of the observation interval by adapting the time waveform may provide a possible approach to estimate the fundamental frequency tracking.

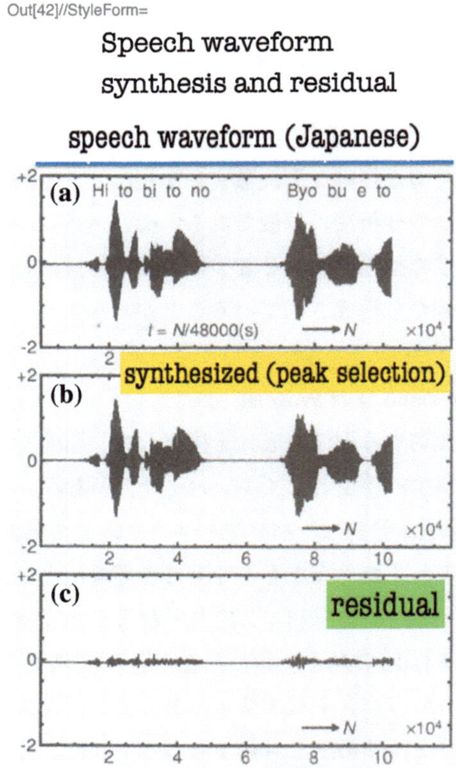

Fig. 7.14 Speech waveform in Japanese synthesized by spectral peak selection from Fig. 4 [9]

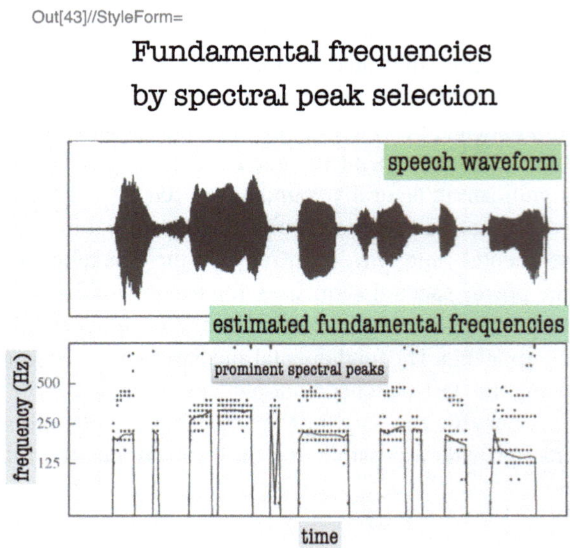

Fig. 7.15 Fundamental frequency tracking for speech waveform using spectral peak selection and auto-correlation analysis from Fig. 6 [11]

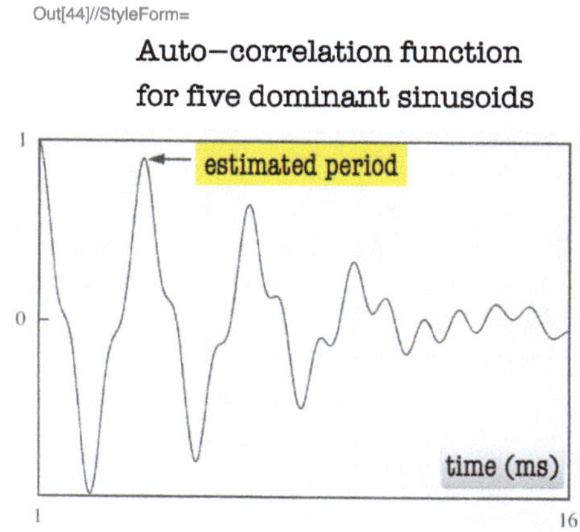

Fig. 7.16 Example of an auto-correlation function using the five prominent sinusoidal components from Fig. 5 [11]

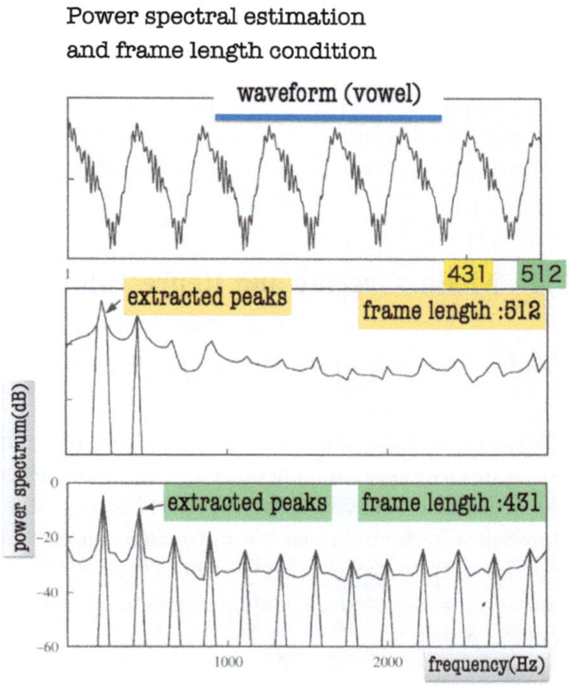

Fig. 7.17 Comparison of power spectral sequence for two window lengths such as 512 and 431 for a vowel portion of the speech, where 431 is suggested from the auto-correlation function from Fig. 7 [11]

Figure 7.18 demonstrates an example of fundamental frequency tracking of a female chorus by adopting the best-matched frame length frame-by-frame [12, 13]. The three vocal parts can be read in the time-dependent fundamental frequencies.

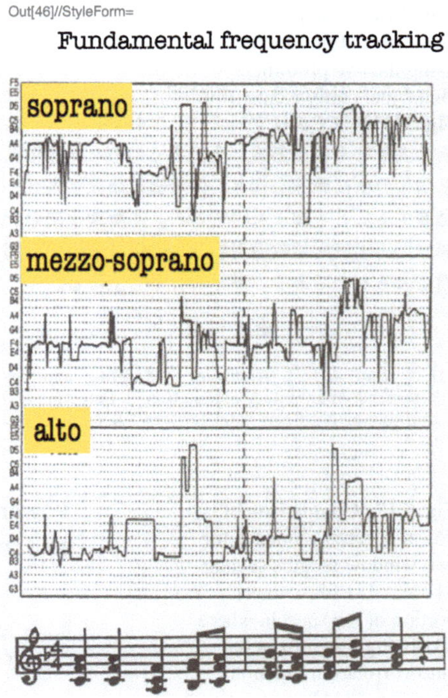

Fig. 7.18 Fundamental frequency tracking of a female vocal chorus (My Old Kentucky Home) from Fig. 6 [13]

7.4 Modal Overlap and Representation of Envelopes

7.4.1 Separation of Auto-correlation Functions

The auto-correlation function is representative of signal signatures in the time domain [14]. The power spectral functions determined by the poles that govern the form of the auto-correlation functions are expressed using spectral peak selection. Figure 7.19 shows power spectral functions in which a pair of poles is seen. The auto-correlation function may be approximately expressed as shown in Fig. 7.20, in which two cosine functions are superposed, and the frequencies of the cosine functions correspond to spectral peaks.

However, if the frequencies of the poles are located too close to identify the spectral peaks distinctly, as shown in Figs. 7.21 and 7.22, it is difficult to properly estimate the auto-correlation function. The responses resulting from the resonance poles can be interpreted as frequency characteristics of a narrow-band filter. If the bandwidth of the filter response is narrow, a sinusoidal wave can be clearly detected even from a compound sound. In contrast, if the responses of two nearby filters corresponding to a pair of poles overlap, then it does not discriminate the individual sinusoids from the compound waves. A way to sharpen the response so that an individual sinusoid might be separated would be to create zeros on both sides of the frequency corresponding

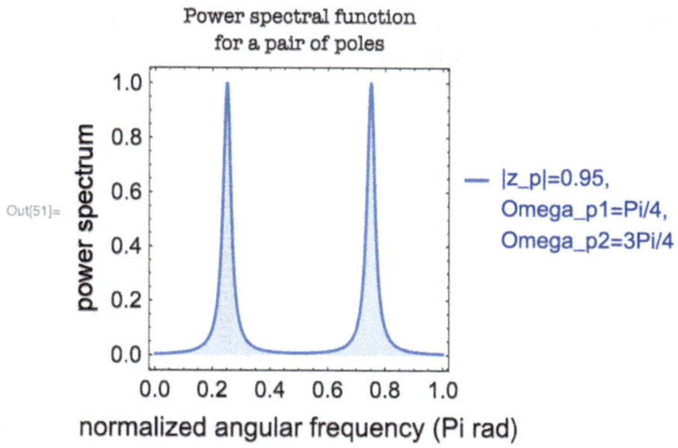

Fig. 7.19 Power spectral functions due to a pair of poles

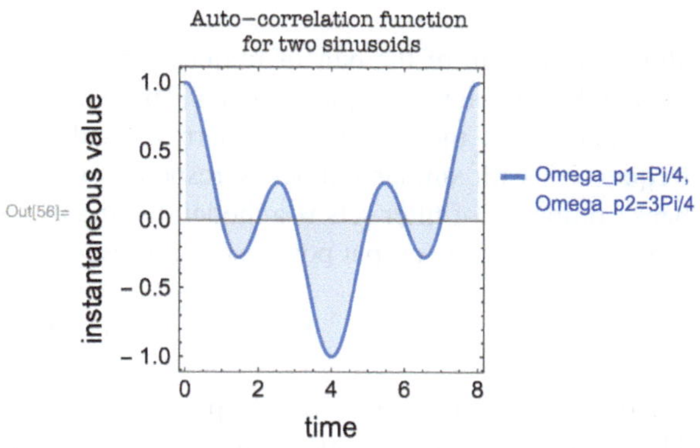

Fig. 7.20 Correlation functions superposed by two cosine functions corresponding to the poles

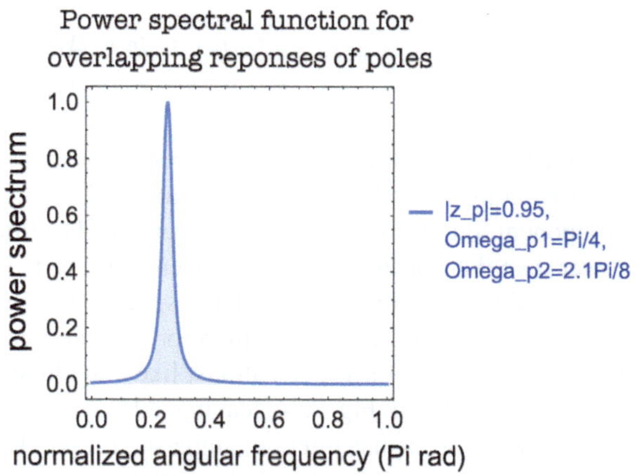

Fig. 7.21 Power spectral functions for overlapping two-pole responses

to the pole [15]. The pair of zeros would be helpful to make the response due to the pole more prominent with the narrower bandwidth. A pair of resonant power spectral responses produces a zero in between the pair of pole frequencies by subtraction,

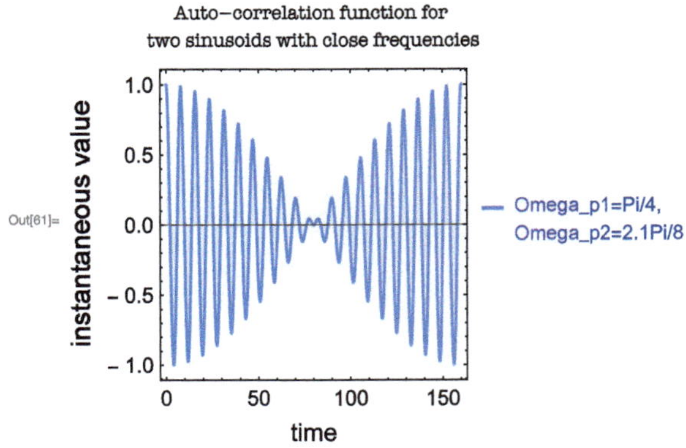

Fig. 7.22 Correlation function for two superposed cosine functions of close frequencies

although the sum of the pair of frequency responses including phases produces a zero in the complex frequency domain [16–18].

Suppose that there is a set of narrow-band filters composed of three central frequencies with similar frequency responses to the single pole response. A filter bank composed of filter sets is a model that draws inspiration from human hearing [15, 19]. Setting the output power spectral response such that

$$\hat{Y}_0 = |2Y_0 - Y_{-1} - Y_{+1}|, \tag{7.36}$$

where Y_{-1}, Y_0, and Y_{+1} represent the power responses for the three component filters. The other central frequencies for the filters Y_{-1} and Y_1 are set to

$$f_{-1} = f_0 - \Delta f \tag{7.37}$$
$$f_1 = f_0 + \Delta f. \tag{7.38}$$

Figure 7.23 illustrates the power spectral responses for the filter set composed of three component filters according to Eq. 7.36, where the response for a single component filter is illustrated for comparison. The filtering bandwidth becomes narrower than that for the single component filter because there is a pair of zeros, one on each side of the main lobe of the response. The response is mostly composed of three prominent spectral peaks, therefore the auto-correlation function for the time response of the filter set looks almost like a modulated cosine function with a long period as depicted in Fig. 7.24, leaving a decaying time envelope.

The response of the filter set suggests a possible means of detecting the sinusoidal components more sharply by comparing that with a single component filter. Figure 7.25 illustrates the power spectral responses for two close filter sets. The overlapping near resonances decreases when using a filter set composed of three component filters compared with that for a single component filter. Although the examples above are ideal or limit cases, the results imply that the filter set may be usable also for monitoring temporal changes in frequencies of a sinusoidal wave.

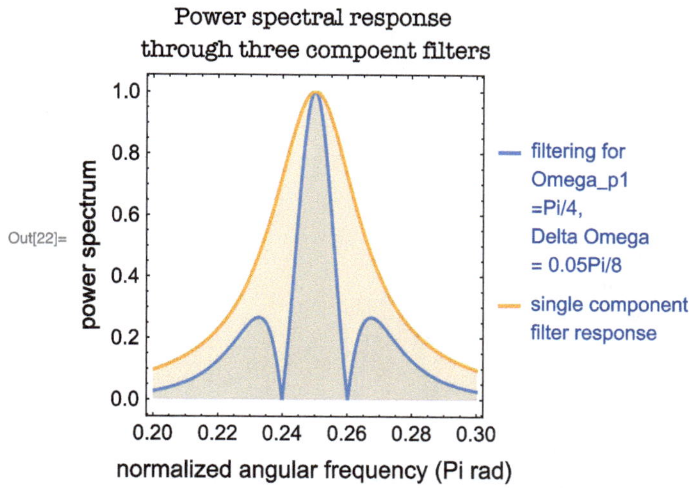

Fig. 7.23 Power spectral response for filter set composed of three component filters according to Eq. 7.36

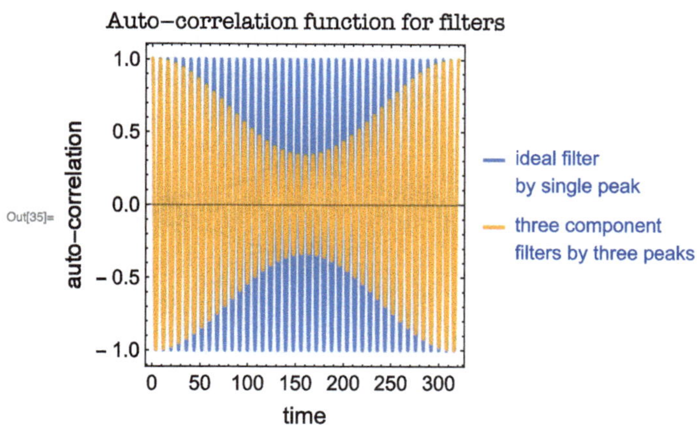

Fig. 7.24 Auto-correlation function representing the filter composed of three component filters

7.4.2 *Estimating Envelopes of Sound in the Time Domain*

The modal response described in the previous subsection assumed some damping of the time-dependent waveform. These decaying waveforms, in principle, cannot be expressed with a single sinusoidal function; an exponential envelope is needed to represent the waveform.

Instead of a single sinusoid [5, 7], an exponential-decaying sinusoid can be expressed as a superposition of several clustered sinusoids in the observation interval, as shown in Fig. 7.26.

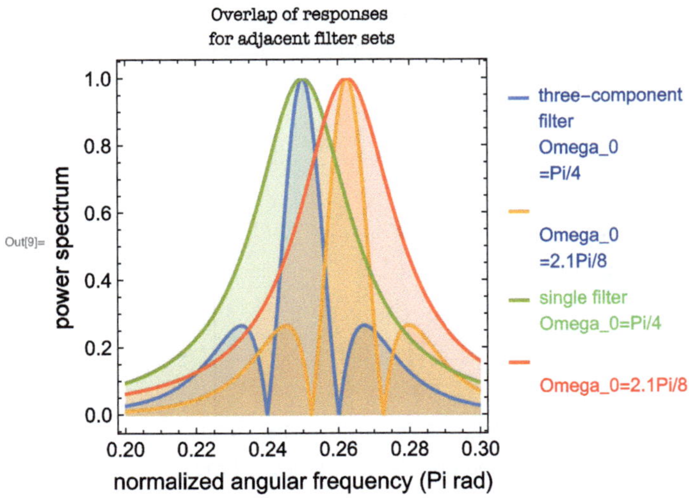

Fig. 7.25 Power spectral responses for adjacent filter sets with close central frequencies

Fig. 7.26 Clustered sinusoids representing exponentially decaying sinusoidal sound

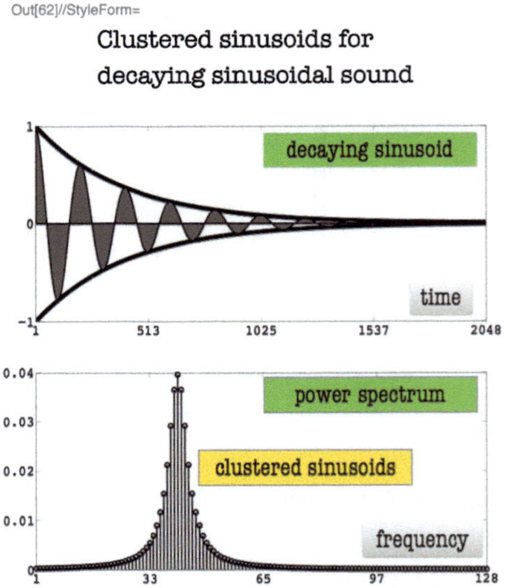

The corresponding sinusoidal components to an exponentially decaying sinusoid can be formally determined as a least-squares-error (LSE) solution of linear equations using clustered line spectral modeling (CLSM) [8, 20]. Hence not only exponentially decaying envelopes but also other envelopes like beats can be represented by CLSM. Finding the LSE solution following the CLSM procedure is similar to decomposing the overlapping responses into corresponding sinusoidal components. Figure 7.27 illustrates a schematic that uses the power spectral presentation instead of the complex functions for the decomposition of the overlapped response to the clustered sinusoids.

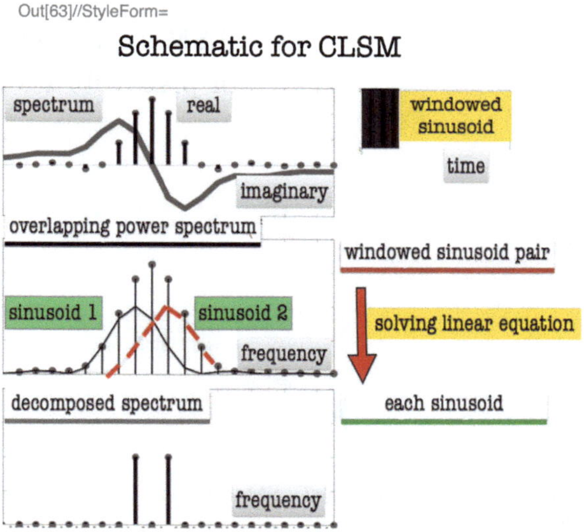

Fig. 7.27 Schematic of CLSM using the power spectral presentation as an example

The decomposition can be formulated as the LSE solution of linear equations that are constructed using the observational spectrum and windowing function [8, 21]. The CLSM is a possible approach to represent a signal with any envelope using several clustered sinusoids. Most of the modulated waveform can be reconstructed, even if the entire or a period of the envelope is not fully observed [5, 7].

7.4.3 Examples of Signal Decomposition into Clustered Sinusoids

CLSM is applicable to transient sounds such as musical instruments. Figure 7.28 shows a schematic of a CLSM experimental setup investigating plate vibration, in which the plate is simply supported at the boundaries and excited by an impulsive hammer.

Figure 7.29 illustrates the CLSM results from a plate vibration. Prominent spectral peaks corresponding to the resonance vibrations of the plate can be observed; however, spectral peak selection does not reproduce the decaying envelope. Assuming three or five sinusoidal components at each spectral peak, the CLSM procedure is repeated iteratively [5, 7, 8] until the prominent peaks disappear. In this manner, the signature of the vibration including the time envelope can be represented as a compound sinusoidal function in the observation interval. Figure 7.30 presents the original, reconstructed, and residual waveforms.

Figures 7.31 and 7.32 show another example of CLSM, this time with the signal of a vibrating piano string. Again the vibration waveform including the envelope is well represented as clustered sinusoids.

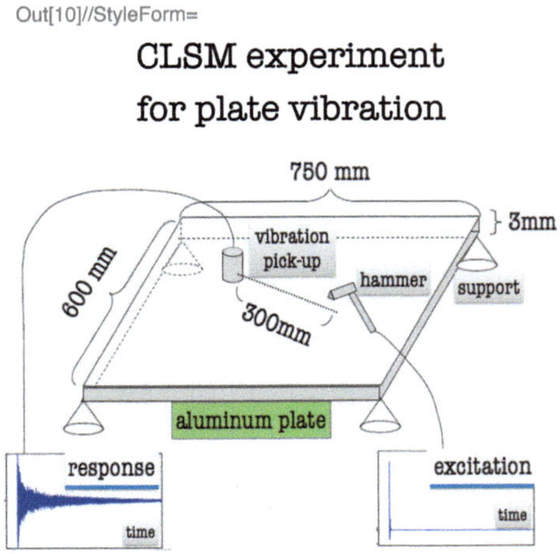

Fig. 7.28 Experiment using CLSM to analyze a simply supported vibrating aluminum plate from Fig. 6 [8]

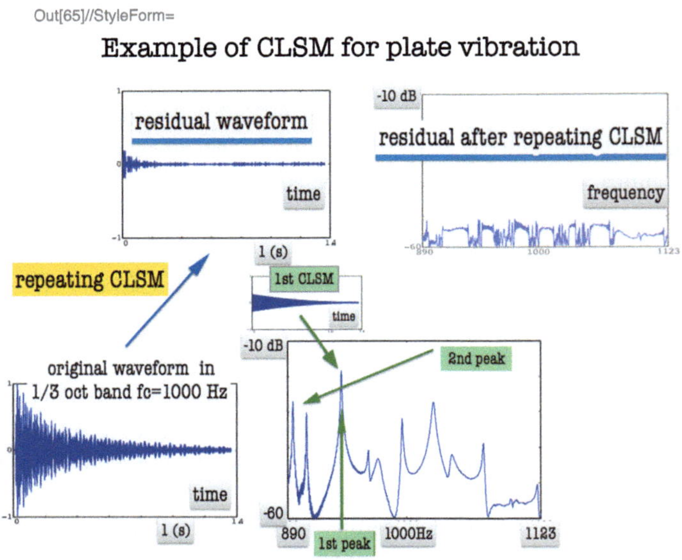

Fig. 7.29 Results from CLSM of plate vibrations from Fig. 5 [8]

7.4.4 Separation of Direct Wave

The observed waveforms of sound can be interpreted approximately as a summation of the direct and reverberant waveforms through a resonant system. The direct sound conveys most of the sound source information, whereas reverberation carries the characteristics of the transmission channels (path information as represented by the transfer function), which is mostly represented by resonances due to the poles

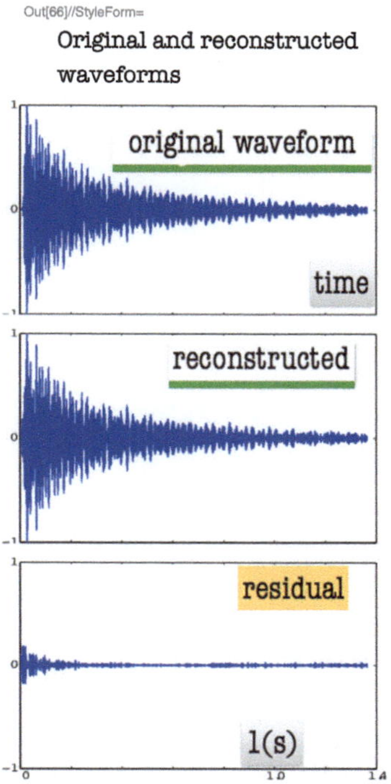

Fig. 7.30 Reconstructed waveform obtained using CLSM for the plate vibration signal along with the residual from Fig. 5 [8]

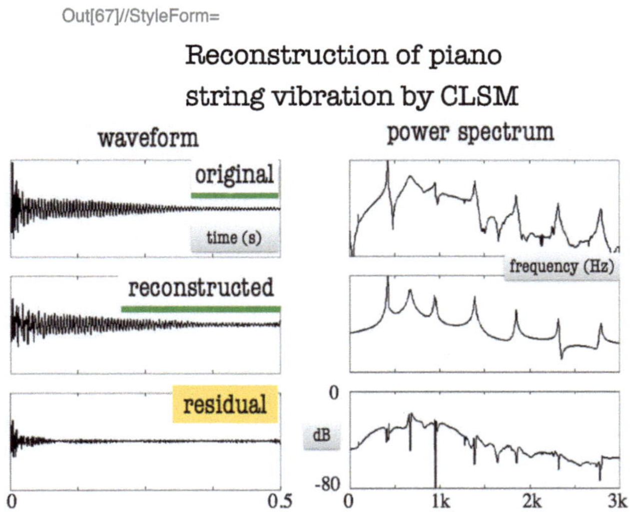

Fig. 7.31 Example of using CLSM to obtain the waveform and spectral representation from the signal of a vibrating piano string from Fig. 5 [22]

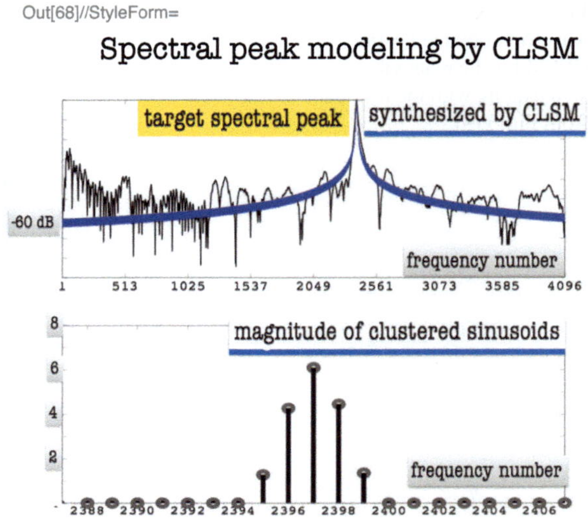

Fig. 7.32 Close-up of spectral peak modeled by clustered sinusoids

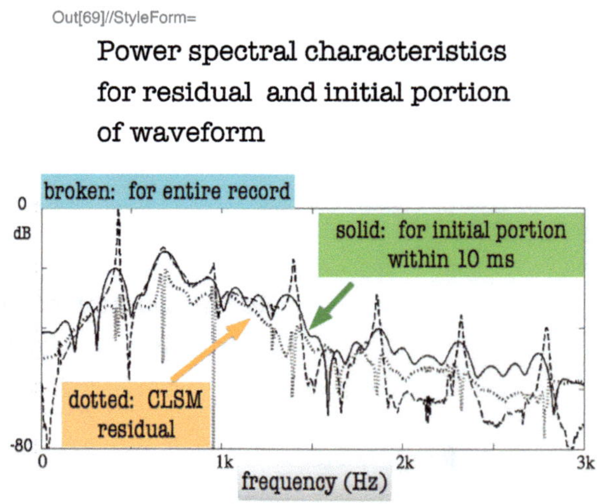

Fig. 7.33 Power spectral functions for the initial 10 (ms) interval of the original waveform, residual component, and the entire original waveform from Fig. 7 [22]

and zeros [5, 7, 16, 17]. The solution of a single-degree-of-freedom system corresponding to a resonant system could be interpreted as a superposition of a particular solution and a so-called transient solution. The transient solution carries path information including resonance properties of the system of interest. Therefore, the residual waveform that is left after CLSM can be interpreted as representing mainly the source information conveyed by the direct wave.

As shown in Fig. 7.31, the residual signal is most apparent in the initial portion of the waveform within a 10(ms) interval. Interestingly, the spectral function for the residual sound mainly follows that for the initial portion as seen in Fig. 7.33. In addition, the initial interval already determines the entire spectral envelope of the original waveform. Hence, the initial interval of the residual still conveys a significant

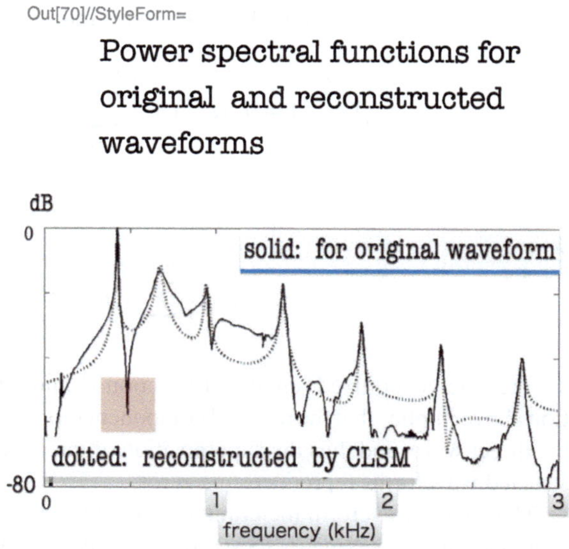

Fig. 7.34 Power spectral comparison between the original and reconstructed waveforms from Fig. 2 [23]

signature of the original waveform. In other words, the residual components, to be superposed onto the reconstructed waveform using CLSM, are invaluable in rendering, in particular, the initial 10 (ms) interval of the waveform.

The residual component, however, does not by itself well represent the first very prominent trough of the original spectral function around 500 (Hz) as confirmed in Fig. 7.34. The trough or zero between an adjacent pair of poles is formulated by solving the equation so that

$$\begin{aligned} H(z^{-1}) &= \frac{A}{1-az^{-1}} + \frac{B}{1-bz^{-1}} + R(z^{-1}) \\ &= H_p(z^{-1}) + R(z^{-1}) \\ &= 0, \end{aligned} \tag{7.39}$$

where a and b denote the pair of adjacent poles, A and B are the residues of the poles, and $R(z^{-1})$ represents the response from the resonance modes other than those associated with the two poles [5, 7, 16, 17]. The remainder function can be generalized to include the source information [23]. The response record $y(n)$ for the source signal $x(n)$ is represented as

$$Y(z^{-1}) = X(z^{-1})H(z^{-1}) \tag{7.40}$$

by taking the z-transforms. The z-transform $X(z^{-1})$ represents the source characteristics so that

$$X(z^{-1}) = \sum_{n=0}^{N-1} x(n)z^{-n}, \tag{7.41}$$

where $x(0) = 1$. Introducing Eq. 7.41 above into Eq. 7.40,

$$\begin{aligned} Y(z^{-1}) &= H(z^{-1}) + \left[\sum_{n=1}^{N-1} x(n)z^{-n}\right] H(z^{-1}) \\ &= H(z^{-1}) + \hat{R}(z^{-n}), \end{aligned} \tag{7.42}$$

where

$$\hat{R}(z^{-n}) = \left[\sum_{n=1}^{N-1} x(n)z^{-n}\right] H(z^{-1})$$

denotes the generalized remainder function including source effects. The formulation above implies that the residual component represents the source effects including the direct wave, and thus the superposition or interference of the residual and reconstructed waveforms reproduces the zeros.

The source waveform including the direct wave can be identified using clustered time sequence modeling (CTSM) from the initial portion of the waveform [22–24]. Specifically, a brief transient time signal can be decomposed into clustered time sequences following the relationship between the time and frequency domains as well as CLSM. Figure 7.35 presents a schematic of the relationship between the time and frequency domains.

Figure 7.36 outlines the CTSM process where a target waveform is represented by the convolution of the impulse response of the band-limiting filter and the clustered time sequence following the relationship between the time and frequency domains,

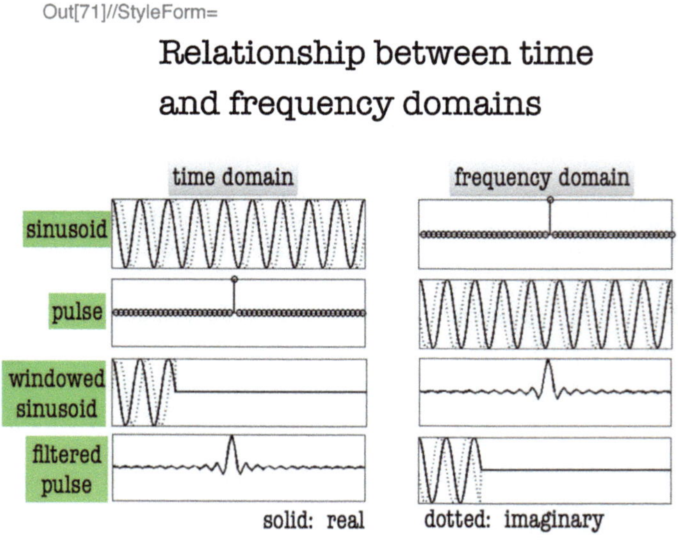

Fig. 7.35 Schematic of the relationship between time and frequency domains from Fig. 8 [22]

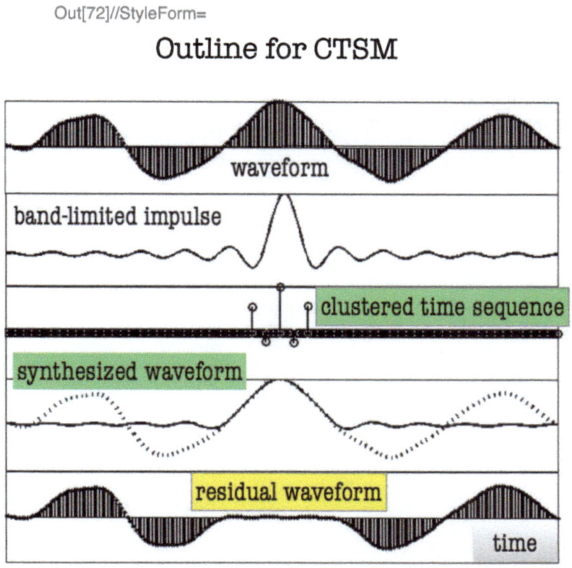

Fig. 7.36 Schematic for CTSM from Fig. 14.42 [5]

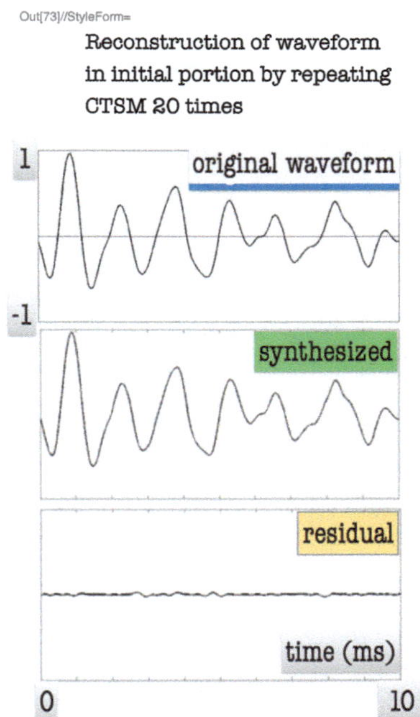

Fig. 7.37 Reconstructed waveform for the initial portion obtained by iterating CTSM 20 times from Figs. 11 and 12 [22]

as sketched in Fig. 7.35. The initial portion of the original waveform can be reconstructed by CTSM as shown in Fig. 7.37.

Figure 7.38 presents the dominant transient signals constructing the synthesized waveform in the time order where each pulse-like wave travels on the string and

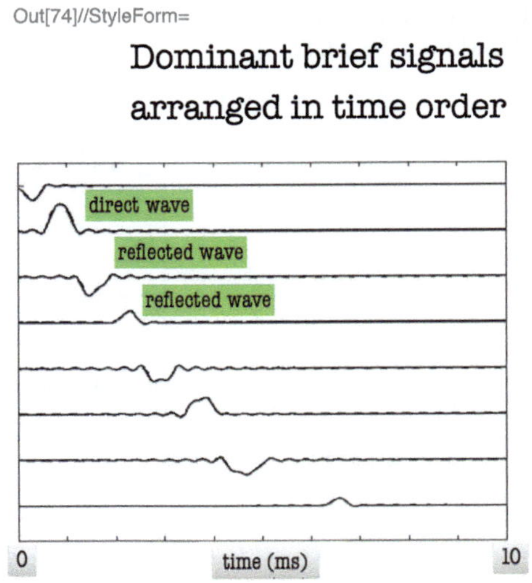

Fig. 7.38 Decomposition of synthesized waveform of initial portion into dominant pulse-like waveforms in time order from Fig. 13 [22]

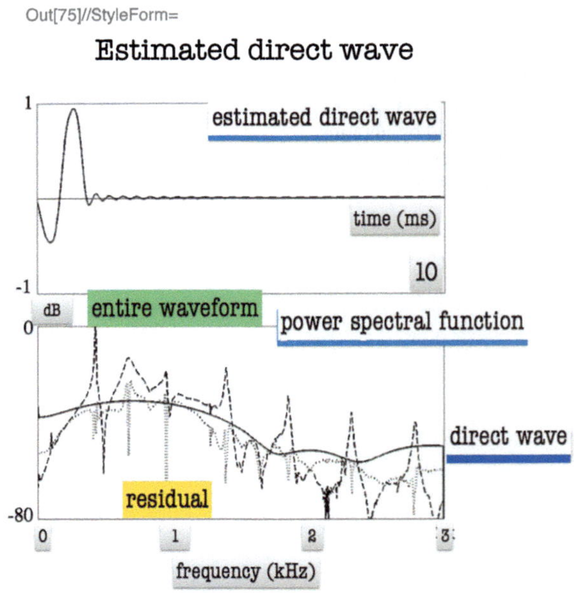

Fig. 7.39 Estimated direct wave and its power spectral function from Fig. 14 [22]

is reflected at the edges. The first and second waveforms can be assumed to be the direct waveform from the source. The third and fourth can then be assumed to be the reflected waves from the boundaries. Figure 7.39 shows the estimated direct wave obtained using CTSM with its power spectral function.

The direct wave reflects the properties of the overall spectral characteristics of the original waveform, and the power spectral property of the residual component is also quite similar to that for the direct wave. However, the direct wave still does

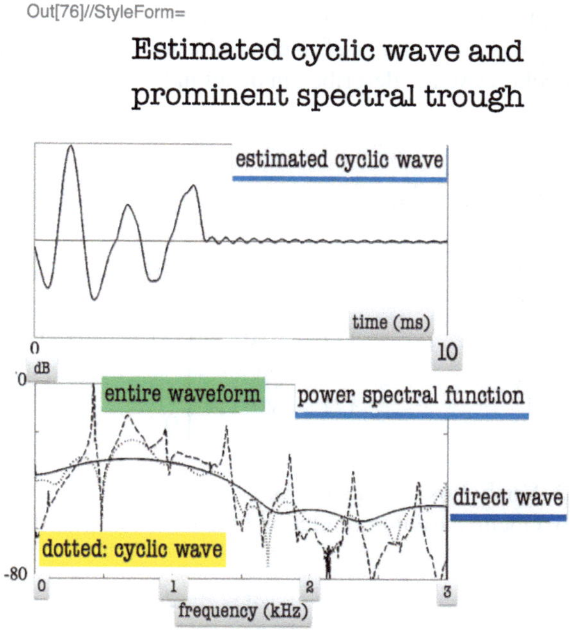

Fig. 7.40 Estimated cyclic waveform and its power spectral function from Fig.15 [22]

not reproduce the prominent zero that can be seen in the original entire waveform around 500 (Hz).

A single cycle of a periodic wave is a significant signature of musical sounds. The sum of the direct and two reflected waves, shown in Fig. 7.38, provides a satisfactory representation of most of the cycle of the waveform for the original sound. The estimated cyclic waveform from the CTSM decomposition is shown in Fig. 7.40. Interestingly, a prominent trough is seen in the power spectral function of the cyclic waveform estimated using CTSM, whereas the overall characteristics are determined by the direct wave.

The residual component is necessary to create the signature represented by the cyclic wave of the original waveform. In particular, the prominent spectral trough seen in the cyclic wave can be reconstructed by the interference (superposition) of the residual and synthesized waveforms using CLSM. The prominent trough can be interpreted as the zero for the single cycle of a periodic waveform that appears in the initial portion. The synthesized waveform obtained from CLSM that reconstructs most of the resonance spectral properties can be interpreted as the transient solution independent of the external source, whereas the residual waveform may be understood as the particular solution that conveys the external source information in the time domain. A brief time sequence such as the initial portion of the waveform is represented by the zeros rather than the resonance poles. The methods for determining the zeros of a brief interval of a waveform such as the prominent trough in the cyclic wave are described in detail in [7, 25] along with CLSM and CTSM.

7.5 Exercises

1. Obtain the discrete Fourier transform for the two following sequences,

$$\begin{aligned} &(1) \quad x(0) = x(1) = \cdots = x(N-1) = 1 \\ &(2) \quad y(n) = \begin{cases} x(n) & 0 \le n \le N-1 \\ 0 & N \le n \le M-1. \end{cases} \end{aligned} \tag{7.43}$$

2. Verify Eqs. 7.26 and 7.35.
3. Obtain the auto-correlation functions from the following power spectral functions,

$$\begin{aligned} &(1) \quad \frac{1}{2}\delta(\Omega \pm \Omega_0) \\ &(2) \quad \delta(\Omega - \Omega_0) + \frac{1}{2}\delta(\Omega - (\Omega_0 \pm \Delta\Omega)) \\ &(3) \quad 1 + \frac{1}{2}\delta(\Omega \pm \Omega_0) \\ &(4) \quad \delta(\Omega) + \frac{1}{2}\delta(\Omega \pm \Omega_0). \end{aligned} \tag{7.44}$$

4. Obtain the discrete Fourier transform of $y(n) = w(n)x(n)$ for $0 \le n \le N-1$.
5. Derive the inverse discrete Fourier transform of $Y(k) = W(k)X(k)$.
6. Suppose the discrete Fourier transforms for $x(n)$ and $y(n)$ are $X(k)$ and $Y(k)$, respectively. Obtain the discrete Fourier transform for each of the following sequences

$$\begin{aligned} &(1) \quad x(n) + y(n) \\ &(2) \quad x(n) - y(n) \\ &(3) \quad x \cdot y(n) \\ &(4) \quad x * y(n) \\ &(5) \quad x^2(n) + y^2(n). \end{aligned} \tag{7.45}$$

7. Suppose once more $x(n)$, $y(n)$, and $X(k)$, $Y(k)$. Get the inverse Fourier transforms for the spectral sequences such that

$$
\begin{array}{ll}
(1) & X(k) + Y(k) \\
(2) & X(k) - Y(k) \\
(3) & X(k) + Y(k) + \mathrm{e}^{-ikm} \\
(4) & [X(k) + Y(k)]\mathrm{e}^{-ikm} \\
(5) & |X(k) + Y(k)|^2 \\
(6) & X(k) \cdot Y(k) \\
(7) & X * Y(k) \\
(8) & |X(k)|^2 + |Y(k)|^2 \\
(9) & X^2(k) + Y^2(k).
\end{array}
\tag{7.46}
$$

References

1. Y. Hirata, A method of eliminating noise in power responses. J. Sound Vib. **84**, 593–595 (1982)
2. Y. Hirata, Estimation of the frequency response of a structure using its non-stationary vibration. J. Sound Vib. **313**, 363–366 (2008)
3. Y. Hirata, M. Tohyama, M. Matsumoto, and S. Gotoh, *Health monitoring for a structure using its non-stationary vibration* Advances in Acoust. and Vib. 2010, Article ID 696512. https://doi org/10.1155/2010/696512 Hindawi Publishing Corporation, (2010)
4. Y. Hirata, S. Goto, *Estimation of the Resonance Frequency of a Structure Using the Natural Force of Vibration*, relevant articles (2017), http://wavesciencestudy.com
5. M. Tohyama, *Sound and Signals* (Springer, Berlin, 2011)
6. S. Gotoh, *Private communication* (2017)
7. M. Tohyama, *Waveform Analysis of Sound* (Springer, Berlin, 2015)
8. M. Kazama, K. Yoshida, M. Tohyama, Signal representation including waveform envelope by clustered line-spectrum modeling. J. Audio Eng. Soc. **51**(3), 123–137 (2003)
9. S. Ushiyama, M. Tohyama, M. Iizuka, Y. Hirata, *Generalized Harmonic Analysis of Non-stationary Waveforms (in Japanese with English Abstract)*, Inst. Elect., Info., Communication Engineers, Japan, Technical Report of IEICE, EA-93-103 (1994-03)
10. T. Terada, H. Nakajima, M. Tohyama, Y. Hirata, Non-stationary waveform analysis and synthesis using generalized harmonic analysis, in *Proceedings of IEEE-SP, International Symposium on Time-Frequency and Time-Scale Analysis* (1994), pp. 429–432
11. M. Kazama, M. Tohyama, Estimation of speech components by ACF analysis in a noisy environment. J. Sound Vib. **241**(1), 41–52 (2001)
12. T. Ohnishi, M. Kazama, M. Tohyama, Acoustic signal processing using multi-windowed STFT and harmonic sieving, in *Proceedings of 5th International Congress Sound and Vibration* (1997), pp. 2167–2174
13. M. Tohyama, T. Koike, High resolution frequency analysis (in Japanese). J. Acoust. Soc. Jpn. **54**(8), 568–574 (1998)
14. Y. Ando, *Auditory and Visual Sensation* (Springer, Berlin, 2009)
15. Y. Hirata, The Frequency Discrimination of the Ear, relevant articles (2017), HP:wavesciencestudy.com
16. R.H. Lyon, Progressive phase trends in multi-degree-of-freedom systems. J. Acoust. Soc. Am. **73**(4), 1223–1228 (1983)
17. R.H. Lyon, Range and frequency dependence of transfer function phase. J. Acoust. Soc. Am. **76**(5), 1435–1437 (1984)
18. M. Tohyama, R.H. Lyon, Zeros of a transfer function in a multi-degree-of-freedom system. J. Acoust. Soc. Am. **86**(5), 1854–1863 (1989)

19. R. Meddis, L. O'Mard, A unitary model of pitch perception. J. Acoust. Soc. Am. **102**(3), 1811–1820 (1997)
20. O. Yasojima, Y. Takahashi, M. Tohyama, Resonant bandwidth estimation of vowels using clustered line-spectrum modeling for pressure speech waveform, in *International Symposium Signal Processing and Information Technology* (IEEE, 2006), pp. 589–593
21. T. Quatieri, R. Danisewicz, An approach to co-channel talker interference suppression using s sinusoidal model for speech. IEEE Trans. Acoust. Speech Signal Process **38**, 56–69 (1990)
22. T. Hasegawa, M. Tohyama, Analysis of spectral and temporal waveforms of piano-string vibration. J. Audio Eng. Soc **60**(4), 237–245 (2012)
23. T. Hasegawa, M. Tohyama, Separation of zeros for source signature identification under reverberant path condition. J. Acoust. Soc. Am. **130**(4) EL271–EL275 (2011)
24. H. Nakajima, M. Tanaka, M. Tohyama, Signal representation and inverse filtering using recursive vector projection (in Japanese with English abstract). J. Inst. Electron. Inf. Commun. Eng. Jpn. **J83-A**(4) 353–360 (2000)
25. T. Hasegawa, M. Tohyama, Source signature identification by using pole/zero modeling of transfer function. Inter Noise **2011**, 431616 (2011)

Chapter 8
Time and Frequency Responses of Spherical Source

Abstract A symmetric spherical wave is a particular representation of sound waves propagating in a medium. A spherical source such as a point source in a limit case provides a fundamental model of sources radiating sound waves. This chapter describes the sound sources for symmetric spherical waves in terms of the radiation impedance, which explains in principle why radiated sound pressure is proportional to the acceleration of the source. Radiation impedance is a basic notion within the theory of linear systems that deal with the sound field including sound sources. The impedance function is defined using both the magnitude and phase of the transfer function. The Fourier transform of the impedance function renders the transient response of the source to an impulsive excitation in the time domain. The impulse response of a loudspeaker system may be interpreted as a representation of radiation impedance functions in the time domain.

Keywords Point source · Transfer impedance · Impulse response · Sinc function · Transient response · Spherical source · Radiation impedance · Sound power output of source · Intensity · Driving point impedance

8.1 Impulse Responses of a Point Source

8.1.1 Pressure Response of a Point Source

Consider a spherical sound wave produced by a point source radiating into an isotropic medium. As the wavefront spreads out over a spherical shell formed of small volume elements as depicted in Fig. 8.1 [1], the force

$$f = -\frac{\partial p}{\partial r} dr 4\pi r^2 \qquad \text{(N)} \tag{8.1}$$

is applied uniformly to the total spherical shell, where p(Pa) denotes the sound pressure radiated by a point source, and r(m) is the distance from the point source. The acceleration, the rate of change of the volume velocity, of the total spherical shell can be written as

M. Tohyama, *Sound in the Time Domain*, Signals and Communication Technology,
https://doi.org/10.1007/978-981-10-5889-9_8

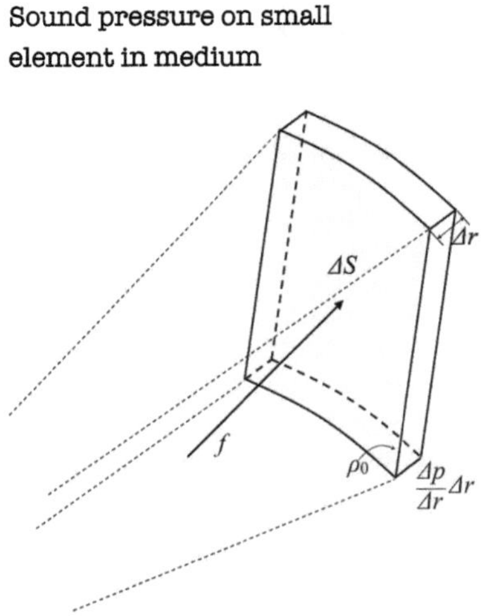

Fig. 8.1 Sound pressure in a small volume element of a medium from Fig. 8.7 [1]

$$\frac{\partial v}{\partial t}4\pi r^2 = \frac{\partial q}{\partial t}(t - r/c), \qquad (\mathrm{m}^3/\mathrm{s}^2) \tag{8.2}$$

where $v \cdot 4\pi r^2(\mathrm{m}^3/\mathrm{s})$ is the volume velocity of the spherical shell $q(t - r/c), v(\mathrm{m/s})$ gives the particle velocity of the shell, and $q(\mathrm{m}^3/\mathrm{s})$ denotes the volume velocity of the point source. The equation of motion becomes

$$\frac{\rho_0}{4\pi r^2} \cdot \frac{\partial q}{\partial t} = -\frac{\partial p}{\partial r}, \qquad (\mathrm{Pa/m}) \tag{8.3}$$

where $\rho_0(\mathrm{kg/m}^3)$ is the density of the medium.

Integration of the equation above yields the sound pressure

$$p = \frac{\rho_0}{4\pi r} \cdot \frac{\partial q}{\partial t}(t - r/c) \qquad (\mathrm{Pa}) \tag{8.4}$$

at a distance $r(\mathrm{m})$ from the point source. The sound pressure is proportional to the acceleration of the volume velocity of the source.

Suppose that the displacement of the source is

$$x = x_0(t/\tau_0)\mathrm{e}^{-t/\tau_0}, \qquad (\mathrm{m}) \tag{8.5}$$

where $\tau_0(\mathrm{s})$ and $x_0(\mathrm{m})$ denote scaling factors. The velocity of vibration of the source $v(t)$ is

$$\begin{aligned} v &= \frac{d}{dt}x \\ &= \frac{x_0}{\tau_0}(1 - t/\tau_0)\mathrm{e}^{-t/\tau_0}, \qquad (\mathrm{m/s}) \end{aligned} \tag{8.6}$$

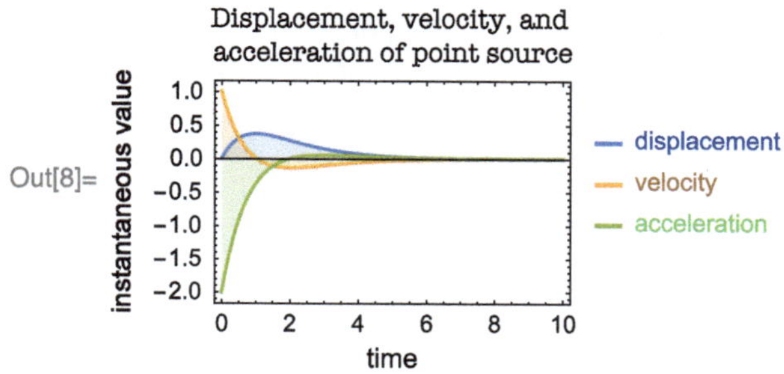

Fig. 8.2 Example of the displacement, velocity, and acceleration of a point source

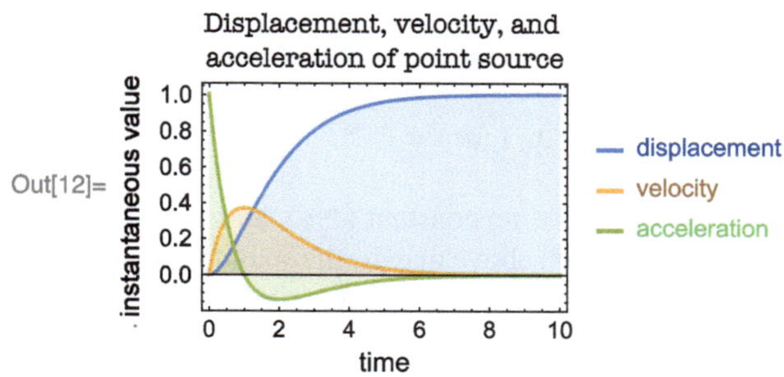

Fig. 8.3 Another example of displacement, velocity, and acceleration of a point source

and the acceleration $a_c(t)$ is

$$a_c = \frac{d}{dt}v \tag{8.7}$$
$$= -\frac{x_0}{\tau_0^2}(2 - t/\tau_0)\mathrm{e}^{-t/\tau_0}. \qquad (\mathrm{m/s^2})$$

Figure 8.2 shows the time dependence of these three functions, the displacement, velocity, and acceleration. The acceleration gives the sound pressure response and indicates a negative initial response.

Take another example. If the velocity $v(t)$ is given by Eq. 8.5 instead of the displacement, the displacement is written as

$$x = \int_0^t v d\tau \tag{8.8}$$
$$= v_0\tau_0\left[1 - (1 + t/\tau_0)\mathrm{e}^{-t/\tau_0}\right], \qquad (\mathrm{m})$$

where the displacement does not return to its initial position. Figure 8.3 displays the three functions. The sound pressure response shows a positive response followed by a negative tail.

Fig. 8.4 Impulse functions for displacement, velocity, and acceleration

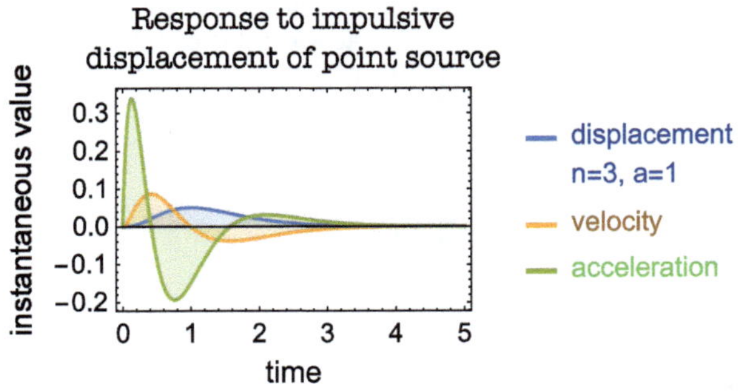

Figure 8.4 displays an example of the impulsive displacement function,

$$x = x_0 \cdot (at)^n \mathrm{e}^{-(n \cdot at)^2} \qquad (\mathrm{m}) \tag{8.9}$$

for $t > 0$, where $n \geq 1$, a is a time constant (1/s), and x_0(m) is a scaling constant. The acceleration of the function above may be the sound pressure response of a small loudspeaker system excited by an impulsive function.

8.1.2 *Steady-State Response of a Point Source*

Suppose that the volume velocity of a point source is

$$q = q_0 \mathrm{e}^{\mathrm{i}\omega t} \quad (\mathrm{m}^3/\mathrm{s}) \tag{8.10}$$

written in complex function form. According to Eq. 8.4, the sound pressure due to a spherical wave spreading out from the source can be written as [1]

$$\begin{aligned} p &= \frac{\rho_0}{4\pi r} \frac{\partial q}{\partial t}(t - r/c) \\ &= \mathrm{i}\omega\rho_0 \frac{q_0}{4\pi r} \mathrm{e}^{\mathrm{i}(\omega t - kr)} \quad (\mathrm{Pa}) \end{aligned} \tag{8.11}$$

at a distance r(m) from the point source, where $k = \omega/c(1/\mathrm{m})$. The magnitude decreases and the phase is delayed as distance increases. Here, ρ_0, c, and k denote the volume density (kg/m^3), speed of sound in the medium (m/s), and wavenumber (1/m), respectively. The phase $\phi = -kr$(rad) is called the propagation phase [2, 3], and $\tau = -\partial\phi/\partial\omega$ gives the group delay in (s) [1].

The particle velocity is formulated according to the equation of motion, which can be written as

$$\rho_0 \frac{\partial v}{\partial t} = -\frac{\partial p}{\partial r}, \quad (\mathrm{Pa/m}) \tag{8.12}$$

where the acceleration is proportional to the gradient of the sound pressure with a negative sign. Following Eqs. 8.4 and 8.12, the particle velocity is expressed as

$$v = -\frac{1}{4\pi}\frac{\partial}{\partial r}\frac{q(t-r/c)}{r}. \quad \text{(m/s)} \tag{8.13}$$

The particle velocity is proportional to the local slope of the volume velocity of the source. Following Eqs. 8.4 and 8.11, the particle velocity can be rewritten as

$$\begin{aligned} v &= \frac{1}{4\pi}\left(\frac{1}{cr}\frac{\partial}{\partial t}+\frac{1}{r^2}\right)q(t-r/c) \\ &= \frac{p(t-r/c)}{\rho_0 c}+\frac{q(t-r/c)}{4\pi r^2} \\ &= \frac{q_0}{4\pi r}\mathrm{e}^{\mathrm{i}k(ct-r)}\left(\frac{\mathrm{i}kr+1}{r}\right) \\ &= \frac{p(t-r/c)}{\mathrm{i}\omega\rho_0}\left(\frac{\mathrm{i}kr+1}{r}\right) \quad \text{(m/s)}, \end{aligned} \tag{8.14}$$

where $\omega = ck$. The first term in the second identity is in phase with the sound pressure, which is proportional to the volume acceleration according to Eq. 8.4. In contrast, the second term which is proportional to the volume velocity of the source is out-of-phase with the sound pressure. In addition, this second term decreases inversely proportional to the square of the distance from the source, whereas the first term decreases with distance. A point source has a singular point where the sound pressure cannot be defined; therefore, from a theoretical point of view, a point source is a virtual source.

8.1.3 Acoustic Transfer Impedance and Impulse Response of a Point Source

According to Eq. 8.4, the sound pressure can be written as

$$\begin{aligned} p &= \frac{\rho_0}{4\pi r}\frac{\partial q}{\partial t}(t-r/c) \\ &= \mathrm{i}\omega\rho_0\frac{1}{4\pi r}q(t-\tau) \quad \text{(Pa)} \end{aligned} \tag{8.15}$$

at distance r (m) from a point source and $\tau = r/c$, assuming that

$$q = q_0\mathrm{e}^{\mathrm{i}\omega t} \quad (\mathrm{m}^3/\mathrm{s}) \tag{8.16}$$

for $r \to 0$. The acoustic transfer impedance is defined as

$$H = \frac{p}{q_0 e^{i\omega t}} = i\omega \frac{\rho_0}{4\pi r} e^{-i\omega\tau} = H_0 e^{-i\omega\tau} \quad (\text{Pa} \cdot \text{s/m}^3) \tag{8.17}$$

between the volume velocity of the source and the sound pressure created at distance r(m) from the source.

The inverse Fourier transform of $H(r, \omega)$ gives the impulse response of the sound pressure at distance r to the impulsive volume velocity of the source. The acoustic transfer impedance H_0 by Eq. 8.17 is pure imaginary. The impulse response derived by the inverse Fourier transformation from the transfer impedance must be an odd function and is assumed noncausal. Visualizing the noncausal response is helpful in the understanding that sound pressure follows the acceleration of a point source.

The frequency-band-limited impulse response is represented in the form

$$\begin{aligned} h &= \frac{1}{2\pi} \frac{1}{2\omega_c} \int_{-\omega_c}^{\omega_c} H e^{i\omega t} d\omega \\ &= \frac{1}{2\pi} \frac{\rho_0}{4\pi r} f * \frac{d}{dt} g(t) \quad (\text{Pa} \cdot \text{s/m}^3), \end{aligned} \tag{8.18}$$

where

$$g = \frac{1}{2\omega_c} \int_{-\omega_c}^{+\omega_c} e^{i\omega t} d\omega = \frac{\sin \omega_c t}{\omega_c t} \tag{8.19}$$

$$f = \delta(t - \tau) \quad (1/\text{s}) \tag{8.20}$$

and ω_c is called the cut-off angular frequency that gives the limit of the frequency interval. The function above $\frac{\sin \omega_c t}{\omega_c t}$ is the so-called sinc function that gives the impulse response of an ideal low-pass filter. Here, it represents the waveform of the band-limited impulsive volume velocity of the point source. Differentiating the sinc function yields

$$\frac{d}{dt} g = \frac{\omega_c t \cos \omega_c t - \sin \omega_c t}{\omega_c t^2}, \quad (1/\text{s}) \tag{8.21}$$

which corresponds to the waveform of the impulsive volume acceleration of the point source. Consequently, the impulse response becomes

$$h = \frac{1}{2\pi} \frac{\rho_0}{4\pi r} \left[\frac{\omega_c(t - \tau) \cos \omega_c(t - \tau) - \sin \omega_c(t - \tau)}{\omega_c(t - \tau)^2} \right]. \quad (\text{Pa} \cdot \text{s/m}^3) \tag{8.22}$$

Figure 8.5 presents an example of $h(r, t)$ at $r = r_0 \neq 0$. The response depends on the first derivative of the sinc function. The main lobe of the pressure response is composed of both positive and negative components corresponding to the acceleration of the impulsive source. The pressure response to the volume velocity of the source propagates in the surrounding medium with an amplitude that decreases in proportion to the inverse of the distance from the source.

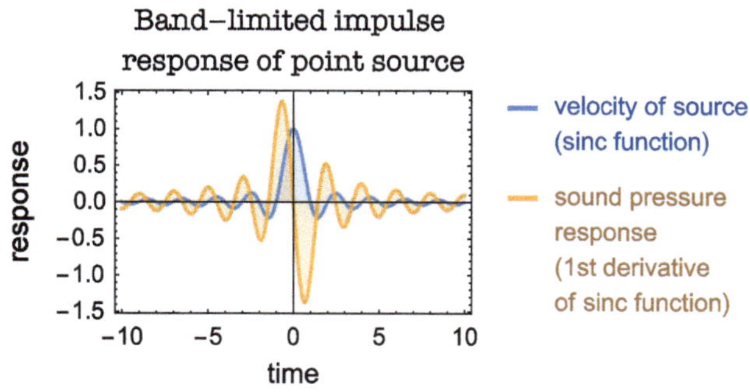

Fig. 8.5 Example of band limiting for the sound pressure impulse response of a point source

8.2 Transient Response of a Spherical Source

8.2.1 *Steady-State Response*

A point source is a virtual source because it has no geometrical size. Indeed, the sound pressure cannot be defined at the point source. To avoid this point singularity of the source, consider a spherical source of radius a. The sound pressure radiated from the source can be written as

$$p = \frac{A}{r}\mathrm{e}^{\mathrm{i}(\omega t - kr)} = \frac{A}{r}\mathrm{e}^{\mathrm{i}\omega(t-\tau)}, \qquad \text{(Pa)} \tag{8.23}$$

where $\tau = r/c$(s) assuming that the surface velocity v_a(m/s) at the surface of the source can be given by

$$v_a = v_0\mathrm{e}^{\mathrm{i}\omega t}. \qquad \text{(m/s)} \tag{8.24}$$

The surface velocity must be equal to the particle velocity along the direction of propagation of the spherical wave [1]. Recalling the relationship between sound pressure and particle velocity given by Eqs. 8.14 and 8.23, the particle velocity at the surface ($r = a$) can be rewritten as

$$\begin{aligned} v_a &= v_0\mathrm{e}^{\mathrm{i}\omega t} \\ &= \frac{q_0}{4\pi a^2}\mathrm{e}^{\mathrm{i}\omega t} \\ &= \frac{p(t-a/c)}{\rho_0 c}\frac{\mathrm{i}\omega a + c}{\mathrm{i}\omega a} \\ &= \frac{A}{a\rho_0 c}\mathrm{e}^{\mathrm{i}\omega(t-\tau_a)}\frac{\mathrm{i}\omega a + c}{\mathrm{i}\omega a}. \qquad \text{(m/s)} \end{aligned} \tag{8.25}$$

Then

$$A = \frac{q_0}{4\pi} e^{i\omega\tau_a} \frac{i\omega\rho_0 c}{i\omega a + c} \qquad (\mathrm{Pa \cdot m}) \tag{8.26}$$

are derived where $\tau_a = a/c(\mathrm{s})$. The variable r in Eq. 8.23 denoting the distance from the center of the source is defined as $r \geq a$. Therefore, the singularity of the source is removed from the radiated sound field or left inside the source, so that the spherical wave can be defined.

8.2.2 *Acoustic Transfer Impedance of Spherical Source*

The acoustic transfer impedance H is defined for the spherical source of radius a (m). Specifically

$$H = \frac{p}{q_0 e^{i\omega t}} = H_0 e^{-i\omega\tau_0}, \qquad (\mathrm{Pa \cdot s/m^3}) \tag{8.27}$$

where

$$q = q_0 e^{i\omega t} = 4\pi a^2 v_0 e^{i\omega t} \qquad (\mathrm{m^3/s}) \tag{8.28}$$

denotes the volume velocity of the spherical source, and

$$H_0 = \frac{1}{4\pi r} \frac{i\omega\rho_0 c}{i\omega a + c} \qquad (\mathrm{Pa \cdot s/m^3}) \tag{8.29}$$

$$\tau_0 = \frac{r - a}{c}. \qquad (\mathrm{s}) \tag{8.30}$$

The acoustic transfer impedance with the propagation phase discarded is not pure imaginary but is represented as a complex function. The sound pressure is no longer proportional to the acceleration of the source. This differs from the point source, in which the acoustic transfer function is pure imaginary, and therefore the radiating sound pressure varies linearly with the acceleration of the source.

Removing the propagation phase, the acoustic transfer function can be rewritten as

$$H_0 = X + iY \tag{8.31}$$

$$X = \frac{\omega\rho_0 c}{4\pi r} \frac{\omega a}{\omega^2 a^2 + c^2} \tag{8.32}$$

$$Y = \frac{\omega\rho_0 c}{4\pi r} \frac{c}{\omega^2 a^2 + c^2}. \qquad (\mathrm{Pa \cdot s/m^3}) \tag{8.33}$$

Taking the limit as a approaches 0

$$X \to 0, \tag{8.34}$$

$$Y \to \frac{\omega\rho_0}{4\pi r}, \qquad (\mathrm{Pa \cdot s/m^3}) \tag{8.35}$$

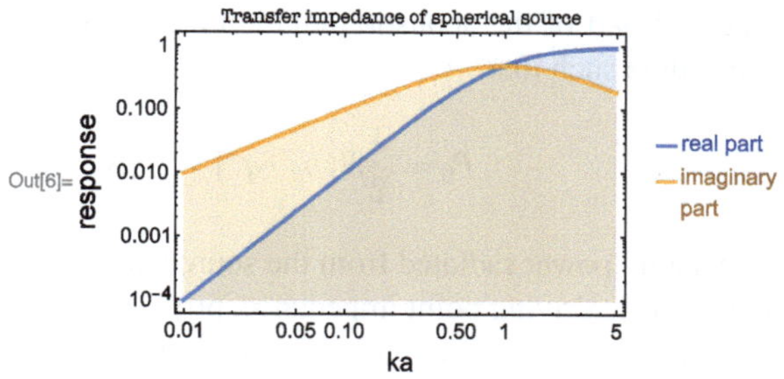

Fig. 8.6 Illustration of the transfer impedance of a spherical source

which uniquely defines the acoustic transfer impedance of a point source.

The acoustic transfer impedance of a spherical source can be rewritten such that

$$X = \frac{\rho_0 c}{4\pi r a} \frac{\chi^2}{\chi^2 + 1} \qquad (\text{Pa} \cdot \text{s/m}^3) \tag{8.36}$$

$$Y = \frac{\rho_0 c}{4\pi r a} \frac{\chi}{\chi^2 + 1}, \tag{8.37}$$

where $\chi = ka$. Figure 8.6 illustrates the acoustic transfer impedance in which the imaginary (real) part dominates the impedance rather than the real (imaginary) part when χ is smaller (larger) than unity. A spherical source of radius a(m) may be represented as an ideal point source of waves, as the wavelengths become suitably longer than radius a.

8.3 Acoustic Radiation Impedance and Pressure Response in the Time Domain

8.3.1 Acoustic Radiation Impedance

The acoustic transfer impedance when r is taken at the surface of the source, i.e., $r = a$ for the acoustic transfer impedance, is called the acoustic driving point impedance or acoustic radiation impedance of the source. The acoustic (driving point or radiation) impedance of the source is given by

$$H_{0a} = X_z + \mathrm{i}Y_z \tag{8.38}$$

$$Y_z = \frac{\rho_0 c}{4\pi a} \frac{\omega c}{\omega^2 a^2 + c^2} \tag{8.39}$$

$$X_z = \frac{\rho_0 c}{4\pi} \frac{\omega^2}{\omega^2 a^2 + c^2}. \qquad (\text{Pa} \cdot \text{s/m}^3) \tag{8.40}$$

The real part of the acoustic radiation impedance gives the sound power output of the source such that

$$P_0 = \frac{1}{2}\Re[p_a \cdot q^*] = \frac{1}{2}[X_z \cdot |q|^2]. \quad \text{(W)} \tag{8.41}$$

The sound power radiated from the source increases in proportion to the real part of the acoustic driving point impedance, provided that the strength of the sound source (volume velocity of the source) is constant. A point source is ideally the limit when the radius of the source a (m) approaches 0. Consequently, the transfer impedance becomes pure imaginary for the point source, and therefore the acoustic driving point impedance is not well defined for the point source.

Nevertheless, taking the limit as a approaches 0, then

$$X_{0_z} = \frac{\rho_0 c}{4\pi}\frac{\omega^2}{c^2}. \quad (\text{Pa} \cdot \text{s/m}^3) \tag{8.42}$$

is formally obtained. This result gives the work done by the radiated spherical wave during propagation. Recalling Eq. 8.14, which relates the sound pressure and the particle velocity at r (m) from the point source,

$$v = p\frac{\mathrm{i}(\omega/c)r + 1}{\mathrm{i}\omega\rho_0 r}. \quad (\text{m/s}) \tag{8.43}$$

The sound intensity across the spherical surface with the radius r centered at the point source is determined by

$$\begin{aligned} I &= \frac{1}{2}\Re[p \cdot v^*] \qquad (8.44) \\ &= \frac{1}{2}\left(\frac{q_0}{4\pi r}\right)^2 \cdot \frac{\omega^2 \rho_0}{c} \quad (\text{W/m}^2) \end{aligned}$$

according to Eqs. 8.11 and 8.14. The energy corresponding to the sound intensity (W/m^2) is consumed or necessary for then propagation of the wave from the source, and hence the energy must be supplied by the sound source. The sound power output is given by

$$P_0 = I \cdot 4\pi r^2 = \frac{\rho_0 \omega^2 q_0^2}{8\pi c}. \quad \text{(W)} \tag{8.45}$$

Stated formally, if the relationship between the sound power output and the real part of the driving point impedance R takes the form

$$P_0 = \frac{1}{2}R q_0^2, \quad \text{(W)} \tag{8.46}$$

then the real part of the driving point acoustic impedance $R(\omega)$ becomes

$$R = \frac{\rho_0 \omega^2}{4\pi c} = X_{0_z}. \qquad (\text{Pa} \cdot \text{s/m}^3) \tag{8.47}$$

The limit of X_z as $a \to 0$ gives the driving point acoustic impedance of a point source in the limit of a spherical source.

The power output of the spherical sound source increases following the real part, which becomes dominant as $\chi = ka$ increases. In contrast, the imaginary part governs the radiation impedance, as ka becomes smaller. In general, the low-frequency components are not radiated well by a small source.

8.3.2 *Radiation Impedance in the Time Domain*

The radiation impedance can be expressed in the time domain. Taking the inverse Fourier transforms of the real and imaginary parts, the responses in the time domain are obtained assuming that the variable χ represents the normalized frequency variable.

Performing the inverse Fourier transform of $\mathrm{i}\hat{Y}_z$ yields [4]

$$\begin{aligned} \hat{h}_Y &= \frac{\mathrm{i}}{2\pi} \int_{-\infty}^{\infty} \hat{Y}_z \mathrm{e}^{\mathrm{i}\chi\xi} d\chi \\ &= \frac{1}{2\pi} \int_{-\infty}^{\infty} \frac{\mathrm{i}\chi}{\chi^2 + 1} \mathrm{e}^{\mathrm{i}\chi\xi} d\chi \\ &= \frac{-1}{2} \mathrm{e}^{-|\xi|}, \end{aligned} \tag{8.48}$$

where the inverse Fourier transform is a real and odd function of the time variable. The inverse Fourier transform of the real part is not well defined in the time domain, and thus the impulse response must be noncausal. One way to define the impulse response is taking only the causal component. The response from the real part of the impedance function can be formally defined from that for the imaginary part so that the impulse response might be causal. Consequently,

$$\hat{h}_z = \hat{h}_X + \hat{h}_Y = \begin{cases} -\mathrm{e}^{-\xi} & \xi > 0 \\ 0 & \xi < 0 \end{cases} \tag{8.49}$$

indicating a causal response. The time response and the frequency characteristics stated above simply demonstrate an example of the relationship between the real and imaginary parts of the Fourier transform of a causal response except when the time variable is zero.

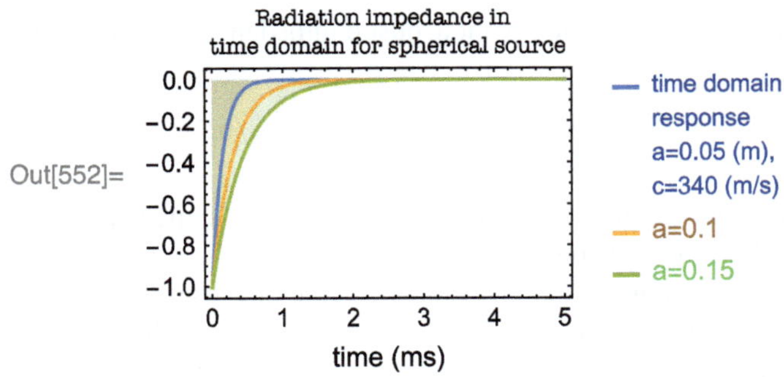

Fig. 8.7 Example of radiation impedance in time domain

The impulse response corresponding to the radiation impedance decays following the form of the exponential function. This decay is a consequence of the work done by the radiating spherical wave during propagation as mentioned in Subsection 8.3.1.

Figure 8.7 provides an example of the radiation impedance of a spherical source in the time domain, where $a = 0.1\,(\mathrm{m})$, $c = 340\,(\mathrm{m/s})$, and $\xi = ct/a$. The solid line gives a waveform of the function $-\mathrm{e}^{-\xi}$. The duration of the response can be seen to be a few milliseconds. The response decays sufficiently fast, as the elapse time extends beyond a few milliseconds.

8.3.3 *Pressure Response in Time Domain*

The causal time response $\hat{h}_z(\xi)$ can be understood as the impulse response of the sound pressure on the surface of the spherical source to the impulsive volume velocity. The sound pressure $p(t)$ on the surface can be expressed as

$$p = h_z * v(t), \qquad (\mathrm{Pa}) \tag{8.50}$$

where $h_z(t)$ denotes the impulse response corresponding to the time response of the acoustic radiation impedance, and $v(t)(\mathrm{m}^3/\mathrm{s})$ denotes the volume velocity function of the source. Consider an example of the volume displacement of a spherical source in dimensionless form,

$$x = \xi \mathrm{e}^{-\xi}, \tag{8.51}$$

where $\xi = ct/a > 0$ and a denotes the radius(m) of the source. The volume velocity $v(\xi)$ is given by

$$v = \frac{d}{d\xi}x(\xi) = (1 - \xi)\mathrm{e}^{-\xi}. \tag{8.52}$$

The sound pressure response at the surface of the spherical source can be written as

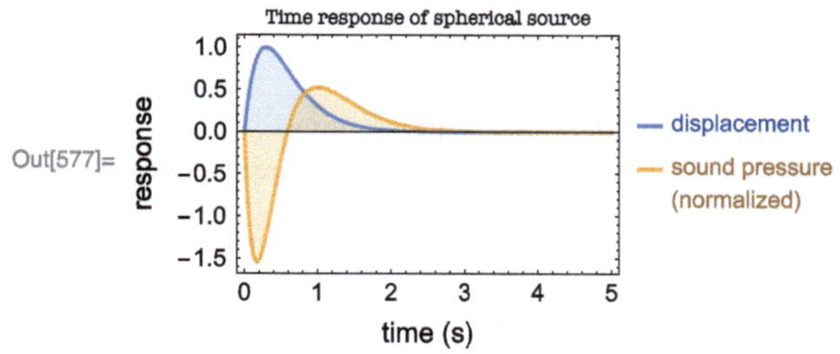

Fig. 8.8 Time response of a spherical source to volume velocity given by Eq. 8.52

$$\begin{aligned} y &= v * h_z \\ &= -\mathrm{e}^{-\xi}\int_0^{\xi}(1-\eta)d\eta \\ &= \frac{\xi}{2}(\xi-2)\mathrm{e}^{-\xi}, \end{aligned} \tag{8.53}$$

in dimensionless form. Figure 8.8 shows the time response to the volume velocity.The duration of the transient response of a small loudspeaker system might be estimated using the radiation impedance in the time domain of the spherical source.

8.4 Exercises

1. Confirm Eq. 8.14.
2. Confirm Eq. 8.25.
3. Confirm Eq. 8.44.
4. The minimum audible sound pressure is estimated to be 2×10^{-5}(Pa) indicating the root mean square (rms) of the minimum audible sound wave (plane wave). Suppose the frequency of a plane wave is 1000 (Hz). Obtain the rms for the displacement of the minimum audible sound wave in the air.
5. The sound pressure level (SPL) is defined as

$$SPL = 10\log_{10}\frac{\overline{p^2}}{(2\times 10^{-5})^2}, \qquad \text{(dB)}$$

where $\overline{p^2}$ denotes the mean of the square pressure in the period. Obtain the root mean square of the sound pressure (Pa) for (1) $SPL = 40$ dB, (2) 0 dB, (3) 60 dB, and (4) 100 dB.
6. Find another example of a response that gives the sinc function.
7. Determine the frequency response for the sinc function in the time domain.
8. Derive the auto-correlation function for the ideal low-pass filter.

References

1. M. Tohyama, *Sound and signals* (Springer, Berlin, 2011)
2. R.H. Lyon, Progressive phase trends in multi-degree-of-freedom systems. J. Acoust. Soc. Am. **73**(4), 1223–1228 (1983)
3. R.H. Lyon, Range and frequency dependence of transfer function phase. J. Acoust. Soc. Am. **76**(5), 1435–1437 (1984)
4. A. Jeffrey, *Handbook of mathematical formulas and integrals*, 3rd edn. (Elsevier Academic Press, Amsterdam, 2004)

Chapter 9
Wave Equation and Its General Solution in the Time Domain

Abstract Classical themes concerning the general solutions of the wave equations in the time domain are briefly summarized in this chapter. The solution of a one-dimensional wave equation or plane waves may be formulated from point of view of the linear system theory using the convolution of the impulse responses and virtual sources converted from the initial excitation. These plane wave solutions may be extended to spherically symmetric waves following the three-dimensional wave equation. The propagation of transient waves rendered by releasing an initial condensation such as a balloon can be formulated using the general solutions for the spherically symmetric waves or the three-dimensional wave equations. Interestingly, after releasing the initial disturbance, the propagating positive condensation wave is followed by a negative condensation wave. Negative condensation waves without followers may arise from an initial condition in, for example, the limit case of a light tube. The difference in the sound perception of an isolated positive pulse and a following negative pulse would be an intriguing topic in the field of acoustics. Power spectral differences in the low-frequency components might partly explain the difference in the perception of the transient waves rendered by a balloon and a light tube.

Keywords Wave equation · General solution of wave equation · Convolution · Impulse response · Virtual source · Spherically symmetric wave · Wave equation for symmetrically spherical wave · Propagation of initial disturbance · Positive condensation · Negative condensation

9.1 Initial Disturbance and One-Dimensional Waves

Wave propagation along a one-dimensional system of infinite extent reveals the essence of progressive waves without the accompanying effects that may arise if the system were of finite length.

M. Tohyama, *Sound in the Time Domain*, Signals and Communication Technology,
https://doi.org/10.1007/978-981-10-5889-9_9

9.1.1 Initial Condition and Impulse Response

The impulse response determines the transient response of a linear system that has been excited by a transient signal. Suppose that $h(t)$ denotes this impulse response, and $x(t)$ denotes the transient source signal. The response $y(t)$ can be written as

$$y = x * h(t)d\tau, \tag{9.1}$$

where * denotes convolution. A free oscillation or reverberation is a typical example of a transient response. Suppose $x(t)$ is a source signal for which $x(t) = 0$ for $t \geq 0$. After this source signal stops, the free oscillation that is observed is given by

$$y = x * h(t) = \int_{\tau=t}^{\infty} x(t-\tau)h(\tau), \tag{9.2}$$

where $x(t-\tau) = 0$ for $\tau \leq t$.

In a vibrational system, there may be two kinds of initial disturbances: velocity or displacement. The transient response is not always expressed by a single impulse response. The transient sound sources described in this chapter may be interpreted as initial disturbances to systems.

9.1.2 Response to Initial Disturbance of One-Dimensional Wave Equation

Take for example a linear system that is governed by a one-dimensional wave equation such that [1]

$$\frac{\partial^2 u}{\partial t^2} = c^2 \frac{\partial^2 u}{\partial x^2}, \qquad (\mathrm{m/s^2}) \tag{9.3}$$

where u represents the displacement (m). The free oscillation observed in the system responding to the initial disturbances $f(x)$ and $g(x)$ satisfying conditions

$$u_0 = f(x) \qquad (\mathrm{m}) \tag{9.4}$$

and

$$v_0 = \left.\frac{\partial u}{\partial t}\right|_{t=0} = g(x) \qquad (\mathrm{m/s}) \tag{9.5}$$

is obtained by introducing the Fourier transform $U(t, k_x)$ [2, 3]

$$U = \frac{1}{2\pi}\int_{-\infty}^{\infty} u\mathrm{e}^{-\mathrm{i}k_x x}dx. \qquad (\mathrm{m^2}) \tag{9.6}$$

The wave equation can be rewritten as

$$\frac{\partial^2 U}{\partial^2 t^2} = c^2 k_x^2 U \qquad (\mathrm{m/s})^2 \tag{9.7}$$

for which the general solution is

$$U = A(k_x)\mathrm{e}^{\mathrm{i}k_x ct} + B(k_x)\mathrm{e}^{-\mathrm{i}k_x ct}. \qquad (\mathrm{m}^2) \tag{9.8}$$

Taking the inverse Fourier transform of $U(t, k_x)$ yields [2]

$$\begin{aligned} u &= h_0^- * a(x) + h_0^+ * b(x) \qquad (9.9)\\ &= a(x+ct) + b(x-ct), \qquad (\mathrm{m}) \end{aligned}$$

showing that u may be derived as convolutions with

$$h_0^- = \int_{-\infty}^{\infty} \mathrm{e}^{\mathrm{i}k_x(x+ct)} dk_x = \delta(x+ct) \qquad (1/\mathrm{m}) \tag{9.10}$$

$$h_0^+ = \int_{-\infty}^{\infty} \mathrm{e}^{\mathrm{i}k_x(x-ct)} dk_x = \delta(x-ct) \qquad (1/\mathrm{m}) \tag{9.11}$$

$$a = \int_{-\infty}^{\infty} A\mathrm{e}^{\mathrm{i}k_x x} dk_x \qquad (\mathrm{m}) \tag{9.12}$$

$$b = \int_{-\infty}^{\infty} B\mathrm{e}^{\mathrm{i}k_x x} dk_x. \qquad (\mathrm{m}) \tag{9.13}$$

Here, $h_0^-(t, x) = \delta(x + ct)$ denotes the impulse response for the progressive wave moving toward the negative (left) side, and similarly $h_0^+(t, x) = \delta(x - ct)$ gives that moving toward the positive (right) side. Therefore, $a(x)$ and $b(x)$ can be interpreted as the virtual sources when $t = 0$ producing the left- and right-moving progressive waves, respectively.

These virtual sources $a(x)$ and $b(x)$ are represented using the initial conditions (or disturbances) when $t = 0$. The $a(x)$ and $b(x)$ must satisfy initial conditions

$$a + b = f(x) \qquad (\mathrm{m}) \tag{9.14}$$

and

$$a' - b' = \frac{g(x)}{c} \tag{9.15}$$

following Eqs. 9.4 and 9.5. Integrating Eq. 9.15 yields

$$a - b = \int_0^x \frac{g(z)}{c} dz, \qquad (\mathrm{m}) \tag{9.16}$$

and by solving the simultaneous pair of equations, Eqs. 9.14 and 9.16, the virtual sources when $t = 0$ are then obtained as

$$a = \frac{1}{2}\left(f(x) + \int_0^x \frac{g(z)}{c}dz\right) \quad \text{(m)} \tag{9.17}$$

$$b = \frac{1}{2}\left(f(x) - \int_0^x \frac{g(z)}{c}dz\right). \quad \text{(m)} \tag{9.18}$$

Composed of the initial displacement $f(x)$ and velocity $g(x)$, the virtual sources generate the system's displacement observed in responding to the initial disturbances. The initial displacement provides the virtual source associated with the displacement $f(x)/2$ that travels without any distortion toward the left- and right-hand sides following the impulse responses h_0^- and h_0^+, respectively. The initial velocity provides the virtual source for the displacement

$$\pm\frac{1}{2}\int_0^x \frac{g(z)}{c}dz \quad \text{(m)} \tag{9.19}$$

creating the waves that move toward the left- and right-hand sides, respectively. The virtual source above shows the virtual displacement transformed by the initial velocity.

Introducing the virtual sources a and b into Eq. 9.9, the responses for the left and right-hand sides are

$$\begin{aligned} u^- &= h_0^- * a(x) \\ &= \frac{1}{2}\int f(y)\delta(x+ct-y)dy \\ &+ \frac{1}{2}\int\int_0^y \frac{g(z)}{c}dz\delta(x+ct-y)dy \quad \text{(m)} \end{aligned} \tag{9.20}$$

and

$$\begin{aligned} u^+ &= h_0^+ * b(x) \\ &= \frac{1}{2}\int f(y)\delta(x-ct-y)dy \\ &- \frac{1}{2}\int\int_0^y \frac{g(z)}{c}dz\delta(x-ct-y)dy. \quad \text{(m)} \end{aligned} \tag{9.21}$$

Both of the responses above, the left and right-moving progressive waves, are determined as a sum of convolutions between the impulse responses and the virtual sources. The total response is given by the sum of the waves moving toward the left- and right-hand sides such that

$$\begin{aligned} u &= u^- + u^+ \\ &= h_0^- * a(x) + h_0^+ * b(x) \\ &= \frac{1}{2}(f(x+ct) + f(x-ct)) + \frac{1}{2}\int_{x-ct}^{x+ct} \frac{g(z)}{c} dz. \quad \text{(m)} \end{aligned} \tag{9.22}$$

Similarly, the velocity response can be written as

$$\begin{aligned} v =& \frac{\partial u}{\partial t} \\ =& \frac{c}{2}(f'(x+ct) - f'(x-ct)) \\ &+ \frac{1}{2}(g(x+ct) + g(x-ct)). \quad \text{(m/s)} \end{aligned} \tag{9.23}$$

Assume an ideal pulse-like initial disturbance such that

$$f = f_0 \delta(x) \quad \text{(m)} \tag{9.24}$$

$$g = g_0 \delta(x) \quad \text{(m/s)} \tag{9.25}$$

for the initial displacement and the velocity where

$$\int \delta(x) dx = 1. \tag{9.26}$$

The left-moving displacement response becomes

$$\begin{aligned} u^- &= \frac{1}{2}\int \left(f_0 \delta(y) + \int_0^y \frac{g_0 \delta(z)}{c} dz \right) \delta(x + ct - y) dy \\ &= \begin{cases} 0 & x + ct < 0 \\ \frac{1}{2} f_0 \delta(x+ct) + \frac{1}{2}\frac{g_0}{c} & x + ct \geq 0. \quad \text{(m)} \end{cases} \end{aligned} \tag{9.27}$$

Similarly, the right-moving response is written as

$$\begin{aligned} u^+ &= \frac{1}{2}\int \left(f_0 \delta(y) + \int_y^0 \frac{g_0 \delta(z)}{c} dz \right) \delta(x - ct - y) dy \\ &= \begin{cases} 0 & x - ct > 0 \\ \frac{1}{2} f_0 \delta(x-ct) + \frac{1}{2}\frac{g_0}{c} & x - ct \leq 0. \quad \text{(m)} \end{cases} \end{aligned} \tag{9.28}$$

The propagation of the displacement wave depends on the initial displacement or velocity. The displacement wave generated by the initial velocity does not disappear, even after the progressive wave has passed. In contrast, the displacement response to the initial displacement goes over along the coming wave. Figures 9.1 and 9.2 display samples of the propagation of the displacement waves excited by the initial displace-

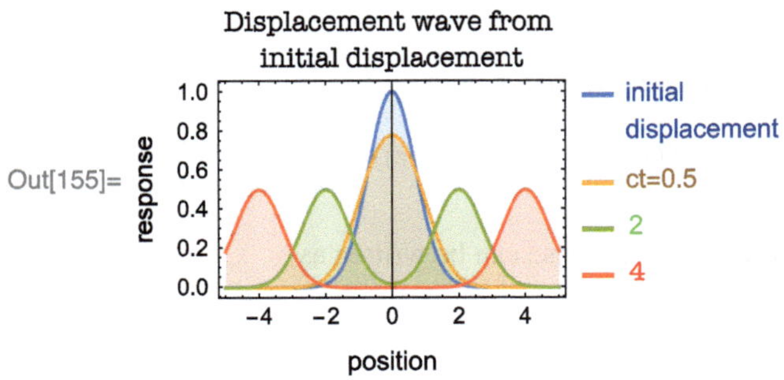

Fig. 9.1 Displacement wave by initial displacement

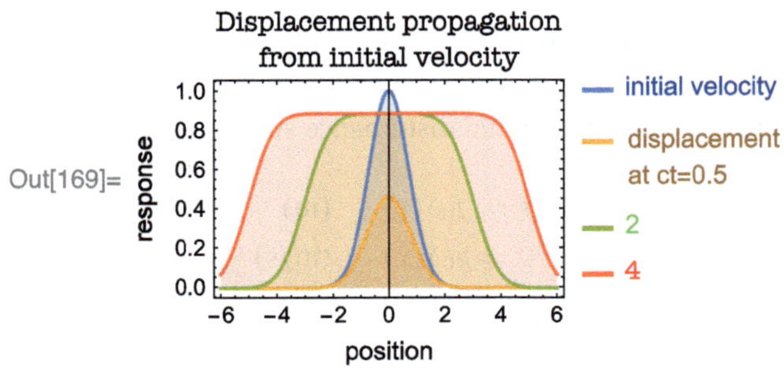

Fig. 9.2 Displacement wave by initial velocity

ment (Fig. 9.1) and velocity (Fig. 9.2), respectively. The differences established from the initial conditions are reconfirmed.

9.2 Wave Propagation of a Spherically Symmetric Wave

9.2.1 Wave Equation for a Spherically Symmetric Wave

The wave equation for a spherically symmetric wave can be formulated as well as that for a plane wave. A spherically symmetric wave depends solely on the distance from a spherical source. The volume velocity of the spherically symmetric wave is expressed as [1]

$$q = v \cdot 4\pi r^2 \qquad (\mathrm{m}^3/\mathrm{s}) \tag{9.29}$$

where v denotes the particle velocity (m/s) at the distance r(m). The small variation of the volume velocity Δq resulting from the extension of the spherical wavefront between r and $r + \Delta r$ can be estimated as

$$\Delta q = \frac{\partial (v \cdot 4\pi r^2)}{\partial r} \cdot \Delta r = \frac{-\partial s}{\partial t} \cdot 4\pi r^2 \cdot \Delta r \qquad (\mathrm{m}^3/\mathrm{s}) \tag{9.30}$$

where s denotes the amount a region of a medium has been compressed and is given by [1]

$$s = \frac{\rho - \rho_0}{\rho_0}, \tag{9.31}$$

and $\rho V = \rho_0 V_0$ (kg) subject to that $|\rho - \rho_0|$ is sufficiently small. Here, ρ represents the volume density of the medium ($\mathrm{kg/m^3}$) and ρ_0 denotes the volume density without the sound wave present. Recalling the relationship between sound pressure and particle velocity is [1]

$$\rho_0 \frac{\partial v}{\partial t} = -\frac{\partial p}{\partial r} \qquad (\mathrm{Pa/m}) \tag{9.32}$$

assuming $\rho \cong \rho_0$ and

$$p = \kappa s \qquad (\mathrm{Pa}) \tag{9.33}$$

then

$$\frac{\partial s}{\partial t} = \frac{-1}{4\pi r^2} \frac{\partial q}{\partial r} = \frac{-1}{4\pi r^2} \frac{\partial}{\partial r} (v \cdot 4\pi r^2) \qquad (1/\mathrm{s}) \tag{9.34}$$

$$\begin{aligned} \rho_0 \frac{\partial^2 s}{\partial t^2} &= \frac{-1}{4\pi r^2} \frac{\partial}{\partial r} \left(\rho_0 \frac{\partial v}{\partial t} \cdot 4\pi r^2 \right) \\ &= \frac{1}{4\pi r^2} \frac{\partial}{\partial r} \left(\frac{\partial p}{\partial r} \cdot 4\pi r^2 \right) \\ &= \frac{1}{c^2} \frac{\partial^2 p}{\partial t^2} \qquad (\mathrm{Pa/m^2}) \end{aligned} \tag{9.35}$$

are derived where $c^2 = \kappa/\rho$ m^2/s^2. Consequently,

$$\frac{\partial^2 p}{\partial t^2} = \frac{c^2}{r^2} \frac{\partial}{\partial r} \left(r^2 \cdot \frac{\partial p}{\partial r} \right) \qquad (\mathrm{Pa/s^2}) \tag{9.36}$$

holds or equivalently

$$\frac{\partial^2 r \cdot p}{\partial t^2} = c^2 \frac{\partial^2 r \cdot p}{\partial r^2}. \qquad (\mathrm{Pa \cdot m/s^2}) \tag{9.37}$$

Interestingly, a spherical wave is governed by an equation that is similar to that for a plane wave but with respect to the function $r \cdot p$.

Wave equations can be formulated for sound pressure and particle velocity, respectively. Introducing into the wave equation, the so-called velocity potential ϕ that determines both the sound pressure and velocity [1],

$$p = \rho_0 \frac{\partial}{\partial t}\phi, \qquad \text{(Pa)} \tag{9.38}$$

$$v_r = -\frac{\partial}{\partial r}\phi, \qquad \text{(m/s)} \tag{9.39}$$

then yields

$$\frac{\partial^2 r \cdot \phi}{\partial t^2} = c^2 \frac{\partial^2 r \cdot \phi}{\partial r^2} \qquad (\text{m}^3/\text{s}^3) \tag{9.40}$$

for spherical waves, and

$$\frac{\partial^2 \phi}{\partial t^2} = c^2 \frac{\partial^2 \phi}{\partial r^2} \qquad (\text{m}^2/\text{s}^3) \tag{9.41}$$

for plane waves are derived.

9.2.2 General Solution for the Spherically Symmetric Wave

The general solution of a spherically symmetric wave is given by

$$r\phi = f(r + ct) + g(r - ct), \qquad (\text{m}^3/\text{s}) \tag{9.42}$$

similar to a plane wave solution. A difference of a spherically symmetric wave from a plane wave is the effect of the source. It is not necessary to identify the sound source for a plane wave, whereas the distance from the source is crucial for a spherical wave.

Suppose that there is an ideal spherical source of small radius a(m) in a space from which a spherically symmetric wave propagates into its surroundings. There can be two types of waves. For the outgoing and returning waves that are radially propagating, even after the source had stopped,

$$-\left.\frac{\partial \phi}{\partial r}\right|_{r=a} \cdot 4\pi a^2 \to 0 \quad (a \to 0) \qquad (\text{m}^3/\text{s}) \tag{9.43}$$

must be imposed on the volume velocity at the source position [1]. Introducing

$$\phi = \frac{f(r + ct) + g(r - ct)}{r} \qquad (\text{m}^2/\text{s}) \tag{9.44}$$

into Eq. 9.43

$$-\frac{\partial \phi}{\partial r} \cdot 4\pi r^2 = 4\pi r^2 \left[\frac{f(r + ct) + g(r - ct)}{r^2} - \frac{r[f'(r + ct) + g'(r - ct)]}{r^2} \right] \tag{9.45}$$

$$\to 0 \quad (r \to a \to 0) \quad (\text{m}^3/\text{s})$$

and hence

$$f(ct) + g(-ct) = 0 \qquad (\mathrm{m}^3/\mathrm{s}) \tag{9.46}$$

holds at $r \to 0$ between the outgoing and returning waves. Consequently, the general solution for a spherically symmetric wave can be rewritten as

$$r\phi = f(ct + r) - f(ct - r). \qquad (\mathrm{m}^3/\mathrm{s}) \tag{9.47}$$

9.2.3 *Free Oscillation of a Spherically Symmetric Wave*

Subject to the initial conditions for $\phi(t, r)$, the spherically symmetric wave can be formulated in a manner such that [3]

$$\phi(0, r) = \phi_0(r) \qquad (\mathrm{m}^2/\mathrm{s}) \tag{9.48}$$

$$\frac{\partial \phi}{\partial t}(0, r) = \phi_1(r). \qquad (\mathrm{m}^2/\mathrm{s}^2) \tag{9.49}$$

Introducing the conditions above into Eq. 9.42

$$r\phi_0(r) = f(r) + g(r) \qquad (\mathrm{m}^3/\mathrm{s}) \tag{9.50}$$

and

$$r\phi_1(r) = c(f'(r) - g'(r)) \qquad (\mathrm{m}^3/\mathrm{s}^2) \tag{9.51}$$

are derived. The latter of the two equations can be integrated to give

$$\frac{1}{c}\int_0^r \xi\phi_1(\xi)d\xi = f(r) - g(r). \qquad (\mathrm{m}^3/\mathrm{s}) \tag{9.52}$$

The simultaneous set of equations obtained from Eqs. 9.50 and 9.52 provides the solutions

$$f(r) = \frac{1}{2}\left[r\phi_0(r) + \frac{1}{c}\int_0^r \xi\phi_1(\xi)d\xi\right] \qquad (\mathrm{m}^3/\mathrm{s}) \tag{9.53}$$

$$g(r) = \frac{1}{2}\left[r\phi_0(r) - \frac{1}{c}\int_0^r \xi\phi_1(\xi)d\xi\right]. \qquad (\mathrm{m}^3/\mathrm{s}) \tag{9.54}$$

Following Eq. 9.42

$$r\phi = \frac{1}{2}(r+ct)\phi_0(r+ct) + \frac{1}{2}(r-ct)\phi_0(r-ct) + \frac{1}{2c}\int_{r-ct}^{r+ct}\xi\phi_1(\xi)d\xi \quad (\mathrm{m}^3/\mathrm{s}) \tag{9.55}$$

can be derived for $r > ct$. Similarly,

$$r\phi = \frac{1}{2}(ct+r)\phi_0(ct+r) - \frac{1}{2}(ct-r)\phi_0(ct-r) + \frac{1}{2c}\int_{ct-r}^{ct+r}\xi\phi_1(\xi)d\xi \quad (\mathrm{m}^3/\mathrm{s}) \tag{9.56}$$

is obtained for $r < ct$.

This solution determines the limit when $r \to 0$. From Eq. 9.47,

$$\phi(t,r)|_{r\to 0} = \lim_{r\to 0}\frac{f(ct+r) - f(ct-r)}{r} = 2f'(ct) \quad (\mathrm{m}^2/\mathrm{s}). \tag{9.57}$$

Recalling Eq. 9.53,

$$\phi(t,r)|_{r\to 0} = 2f'(ct) = \frac{\partial}{\partial t}[t\phi_0(ct)] + t\phi_1(ct) \quad (\mathrm{m}^2/\mathrm{s}) \tag{9.58}$$

can be derived.

9.2.4 Propagation of the Initial Disturbance

Following the general solution in the previous subsection, the spherically symmetric wave $r\phi$ creates the initial disturbance along the distance r (m). Consider the example of an inflated balloon filled with a gas. Suppose that there is a portion of the medium $r < r_B$ with condensation partitioned off by a virtual wall (Fig. 9.3) [1].

If this wall is removed, a pressure wave moves into the surrounding medium. Given the initial conditions, the initial sound pressure $p(t, r)$ and the velocity potential $\phi(t, r)$ at $t = 0$ can be given in the small portion $(r < r_B)$,

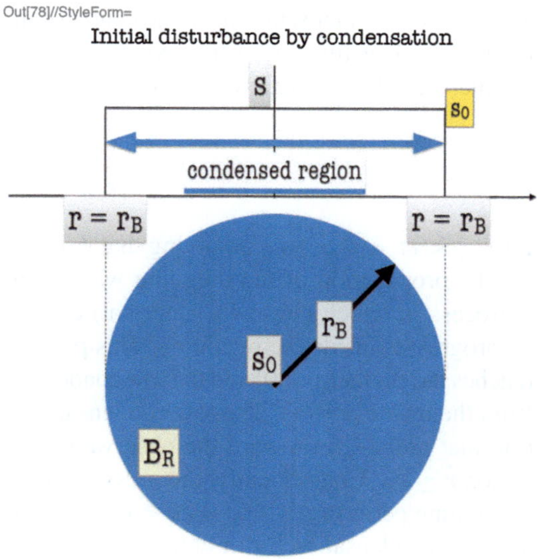

Fig. 9.3 Initial disturbance represented by condensation from Fig. 9.1 [1]

$$p(r) = \kappa s(r) = \kappa s_0 \tag{9.59}$$
$$= \rho_0 \frac{\partial \phi_0}{\partial t}(r) = \rho_0 \phi_1(r) \qquad \text{(Pa)}$$
$$\phi_0(r) = 0, \qquad (\text{m}^2/\text{s}) \tag{9.60}$$

where the sound pressure should be 0 for $r > r_B$ at $t = 0$.

Introducing these initial conditions into the general solution by Eq. 9.55, the pressure wave can be expressed as

$$\frac{\partial \phi}{\partial t} = \frac{(r+ct)\phi_1(r+ct) + (r-ct)\phi_1(r-ct)}{2r} \tag{9.61}$$
$$= c^2 s(t, r). \qquad (\text{m}^2/\text{s}^2)$$

After removing the partition at $t = 0$, a pressure wave $s(t, r)$ propagates outside the initial region $r < r_B$. Setting $r_{out} > r_B$ or taking the observation point on the circle of radius r_{out} outside the initial region $r \leq r_B$, then

$$\phi_1(r_{out} + ct) = 0 \qquad (\text{m/s}^2) \tag{9.62}$$

because $r_{out} + ct > r_B$. The propagating wave can be observed at $r = r_{out}$ as

$$s(t, r_{out}) = \frac{1}{2r_{out}}(r_{out} - ct)s_0 > 0 \tag{9.63}$$

in the time interval corresponding to $r_{out} - r_B < ct$. This positive increase in pressure occurs because of the approaching wave of high condensation from inside $r < r_B$ toward the circle r_{out} outside the initial region. Positive pressure is observed until the arrival of the condensation from the center of the region $r < r_B$ reaches the circle of radius r_{out}. This time interval is expressed as $r_{out} - r_B < ct < r_{out}$.

Interestingly, the negative condensation (or rarefaction) wave obeys [4]

$$s(t, r_{out}) = \frac{1}{2r_{out}}(r_{out} - ct)s_0 < 0 \tag{9.64}$$

for $r_{out} < ct < r_{out} + r_B$ assuming that $\phi_1(r) = \phi_1(-r)$.

The propagation of the outgoing wave of high condensation is simultaneously a process in which the low (or normal) condensation observed outside the region r_B progresses inside the region r_B. This progression into the circle r_B ends when it reaches the center ($r = 0$) and the low condensation part is left without compensation from the inside portion. The low condensation yields a wave of negative condensation that propagates toward the observation position from the center of the initial region $r \leq r_B$. Consequently, the pressure wave expressed by $s(t, r_{out})$ is observed in the time period $r_{out} - r_B < ct < r_{out} + r_B$ as positive condensation followed by negative condensation (or rarefaction).

In summary, the initial disturbance propagates along r as time elapses. For $r_{out} > r_B$

$$s(t, r_{out}) = \begin{cases} 0 & ct < r_{out} - r_B \\ \frac{1}{2r_{out}}(r_{out} - ct)s_0 > 0 & r_{out} - r_B < ct < r_{out} \\ \frac{1}{2r_{out}}(r_{out} - ct)s_0 < 0 & r_{out} < ct < r_{out} + r_B \\ 0 & ct > r_{out} + r_B. \end{cases} \tag{9.65}$$

Figure 9.4 displays the image of a propagating wave expanding. When the initial condensation is negative, the propagating condensation stated above changes the sign. However, if the initial condensation approaches a vacuum as in a light tube, then the speed of propagation in the region r_B is extremely fast, and hence there are no sound waves. Consequently, the outgoing wave is composed of only a negative component without a following component [4].

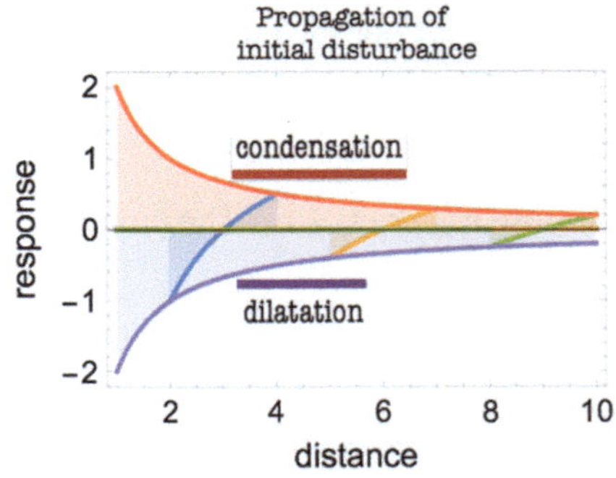

Fig. 9.4 Image of the propagation of the initial disturbance shown in Fig. 9.3 from Fig. 9.2 [1]

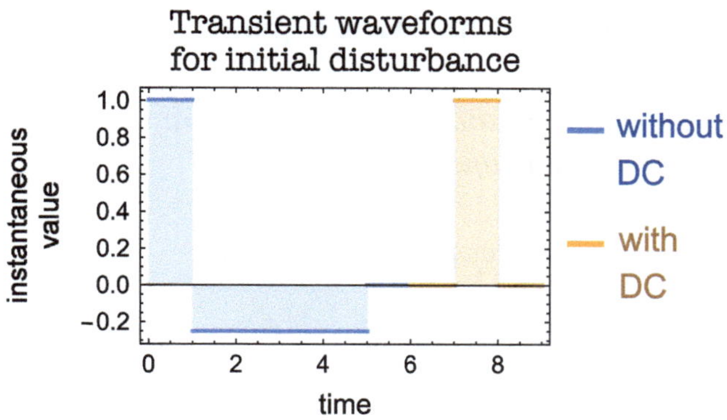

Fig. 9.5 Samples of transient waveforms with and without the following negative component

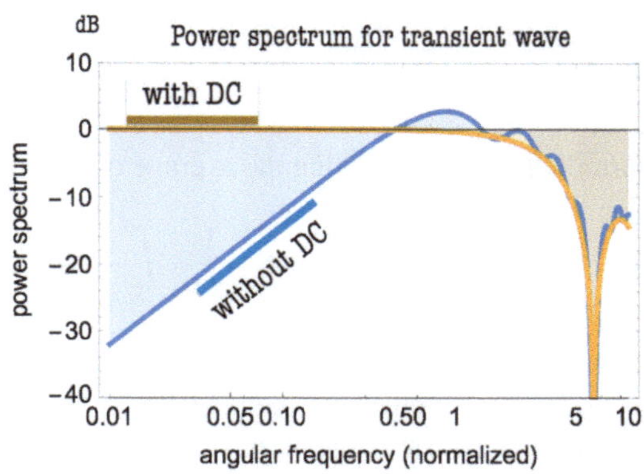

Fig. 9.6 Power spectral characteristics for pulses shown in Fig. 9.5

The difference in sound perception between positive pulses with and without negative followers such as that displayed in Fig. 9.5 is explained by the power spectral characteristics [4]. Figure 9.6 shows the power spectral properties for two pulses, one followed by a negative component and the other not.

A pulse followed by a negative tail does not include a DC component corresponding to one with a time average of 0. In contrast, an isolated pulse without a negative follower contains a DC component. It makes a difference in the low frequency range as indicated in Fig. 9.6. Although detecting the difference between positive and negative isolated pulses may be inaudible, the difference between the explosive sound of a balloon and a light tube may be audible or detectable in spectral characteristics of the power.

9.3 General Solution of the Wave Equation

9.3.1 Initial Condition and Wave Propagation in Three-Dimensional Space

Consider the wave equation in three-dimensional space,

$$\frac{\partial^2 \phi}{\partial t^2} = c^2 \left(\frac{\partial^2 \phi}{\partial x^2} + \frac{\partial^2 \phi}{\partial y^2} + \frac{\partial^2 \phi}{\partial z^2} \right) \quad (\mathrm{m}^2/\mathrm{s}^3) \tag{9.66}$$

with initial conditions at $t = 0$

$$\phi = \phi_0 \quad (\mathrm{m}^2/\mathrm{s}) \tag{9.67}$$

$$\frac{\partial \phi}{\partial t} = \phi_1. \quad (\mathrm{m}^2/\mathrm{s}^2) \tag{9.68}$$

Introducing $\tilde{\phi}$ representing the average of ϕ on the surface $S_r(\mathbf{x_0})$, specifically

$$\tilde{\phi}(t, \mathbf{x_0}) = \frac{1}{4\pi r^2} \int_{S_{r(\mathbf{x_0})}} \phi(t, \mathbf{x}) dS, \qquad (\mathrm{m}^2/\mathrm{s}) \tag{9.69}$$

where $S_r(\mathbf{x_0})$ denotes the surface of the sphere of radius r from the center at $\mathbf{x_0}$, and $\mathbf{x_0} = (x, y, z)$ represents a point taken in the three-dimensional space [3, 5]. The above wave equation can be rewritten as

$$\frac{\partial^2 \phi}{\partial t^2} = c^2 \left[\frac{\partial^2 \phi}{\partial r^2} + \frac{2}{r}\frac{\partial \phi}{\partial r} + \frac{1}{r^2} \left\{ \frac{1}{\sin\theta} \frac{\partial}{\partial \theta} (\sin\theta \frac{\partial \phi}{\partial \theta}) + \frac{1}{\sin^2\theta} \frac{\partial^2 \phi}{\partial \varphi^2} \right\} \right] \tag{9.70}$$

in the spherical coordinate system. Substituting $\tilde{\phi}$ for ϕ, this wave equation becomes

$$\frac{\partial^2 \tilde{\phi}}{\partial t^2} = c^2 \left[\frac{\partial^2 \tilde{\phi}}{\partial r^2} + \frac{2}{r}\frac{\partial \tilde{\phi}}{\partial r} \right] \tag{9.71}$$

$$= \frac{c^2}{r^2}\frac{\partial}{\partial r}\left(r^2 \frac{\partial \tilde{\phi}}{\partial r} \right) \quad (\mathrm{m}^2/\mathrm{s}^3)$$

or simply

$$\frac{\partial^2 \hat{\phi}}{\partial t^2} = c^2 \frac{\partial^2 \hat{\phi}}{\partial r^2} \quad (\mathrm{m}^3/\mathrm{s}^3) \tag{9.72}$$

by setting $\hat{\phi} = r\tilde{\phi}$ [3, 5]. This equation is identical to that for a spherically symmetric wave given by Eq. 9.37, if the average of ϕ is substituted for the wave function ϕ.

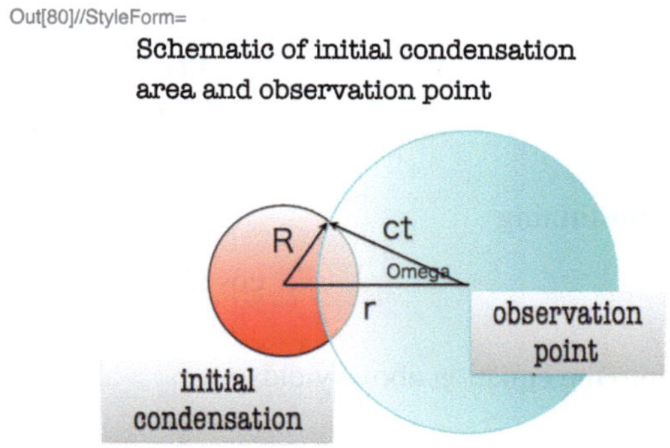

Fig. 9.7 Schematic of an initial area of condensation and a spherical surface surrounding the point of observation

Analogous to the expression for the spherically symmetric wave given at $r \to 0$ in Eq. 9.58, the wave propagation can be expressed as

$$\tilde{\phi}(t, \mathbf{x_0}) = \frac{\partial}{\partial t}[t\tilde{\phi}_0(ct)] + t\tilde{\phi}_1(ct) \qquad (\mathrm{m^2/s}) \tag{9.73}$$

subject to the initial conditions at $t = 0$ of

$$\tilde{\phi} = \tilde{\phi}_0(r) \qquad (\mathrm{m^2/s}) \tag{9.74}$$

$$\frac{\partial}{\partial t}\tilde{\phi} = \tilde{\phi}_1(r). \qquad (\mathrm{m^2/s^2}) \tag{9.75}$$

Similar to the definition of $\tilde{\phi}$ given in Eq. 9.69, the expression in Eq. 9.73 can be rewritten as

$$\phi(t, \mathbf{x_0}) = \frac{1}{4\pi c^2}\left[\frac{\partial}{\partial t}\int_{S_{ct(\mathbf{x_0})}} \frac{\phi_0(\mathbf{x})}{t} dS + \int_{S_{ct(\mathbf{x_0})}} \frac{\phi_1(\mathbf{x})}{t} dS\right]. \quad (\mathrm{m^2/s}) \tag{9.76}$$

In accordance with the solution above, the propagation of the initial condensation described in subsection 9.2.4 can be understood intuitively [3, 5, 6]. Figure 9.7 displays the sphere in which the initial condensation is given and the spherical surface area of radius ct from the center (observational point) where the integration is performed in Eq. 9.76 at time t.

The area of the integrated portion (included within the sphere) can be expressed as

$$\int_0^{\Omega} 2\pi(ct)^2 \sin\theta d\theta = 2\pi(ct)^2(1 - \cos\Omega). \qquad (\mathrm{m^2}) \tag{9.77}$$

Introducing the initial condition $\phi_0 = 0$ into the solution Eq. 9.76, the velocity potential $\phi(t, r)$ becomes

$$\phi(t,r) = \frac{c^2 s_0}{4\pi c^2} \cdot \frac{2\pi (ct)^2}{t}(1-\cos\Omega) \qquad (9.78)$$
$$= \frac{s_0}{2}(1-\cos\Omega)c^2 t. \quad (\mathrm{m}^2/\mathrm{s})$$

Substituting

$$\cos\Omega = \frac{r^2+(ct)^2-R^2}{2rct} \qquad (9.79)$$

into the equation above yields

$$\phi(t,r) = \frac{cs_0}{4} \cdot \frac{R^2-(r-ct)^2}{r}. \quad (\mathrm{m}^2/\mathrm{s}) \qquad (9.80)$$

Consequently, the propagating condensation wave $s(t,r)$ can be expressed as

$$s = \frac{1}{c^2} \cdot \frac{\partial\phi}{\partial t} = \frac{s_0(r-ct)}{2r}. \qquad (9.81)$$

9.4 Exercises

1. Confirm that the following functions satisfy the wave equation.

$$\begin{aligned}
&(1) \quad f(x-ct), \quad (2) f(x-ct)+f(x+ct) \\
&(3) \quad \sin(x-ct), \quad (4)\cos(x-ct) \\
&(5) \quad \sin(x-ct)+\cos(x-ct) \\
&(6) \quad \exp^{-a(x-ct)^2} \quad (7)\exp^{-a(x+ct)^2}.
\end{aligned} \qquad (9.82)$$

2. Show the function

$$r\phi = f(r-ct)+f(r+ct)$$

satisfies the "radial" wave equation

$$\frac{\partial^2 r\phi}{\partial t^2} = c^2\frac{\partial^2 r\phi}{\partial r^2}.$$

3. Derive the above radial wave equation.
4. Show that the function

$$p = \frac{1}{r}\mathrm{e}^{\mathrm{i}(\omega t-kr)}$$

satisfies the radial wave equation above.
5. Derive Eqs. 9.22 and 9.23. Note how the initial displacement (velocity) determines the velocity (displacement) wave.

6. Verify Eq. 9.61.
7. Verify Eq. 9.81.

References

1. M. Tohyama, *Sound and Signals* (Springer, Berlin, 2011)
2. T. Mori, *Gendai no koten kaiseki (in Japanese, Modern Classical-Analysis)* (Chikuma Shobo Publishing, 2006)
3. S.J. Farlow, *Partial Differential Equations for Scientists and Engineers* (Dover Publications Inc., New York, 1993)
4. Y. Hirata, *private communication* (2017)
5. H. Matano, M. Jimbo, *Netsu, hadou to bibun-houteishiki (in Japanese)*, Thermodynamics, waves, and differential equations, Iwanami Shoten (2007)
6. H. Lamb, *Hydrodynamics* (The Cambridge University Press, Cambridge, 1924)

Chapter 10
Sound Traveling in One-Dimensional Space

Abstract Because of the harmonic structure of waves, sound traveling in an acoustic tube or vibration of a string is a simple system providing an informative model to describe sound generation in musical instruments. This chapter deals with sound waves traveling in a space of one-dimensional from the point of view of the periodic structure in the time domain. According to the $z-$transform of the impulse response, the difference is well expressed by a pair of boundary conditions referred to as open–open and open–closed conditions. Allowing the coefficient of reflection to be represented as a complex number rather than solely real, the harmonic structure then varies. In addition, introducing a pulse sequence instead of a single number for the coefficient of reflection, the spectral characteristics are found to change considerably. Not only the transfer functions but also the driving-point impedance may be derived through geometrical modeling in the time domain. For one-dimensional systems, the phase response has a minimum phase with a propagation phase delay. The propagation phase delay is determined by the poles and zeros of the transfer function. Interestingly, the propagation phase delay can be estimated by the difference in the number of poles and zeros, even after the propagation phase is discarded. The geometrical view of the radiation impedance helps in providing an understanding of the schematics of the radiation condition from vibration.

Keywords Acoustic tube · Period and fundamental frequency · Boundary conditions · Time-delayed reflection · Driving impedance · Phase response · Minimum phase · Poles and zeros · Propagation phase · Accumulated phase · Radiation impedance · Coincidence frequency · Radiation from vibration

10.1 Time-Domain Approach to Pulse-Like Sound Traveling in Acoustic Tube

10.1.1 Sound Traveling in a Tube with Open Ends

One fundamental scheme by which to analyze wave phenomena in musical instruments, as well as vibration of strings, is sound traveling in a one-dimensional tube.

M. Tohyama, *Sound in the Time Domain*, Signals and Communication Technology,
https://doi.org/10.1007/978-981-10-5889-9_10

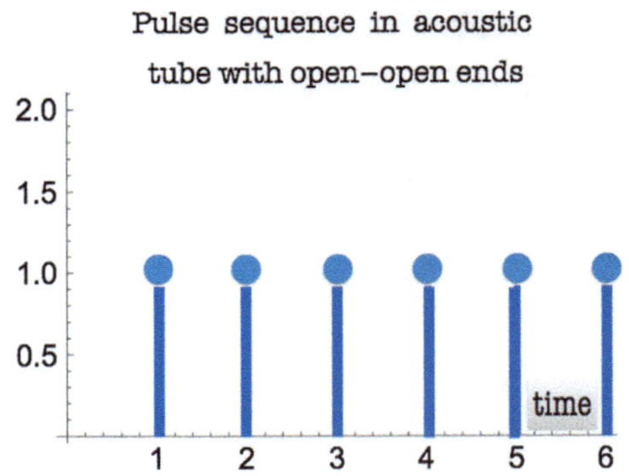

Fig. 10.1 Virtual pulse train inside an open tube observed at $x \to +0$

Consider a thin tube of finite length with open ends. If sound enters the tube, the sound wave travels back and forth between both its ends. The reflection of sound at each end generates a periodic traveling wave with period of $2L/c$(s), where L(m) denotes the length of the tube and c(m/s) gives the speed of sound inside the tube.

Figure 10.1 presents a schematic view of a virtual pulse train observed in time at $x \to +0$; the open ends of the tube are at $x = 0$ and $x = L$. The period of the pulse train, or the spacing between adjacent pair of pulses, is $2L/c$(s). The outgoing waves reflected at both of the open ends have opposite signs to the incoming waves. At the open ends of the tube—the boundaries of the tube—the sound pressure must be zero. The reflected sound pressure waves have opposite signs so that the boundary condition for the pressure wave may be satisfied by a superposition of the coming and reflected waves at the open ends.

10.1.2 Sound in an Open–Closed Tube

Like Figs. 10.1, 10.2 depicts a virtual pulse train but with boundary conditions corresponding to one open end and one closed end.

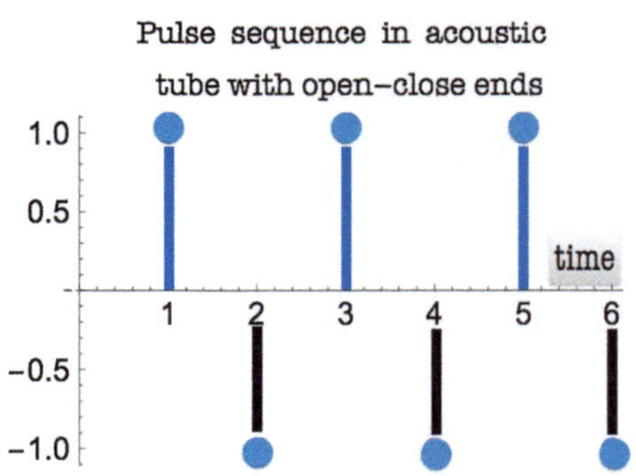

Fig. 10.2 Virtual pulse train open–closed tube similar to Fig. 10.1

At the closed end, the incident and reflected waves are superposed so that the sound pressure increases to twice the amplitude of the incoming wave. Consequently, as shown in Fig. 10.2, the single interval of the periodic wave is composed of a pair of sound waves with relative positive and negative pressure, and thus the period for the traveling wave is $4L/c$(s).

10.1.3 Period and Fundamental Frequency

The period of the sound traveling in the tube depends on the boundary conditions, being $2L/c$(s) for the open–open end condition and $4L/c$(s) for the open–closed ends. Thus the fundamental frequency or pitch of the sound is different for these two instances. The fundamental frequency, defined as the inverse of the period of the sound in the time domain, is a 1/1-octave lower for the open–closed condition than that for the open–open condition. This difference explains why the fundamental frequency of a clarinet could be lower by a 1/1-octave than that for a flute, although the lengths of the clarinet and flute are comparable.

Let us introduce the reflection coefficient r defined for the reflected wave to the incident wave. Its value $r = 1(-1)$ represents the reflection of the wave at the closed (open) end of the tube. In essence, the value of r succinctly determines the boundary condition at the corresponding end.

Consider again an open–open tube for which the boundary conditions are now $r_0 = -1$ at $x = 0$ and $-1 < r_L < 0$ at $x = L$. Figure 10.3 illustrates the virtual pulse train similar to Fig. 10.1 but with $-1 < r_L < 0$. The pulse train is no longer periodic from a mathematical point of view; however, the pulses are equally spaced at $2L/c$(s). The magnitude of each successive pulse decreases in keeping with a geometric series. This decrease in magnitude is intuitively understood as resulting from energy dissipation at the ends through sound waves being transmitted from the tube into the surrounding space.

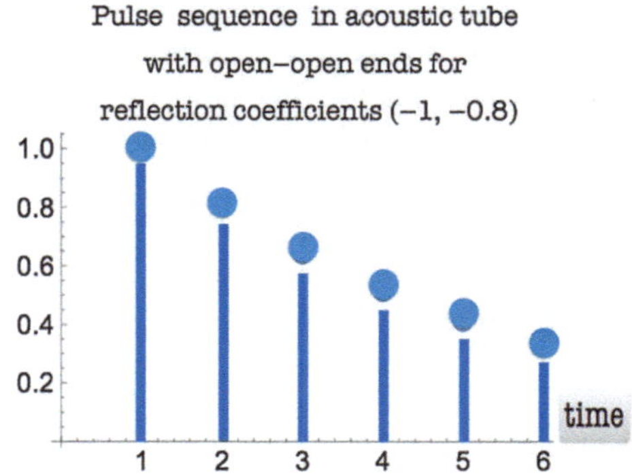

Fig. 10.3 Virtual pulse train open–open tube similar to Fig. 10.1, but assuming the reflection coefficient $r_L = -0.8$ at $x = L$ and $r_0 = -1$ at $x = 0$

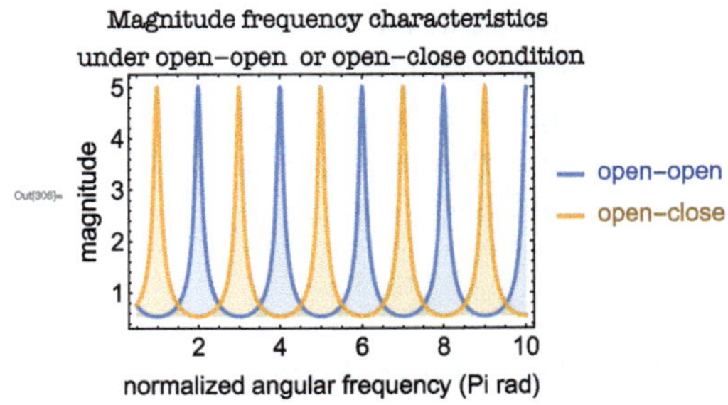

Fig. 10.4 Magnitude frequency characteristics for the pulse trains under open–open ($r_0 = -1, r_L = -0.8$) and open—closed ($r_0 = -1, r_L = 0.8$) boundary conditions

10.1.4 Magnitude Frequency Characteristics of Pulse Train

Assume for simplicity that the interval of the pulse train is one single sample; the z-transform of the virtual pulse train, depicted in Fig. 10.3 at $x \to +0$ (open end), is

$$H(z^{-1}) = \frac{1}{1 + r_L z^{-1}}, \tag{10.1}$$

where r_L denotes the reflection coefficient at $x = L$. The angular frequencies of the poles are located at $\Omega_{poo} = 2m\pi$ for open–open boundary conditions where $r_0 = r_L = -1$ and m is an integer. In contrast, the poles are found at $\Omega_{poc} = (2m - 1)\pi$ for the open–closed boundary conditions where $r_0 = -1$ and $r_L = 1$.

Figure 10.4 illustrates the magnitude frequency characteristics of a pulse train under open–open ($r_0 = -1,\ r_L = -0.8$) and open–closed ($r_0 = -1, r_L = 0.8$) boundary conditions. These characteristics are understood as changes in the magnitude and phase from the original characteristics of the sinusoidal sequence fed into the tube [1, 2]. The angular frequency corresponding to the local maxima in the magnitude frequency characteristics determines the pole.

The local maxima observed in Fig. 10.4 or the angular frequencies of the poles identify the difference in the fundamental frequencies of the two situations. The pitch sensation associated with sound is possible even if its fundamental frequency is removed. This condition is referred to as the missing fundamental in terms of perception of sound. The sensation of pitch arises basically from the characteristics of sound waves in the time domain. The period of a sound in the time domain may be identified irrespective of whether the fundamental component is contained in the sound.

Suppose that a mixture of pulse sequences $h_{oo}(n)$ for open–open conditions ($r_0 = -1, r_L = -|r|$) and $h_{oc}(n)$ ($r_0 = -1, r_L = |r|$) for open–closed conditions, as depicted in Fig. 10.5, such that

$$h(n) = h_{oo}(n) + h_{oc}(n). \tag{10.2}$$

The z-transform becomes

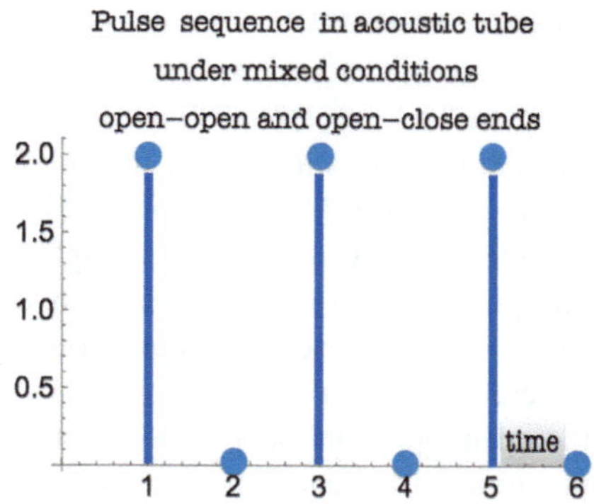

Fig. 10.5 Mixture of pulse trains for open–open and open–closed conditions with $|r| = 1$

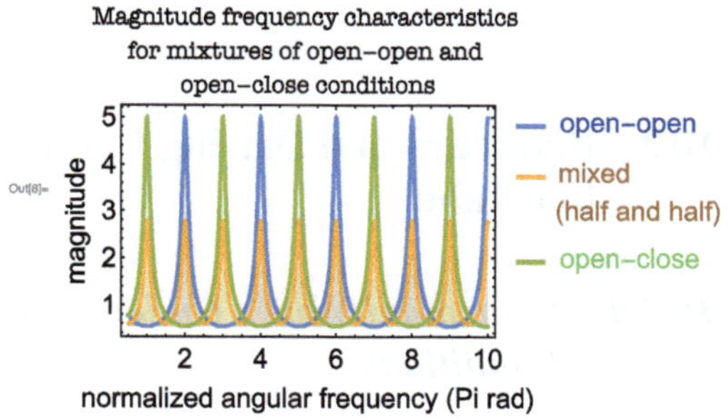

Fig. 10.6 Magnitude frequency characteristics for mixture of the pulse trains shown in Fig. 10.5 with 1/2 magnitudes

$$
\begin{aligned}
H(z^{-1}) &= H_{oo}(z^{-1}) + H_{oc}(z^{-1}) \\
&= \frac{1}{1-|r|z^{-1}} + \frac{1}{1+|r|z^{-1}} \\
&= \frac{2}{(1-|r|z^{-1})(1+|r|z^{-1})}.
\end{aligned}
\tag{10.3}
$$

Figure 10.6 presents the magnitude frequency characteristics for

$$
H(\mathrm{e}^{-\mathrm{i}\Omega}) = \frac{2}{(1-|r|\mathrm{e}^{-\mathrm{i}\Omega})(1+|r|\mathrm{e}^{-\mathrm{i}\Omega})}.
\tag{10.4}
$$

The period of $h(n)$ is equal to $h_{oc}(n)$ as shown in Fig. 10.5. Take the z-transform of $h_{oc}(n)$ once again. The z-transform can be rewritten as

$$
\begin{aligned}
H_{oc}(z^{-1}) &= \frac{(1-|r|z^{-1})}{(1-|r|z^{-1})(1+|r|z^{-1})} \\
&= S(z^{-1})\frac{1}{2}H(z^{-1}),
\end{aligned}
\tag{10.5}
$$

where

$$S(z^{-1}) = 1 - |r|z^{-1}. \tag{10.6}$$

The expression is intuitively understood as the sequence $h_{oc}(n)$ shown in Fig. 10.2 where $h_{oc}(n) = (1/2)s * h(n)$ and $s(n)$ relates to the sequence for conditions $(1, -|r|)$. The waveform in a single period of the mixture changes from a single unit pulse of $h_{oc}(n)$ into the sequence $(1, -|r|)$ for $S(z^{-1})$. Here, the sequence for $S(z^{-1})$ may represent the sound source characteristics under mixed conditions. The difference in the sound sources of flutes and clarinets is governed by the difference in the open–open and open–closed boundary conditions. The mixture model produces magnitude frequency characteristics of an open–open condition combined with the fundamental frequency for an open–closed condition.

10.2 Boundary and Driving Conditions of Sound Traveling in a Tube

10.2.1 *Sinusoidal Waves Under Generalized Boundary Conditions*

The open or closed end condition is a simplified model for the boundary condition under which the reflected waves are produced. However, these conditions are too simplifying to understand properly wave propagation inside tubes such as musical instruments. Allowing complex numbers for the reflection coefficients to represent the magnitude and phase of the reflected sinusoidal waves, the boundary conditions can be extended to account for other physical aspects occurring at open and closed ends. Suppose the reflection coefficient of the end at $x = L$ be $r_L = |r_L|e^{i\theta}$. Introducing the complex reflection coefficient into the z-transform expressed by Eq. 10.1, the poles are located such that

$$1 + r_L z_p^{-1} = 0 \tag{10.7}$$

$$z_p = |z_p|e^{i\Omega_p} = -|r_L|e^{i\theta} \tag{10.8}$$

$$e^{i\Omega_p} = -e^{i\theta} = e^{i(\theta-\pi)} \tag{10.9}$$

$$\Omega_p = \theta + (2m - 1)\pi, \tag{10.10}$$

where m is an integer. A generalized boundary condition is represented by a phase change during reflection.

An example of a generalized reflection coefficient is given by $r_L = |r_L|i$, which produces a $\pi/2$-phase shift between the incident and reflected waves. The frequency characteristics for the case are written as

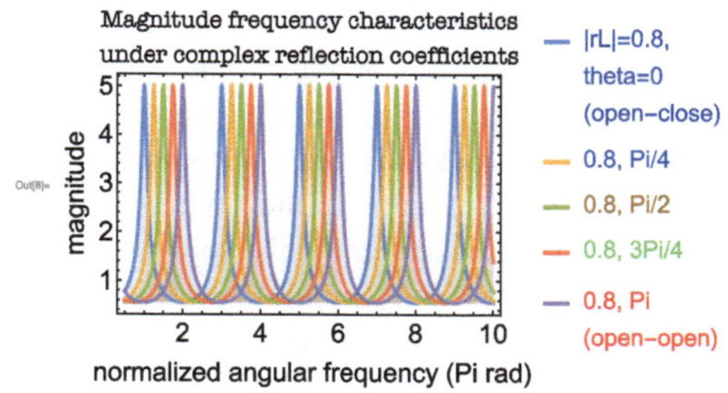

Fig. 10.7 Magnitude frequency characteristics for complex reflection coefficients

$$H(\mathrm{e}^{-\mathrm{i}\Omega}) = \frac{1}{1+\mathrm{i}|r_L|\mathrm{e}^{-\mathrm{i}\Omega}}. \tag{10.11}$$

The magnitude frequency characteristics is displayed in Fig. 10.7.

The poles are located at

$$\Omega_{poi} = \theta - \pi + 2m\pi = \left(2m - \frac{1}{2}\right)\pi, \tag{10.12}$$

where $\theta = \pi/2$. Comparing $\Omega_{poc} = (2m-1)\pi$ for the open–closed and $\Omega_{poo} = 2m\pi$ for the open–open conditions, Ω_{poi} is understood as in-between both of these conditions. Consequently, the harmonic structure in which all the ratios of the fundamental to the harmonics are integers is lost for Ω_{poi}. That is, whether the end is open or closed is very important for the sound characteristics of musical instruments.

10.2.2 *Generalized Boundary Conditions in Time-Delayed Reflection*

The complex representation of the reflection coefficients in terms of the magnitude and phase may not be appropriate to represent the time sequence of sound pulses traveling in the tube. Consider then a virtual sequence $h(n)$ observed at $x = 0$ (open end) that instead of being a single unit sample is periodic with period N. Moreover, this virtual time sequence may be assumed to composed of successive convolutions with the sequence $-g(n)$ where $g(n)$ represents the reflected pulse sequence generated at the other end ($x = L$). The z-transform of the sequence $h(n)$ can be written as

$$H(z^{-1}) = \frac{1}{1+G(z^{-1})z^{-N}}, \tag{10.13}$$

where $G(z^{-1})$ denotes the z-transform of $g(n)$.

The reflection coefficient corresponding to this sequence $g(n)$ in the time domain or $G(z^{-1})$ is largely different from the open or closed end condition. Thus the sound quality for an acoustic tube could be determined by $g(n)$ or $G(z^{-1})$ representing the frequency characteristics of the reflected wave. The function $G(z^{-1})$ is a polynomial that has zeros in the complex frequency domain. If the zeros are located close to the unit circle in the z-plane, the components of the incoming sound associated with these zeros are not reflected back into the tube as the reflected sound, but instead propagate forward outside the tube without creating any resonance effect in the acoustic tube. Consequently, the sound source signal may be transmitted away from the acoustic tube. The denominator of Eq. 10.13 determines the resonance, whereas $G(z^{-1})$ governs the transmitted components in terms of its zeros. Equation 10.13 is the same type of transfer function as that for a closed-loop system. This implies that the sound produced by musical instruments based on resonance may be understood as a howling tone through a closed-loop.

An example of time-delayed reflection is the single-unit sample delay for a reflected wave. Take for example $G(z^{-1})$ such that

$$G(z^{-1}) = r(1 + az^{-1}). \tag{10.14}$$

Substituting Eq. 10.14 into Eq. 10.13, then

$$H(\mathrm{e}^{-\mathrm{i}\Omega}) = \frac{1}{1 + r(1 + a\mathrm{e}^{-\mathrm{i}\Omega})\mathrm{e}^{-\mathrm{i}\Omega N}}. \tag{10.15}$$

Figure 10.8 illustrates examples of the magnitude frequency characteristics for the reflection waves according to Eq. 10.14. A variety of frequency characteristics could be produced by increasing the order of the polynomial for $G(z^{-1})$ as exemplified in Fig. 10.9. The magnitude response as given by Eq. 10.15 is illustrated in Fig. 10.10.

The strength of the harmonic components of the acoustic tube could be modified by the sequence of a reflected sound, as depicted in Fig. 10.10. The reflection sequence $g(n)$ emphasizes lower or higher components than that for the open–open condition in the examples in the figure. The poles of Eq. 10.13 including $G(z^{-1})$ produce such spectral modifications. Figure 10.11 presents another sample of a spectral modification, this time by the reflection sequence shown in Fig. 10.9.

Fig. 10.8 Examples of magnitude frequency characteristics for a reflected wave according to Eq. 10.14

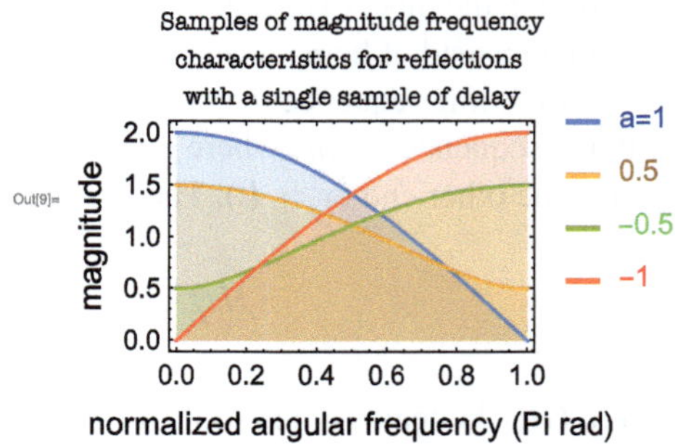

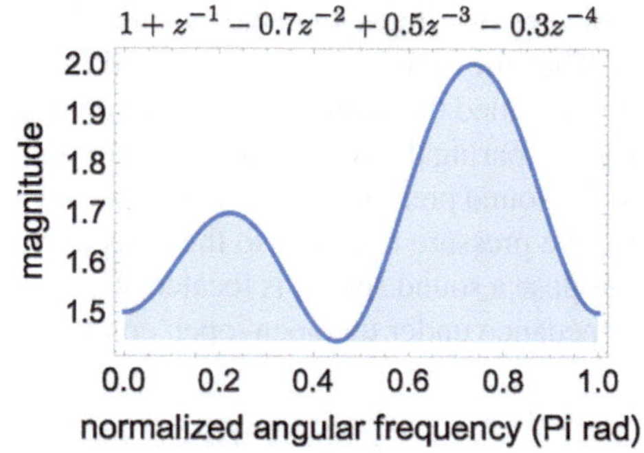

Fig. 10.9 Another example of the magnitude frequency characteristics for a reflected wave

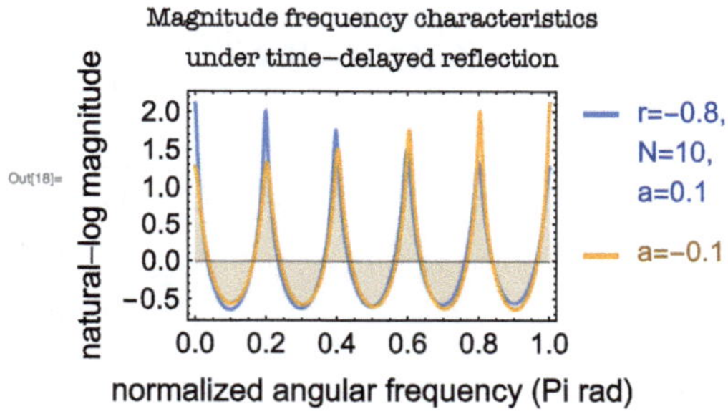

Fig. 10.10 Examples of the magnitude response obtained by Eq. 10.15

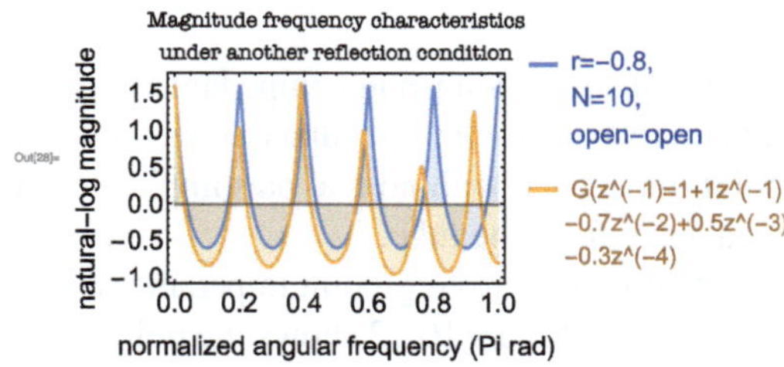

Fig. 10.11 Another sample of the magnitude spectral modification obtained applying a time-delayed reflection condition shown in Fig. 10.9

10.2.3 Driving Condition for the Acoustic Tube

Boundary conditions such as the open–closed conditions developed in the previous sections can also be interpreted as including driving conditions for the acoustic tube. Take again the z-transform setting $r_L = -1$ at the open end ($x = L$) such that

$$H_{oo}(z^{-1}) = \frac{1}{1 - z^{-1}}. \tag{10.16}$$

The open–end condition implies that the sound pressure must vanish at $x = L$. In contrast, the open–closed condition

$$H_{oc}(z^{-1}) = \frac{1}{1 + z^{-1}} \tag{10.17}$$

indicates that the sound velocity must be zero at $x = L$.

The ratio of the sound pressure to the volume velocity responses at the driving position is called the acoustic driving-point impedance, or acoustic radiation impedance. By comparing the two boundary conditions, the poles of the sound pressure response to the sound pressure source under the open–closed condition can be viewed as those for the pressure response to the velocity source under the open–open condition [3]. Suppose a sound source is located at $x = 0$. The poles and zeros of the driving-point impedance under the open–open condition correspond to those for a ratio such that

$$Z_{oo}(x = +0, z^{-1}) = \frac{H_{oc}(z^{-1})}{H_{oo}(z^{-1})} = \left.\frac{1 - |r|z^{-1}}{1 + |r|z^{-1}}\right|_{r=1} \tag{10.18}$$

$$Z_{oo}(x = +0, \mathrm{e}^{-\mathrm{i}\Omega}) = \left.\frac{1 - |r|\mathrm{e}^{-\mathrm{i}\Omega}}{1 + |r|\mathrm{e}^{-\mathrm{i}\Omega}}\right|_{r=1} \tag{10.19}$$

$$= \frac{1 + \mathrm{e}^{\mathrm{i}\Omega} - \mathrm{e}^{-\mathrm{i}\Omega} - 1}{(1 + \cos\Omega)^2 + \sin^2\Omega}$$

$$= \mathrm{i}\frac{2\sin\Omega}{4\cos^2\frac{\Omega}{2}} = \mathrm{i}\frac{\sin\frac{\Omega}{2}}{\cos\frac{\Omega}{2}}, \tag{10.20}$$

where

$$\Omega = \omega T_s = \omega 2L/c = 2kL. \tag{10.21}$$

Note that the driving-point impedance represented by the poles and zeros is pure imaginary because $|r| = 1$; that is, no losses are assumed at the boundaries. A driving-point impedance with no real part implies no sound is radiated into the surrounding area away from the tube.

The complex driving-point impedance can also be defined by Eq. 10.20 assuming $0 < |r| < 1$. Figure 10.12 shows the poles and zeros of the driving-point impedance for open–open condition Z_{oo} assuming $r = -0.8$. The zeros are determined by the numerator of Eq. 10.20; that is,

$$1 - |r|z_0^{-1} = 0. \tag{10.22}$$

Similarly, the poles are given by the roots of the denominator,

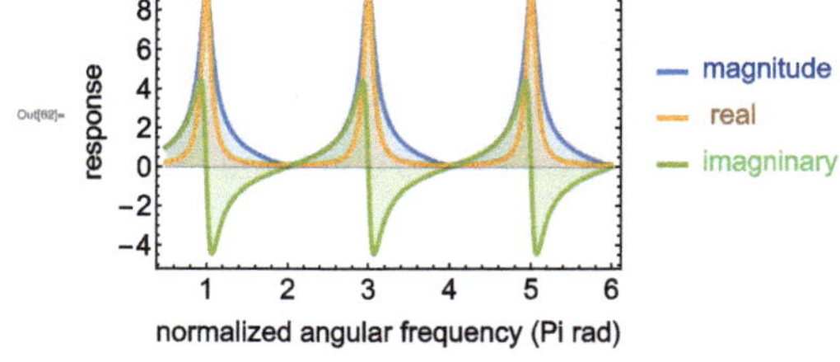

Fig. 10.12 Poles and zeros of driving-point impedance at $x = 0$ for open–open tube

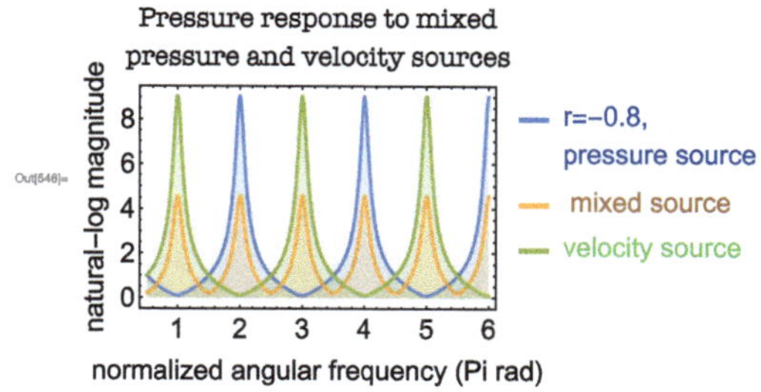

Fig. 10.13 Pressure response to mixed pressure and velocity sources in open–open tube

$$1 + |r|z_p^{-1} = 0. \tag{10.23}$$

Poles are interlaced with zeros. This interlacing is a typical characteristic of the driving-point impedance [1, 2]. Figure 10.12 shows a complex function indicating how the sound energy traveling in the tube is dissipated as sound transmitted into the surroundings because $|r| < 1$.

A sound source is ideally classified as either a pressure or a velocity source; however, from a practical perspective, a sound source may be modeled as a mixture between a pressure and velocity source. According to that H_{oc} can be read as the pressure response to a velocity source in an open–open tube, Fig. 10.13 may be interpreted as pressure responses to a mixed source of pressure and velocity in an open–open tube as these plots seem to reflect to a certain degree of realistic characteristics of musical instruments such as clarinets and flutes.

10.3 Phase Trend of a Periodic Wave Along a Finite Length of String

10.3.1 Periodically Repeated Waves Traveling Between both Ends

As another example of a one-dimensional system, consider a finite length of string fixed at both ends, along which a sound wave travels. As for waves in an acoustic tube investigated in the previous sections, the time-domain approach is taken once again. Figure 10.14 displays an arrangement of source and receiving positions.

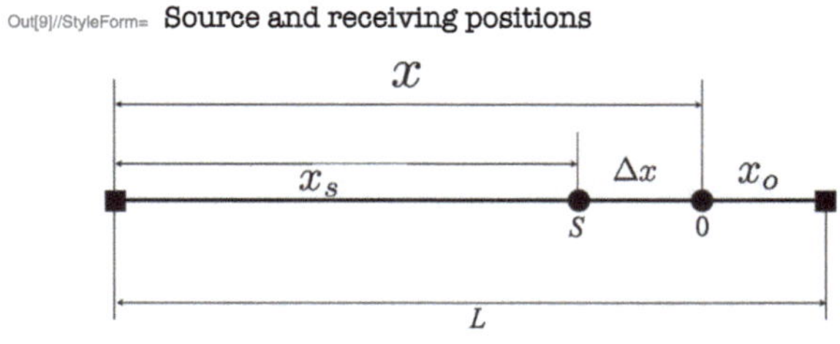

Fig. 10.14 Source and receiving positions along a finite length of string from Fig. 5.11 [3]

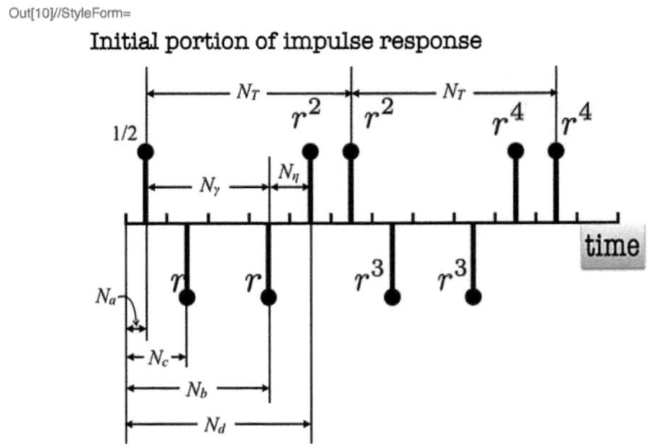

Fig. 10.15 Initial part of the impulse response observed at the receiving position from Fig. 5.12 [3]

An impulse-like wave leaves the source position at $t = 0$ and moves toward both fixed ends. Figure 10.15 illustrates the initial part of the response observed at the observation point, assuming a negative reflection coefficient $-1 < r < 0$. The observed response, called the impulse response, is constructed using an infinitely-long sequence similar to a geometric series, in which each cyclic sequence is composed of four pulses. The length of a single cycle of the wave is N_T indicating the period of the response [3–5].

Take the z-transform of the response record (transfer function) $h(n)$. The z-transform is written as

$$H(z^{-1}) = \sum_n h(n)z^{-n} \tag{10.24}$$

$$= \frac{1}{2}\frac{z^{-N_a} - |r|z^{-N_b} - |r|z^{-N_c} + |r|^2 z^{-N_d}}{1 - |r|^2 z^{-N_T}},$$

where

$$N_a T_s = \frac{x - x_s}{c} = \frac{\Delta x}{c} \tag{10.25}$$

$$N_b T_s = \frac{2x_s + \Delta x}{c} \tag{10.26}$$

$$N_c T_s = \frac{\Delta x + 2x_o}{c} \tag{10.27}$$

$$N_d T_s = \frac{2x_s + \Delta x + 2x_o}{c} \tag{10.28}$$

$$N_T T_s = \frac{2x + 2x_o}{c} = \frac{2L}{c}, \tag{10.29}$$

where L denotes the length of the string and T_s gives the sampling period in seconds. Here, $N_a T_s$ represents the time delay required for the direct wave to reach the observation position.

10.3.2 Poles and Zeros of a Cyclic Wave on a Finite String

The transfer function stated above can be rewritten as

$$\begin{aligned} H(z^{-1}) &= \frac{1}{2}\frac{z^{-N_a}}{1-|r|^2 z^{-N_T}}\left[1-|r|z^{-(N_b-N_a)}\right. \\ &\left. -|r|z^{-(N_c-N_a)}+|r|^2 z^{-(N_d-N_a)}\right] \\ &= \frac{1}{2}z^{-N_a}\hat{H}(z^{-1}) \\ &= \frac{1}{2}z^{-N_a}\frac{I(z^{-1})}{J(z^{-1})}, \end{aligned} \tag{10.30}$$

where

$$I(z^{-1}) = \left(1-|r|z^{-N_{ba}}\right)\left(1-|r|z^{-N_{ca}}\right) \tag{10.31}$$

$$J(z^{-1}) = 1-|r|^2 z^{-N_T} \tag{10.32}$$

$$N_{ba} = N_b - N_a = \frac{2x_s}{cT_s} \tag{10.33}$$

$$N_{ca} = N_c - N_a = \frac{2x_o}{cT_s} \tag{10.34}$$

$$N_{da} = N_d - N_a = \frac{2(x_s+x_o)}{cT_s} \tag{10.35}$$

$$N_{ca} - N_{da} = N_c - N_d = -N_{ba}. \tag{10.36}$$

The numerator represents the transfer function for a single cycle of the response [4, 5], excluding the pure delay between the source and observation positions, and determines the zeros of the transfer function $H(z^{-1})$. This single cycle mostly determines the transient (on-set) response after a source signal is fed into the system. The on-set response can be understood as the process in which zeros are produced rather than poles corresponding to resonances. The resonances defined by the poles mostly represent characteristics of the off-set or reverberation conditions. The zeros z_0 of a single cycle of the wave depend on the locations of the source and observation positions. For the source position,

$$|z_{0_s}| = |r| \tag{10.37}$$

$$e^{-iN_{ba}\Omega_{0_s}} = 1. \tag{10.38}$$

The number of zeros is

$$N_{0_s} = N_{ba}. \tag{10.39}$$

Similarly, the zeros for the observation position are located such that

$$|z_{0_o}| = |r| \tag{10.40}$$

$$e^{-iN_{ca}\Omega_{0_o}} = 1. \tag{10.41}$$

The number of the zeros is

$$N_{0_o} = N_{ca}. \tag{10.42}$$

Consequently, the number of zeros N_0 can be estimated as

$$N_0 = N_{0_s} + N_{0_o} = N_{ba} + N_{ca}, \tag{10.43}$$

all of which are located inside the unit circle assuming that $-1 < r < 0$.

In contrast, the zeros of the denominator or $J(z^{-1})$ determine the poles of the transfer function. The poles are determined from

$$|z_p| = |r|^2 \tag{10.44}$$

$$e^{-iN_T\Omega_p} = 1, \tag{10.45}$$

independently of the source and observation positions. The number of poles is

$$N_p = N_T, \tag{10.46}$$

all of which are also inside the unit circle.

10.4 Minimum-Phase Property and Propagation Phase

10.4.1 Poles and Zeros for Minimum Phase

The zeros of the transfer function described above are all located inside the unit disc. Such a system is referred to as minimum phase. The transfer function $\hat{H}(z^{-1})$ have zeros also at the origin $z = 0$. Their number N_{00} is given by N_T as fixed by the denominator $J(z^{-1})$ given in Eq. 10.32. In addition, the poles are located at the origin, their number being

$$N_{p0} = N_{ba} + N_{ca} \tag{10.47}$$

established by the numerator $I(z^{-1})$ in Eq. 10.31. Therefore, the zeros at the origin are partly canceled by the poles, and hence the number of zeros at the origin becomes

$$\begin{aligned} \hat{N}_{00} &= N_{00} - N_{p0} \\ &= N_T - (N_{ba} + N_{ca}) \end{aligned} \tag{10.48}$$

$$= \frac{2\Delta x}{cT_s}$$
$$= 2N_a.$$

Consequently, there are

$$\begin{aligned} N_0 &= N_{0s} + N_{0o} \\ &= N_{ba} + N_{ca} \\ &= N_b + N_c - 2N_a \end{aligned} \tag{10.49}$$

zeros inside the disc (not at the origin), and

$$\begin{aligned} \hat{N}_{00} &= N_{00} - N_{p0} \\ &= 2N_a \end{aligned} \tag{10.50}$$

zeros at the origin. Therefore, the total number of the zeros is

$$\begin{aligned} N_z &= \hat{N}_{00} + N_0 = 2N_a + N_b + N_c - 2N_a \\ &= N_b + N_c. \end{aligned} \tag{10.51}$$

In addition there are N_T poles inside the disc. Therefore, the total accumulation in phase due to poles and zeros is estimated using

$$\begin{aligned} \Phi &= -2\pi(N_p - N_z) \\ &= -2\pi(N_T - N_b - N_c) \\ &= 0. \end{aligned} \tag{10.52}$$

Excluding direct sound representing the propagation phase, the transfer function is referred to as the minimum phase when it does not produce an accumulation other than the propagation phase. A one-dimensional wave-propagating system is a typical example of a minimum-phase system.

The behavior in phase is not produced in minimum-phase systems if the propagation phase is removed. However, it would be desirable to intuitively understand the propagation phase for one-dimensional systems, if the phase behavior can be observed in accordance with the difference between the number of poles and zeros inside the unit disc, apart even after removing the direct sound [1, 2].

10.4.2 Accumulated Phase for One-Dimensional Systems

Figure 10.16 illustrates examples of the accumulation of phase that can be obtained by integration of the group delay function (see Sects. 3.4 and 6.2).

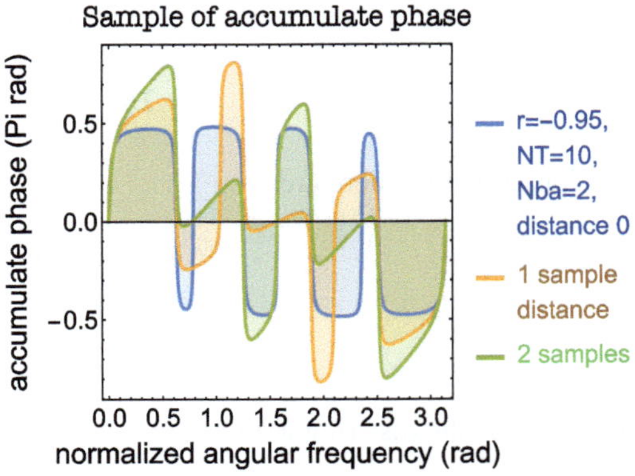

Fig. 10.16 Example of the accumulation of phase

The panel shows cases when the distance from the source position is 0, 1, and 2 samples, respectively. No accumulation phase can be seen for any case, because the propagation phase is removed.

The propagation phase, however, is produced by the difference in the numbers of the poles and zeros, even if the propagation phase is discarded. For the positive frequency region ($0 < \Omega < \pi$) corresponding to physical frequencies, the numbers of poles and zeros inside the disc are estimated to be

$$N_0^+ = N_0/2 = \frac{1}{2}(N_{ba} + N_{ca}) \tag{10.53}$$

$$N_p^+ = N_T/2. \tag{10.54}$$

The difference corresponding to the number of un-canceled poles by the zeros is

$$\begin{aligned} N_p^+ - N_0^+ &= \frac{1}{2}(N_T - N_{ba} - N_{ca}) \\ &= N_a \\ &= \frac{1}{2}\hat{N}_{00}. \end{aligned} \tag{10.55}$$

The zeros found at the origin cancel the accumulation phase via the residual poles. Therefore, by counting the number of the residual poles inside the disc instead of the accumulation phase, the propagation phase then can be estimated even if the propagation component is removed.

Figure 10.17 illustrates group delay with samples in which the positive (negative) pulses are due to poles (zeros). The difference in the number of positive and negative pulses gives the number of the residual poles that produce the propagation phase or the time delay in the impulse response. The poles are independent of the source and receiving positions. In contrast, the zeros are determined by the source and receiving positions [1, 2]. When the position of the sound source is fixed, the number of zeros varies depending on only the receiving position or the distance from the source

Out[11]//StyleForm=

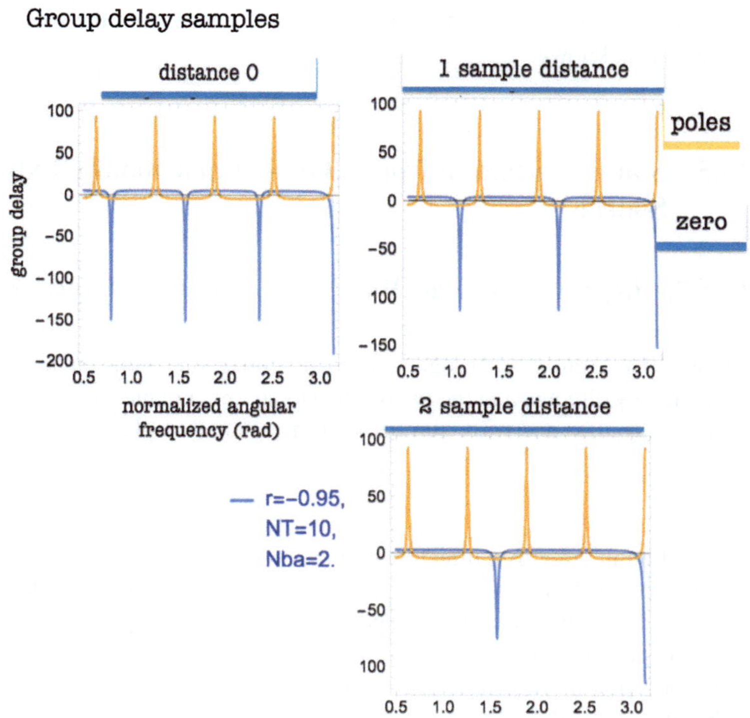

Fig. 10.17 Samples of group delay

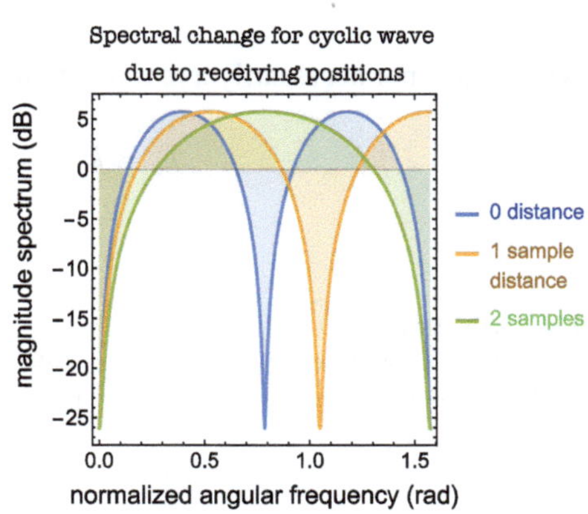

Fig. 10.18 Magnitude spectral characteristics for single cycle of response due to receiving positions

position. Moving the receiving position away from the source, the zeros then move toward high frequencies. This movement in the zeros changes the numbers of poles and zeros. Consequently, a propagation phase is produced [1, 2].

Figure 10.18 presents the magnitude spectral characteristics for the single cycle of response [4, 5] composed of the zeros due to the receiving positions. The change in the magnitude spectrum or waveform resulting from a change in distance from the source implies a possibility of perceiving distance from the source, even if the distance is unknown.

10.5 Sound Radiation from a One-Dimensionally Extended Source

10.5.1 Representation of Waves and Radiation Impedance

Suppose that a transversal progressive wave propagating along an infinitely extended structure ($x-$direction) in the two-dimensional (x, z) plane. A wave of angular frequency ω_0(rad/s) and wave number k_0(1/m) is an example of a velocity wave

$$v_0 = |v_0|e^{-ik_x x}e^{i\omega_0 t} \quad (\text{m/s}) \tag{10.56}$$

propagating in the x-axis direction of a $(x, z)-$plane representing a vibrating structure. Figure 10.19 shows an image of the sound pressure field generated by the structural vibration in the surrounding medium. The radiation impedance for the structural vibration can be formulated according to the boundary condition that requires the vibration velocity which must be continuous at $z = 0$.

Suppose the radiated pressure wave to be

$$p_0 = P_0 e^{-ik_x x - ik_z z}e^{i\omega_0 t}, \quad (\text{Pa}) \tag{10.57}$$

where P_0 is not necessarily a real function and

$$k_0^2 = k_x^2 + k_z^2. \quad (1/\text{m}^2) \tag{10.58}$$

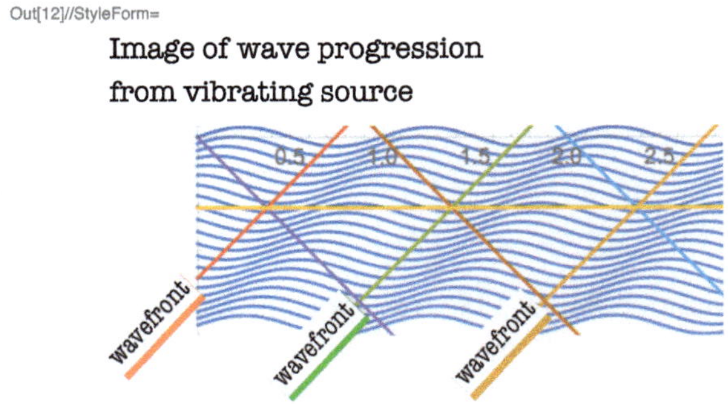

Fig. 10.19 Image of sound pressure waves produced by a vibrating structure

From the boundary condition

$$-B_0 \frac{\partial}{\partial z} p_0 \bigg|_{z=0} = B_0 \rho_0 \frac{\partial}{\partial t} v_0, \quad \text{(N)} \tag{10.59}$$

where $B_0(\mathrm{m}^3)$ denotes the unit volumes,

$$P_0 = \rho_0 c_0 |v_0| \frac{k_0}{\sqrt{k_0^2 - k_x^2}} \quad \text{(Pa)} \tag{10.60}$$

can be obtained. Here, k_0 and c_0 denote the wave number (1/m) and the speed of sound (m/s) in the radiated field, respectively. The radiated sound pressure is written as

$$p_0 = \rho_0 c_0 |v_0| \frac{k_0}{\sqrt{k_0^2 - k_x^2}} \mathrm{e}^{-\mathrm{i}k_x x - \mathrm{i}k_z z} \mathrm{e}^{\mathrm{i}\omega_0 t}, \quad \text{(Pa)} \tag{10.61}$$

where

$$v_0 = |v_0| \mathrm{e}^{-\mathrm{i}k_x x} \mathrm{e}^{\mathrm{i}\omega_0 t} \quad \text{(m/s)} \tag{10.62}$$

by assuming that the one-dimensional vibrating structure is a strip of unit width.

Setting the volume velocity of the unit area $S_0(\mathrm{m}^2)$ at the structure to be

$$q_0 = S_0 v_0, \quad (\mathrm{m}^3/\mathrm{s}) \tag{10.63}$$

the acoustic radiation impedance of the vibrating structure is defined as

$$\begin{aligned} Z_{rad_A} &= \frac{p_0}{q_0} \bigg|_{z=0} \\ &= \frac{\rho_0 c_0}{S_0} \frac{k_0}{\sqrt{k_0^2 - k_x^2}}. \quad (\mathrm{Pa} \cdot \mathrm{s/m}^3) \end{aligned} \tag{10.64}$$

Similarly, the mechanical radiation impedance becomes

$$\begin{aligned} Z_{rad_M} &= \frac{f_0}{v_0} \\ &= S_0 \rho_0 c_0 \frac{k_0}{\sqrt{k_0^2 - k_x^2}}, \quad (\mathrm{N} \cdot \mathrm{s/m}) \end{aligned} \tag{10.65}$$

where $f_0(\mathrm{N})$ means the force working on the area $S_0(\mathrm{m}^2)$ of the structure (strip with a unit width).

Out[13]//StyleForm=

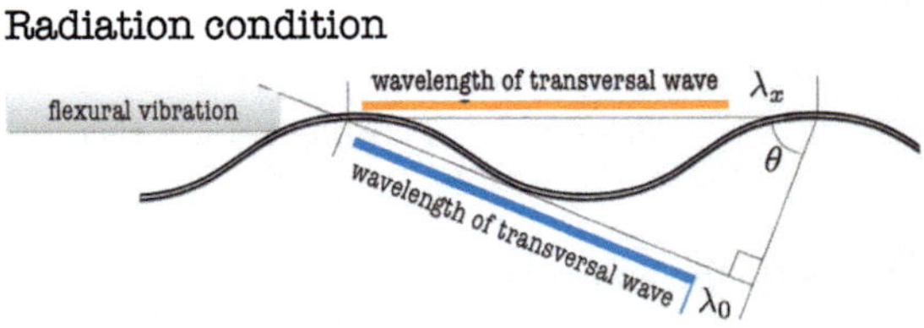

Fig. 10.20 Right-angle triangle representing the radiation condition from Fig. 6.4 [3]

10.5.2 Geometric Interpretation of Radiation Impedance

The image of wave propagation in a vibrating structure (Fig. 10.19) shows a right-angle triangle construction illustrated in Fig. 10.20. From this geometrical schematic, the acoustic radiation impedance can be interpreted as a function of k_x such that

$$\begin{aligned} Z_{rad_A}(k_x) &= \frac{\rho_0 c_0}{S_0} \frac{k_0}{\sqrt{k_0^2 - k_x^2}} \\ &= \frac{\rho_0 c_0}{S_0} \frac{\lambda_x}{\sqrt{\lambda_x^2 - \lambda_0^2}} \\ &= \frac{\rho_0 c_0}{S_0} \frac{c_x}{\sqrt{c_x^2 - c_0^2}} \\ &= \frac{\rho_0 c_0}{S_0} \frac{1}{\cos\theta}, \quad (\text{Pa} \cdot \text{s/m}^3) \end{aligned} \tag{10.66}$$

where θ denotes the radiation angle of the radiated wave, and λ_x and λ_0 (m) are the wavelengths of the waves propagating on the structure and that in the surrounding medium, respectively, and similarly, c_x and c_0 (m/s) indicate the speed of sound.

The geometric schematic is useful in understanding the radiation condition enabling the structural vibration to generate a sound wave traveling in the surrounding medium. The right-angle triangle condition must occur to produce a transmitted wave into the medium. For that purpose, the condition $c_x > c_0$ must hold between the structural vibration and the traveling wave in the medium. In other words, the radiation impedance is real for the condition above; otherwise, it is imaginary.

10.5.3 Coincidence Frequency and Radiation Condition

Suppose again

$$v_0 = |v_0| e^{-ik_x x} e^{i\omega_0 t} \quad (\text{m/s}) \tag{10.67}$$

with angular frequency ω_0(rad/s) along the structure. The propagation speed of the vibration is in general dependent on the vibration frequency [6]. Thus the wavelength of the structural wave is a function of frequency. If the wavelength is longer than that for the radiated wave into the surrounding medium at the angular frequency ω_0(rad/s), the radiated sound propagates into the medium. In contrast, if the wavelength of the structural vibration is shorter than that for the sound wave, then the radiation sound field does not excite the medium. The frequency at which the wavelength of the vibration along the structure coincides with that for sound is often called the coincidence frequency; the radiation impedance is, however, not well defined at this frequency.

The radiation condition implies that reducing the vibration of a source is not always the best way to control noise radiated from the vibrating object. Instead, the vibration component for which the radiation impedance is real should be well controlled. In terms of the coincidence frequency, the vibration that has higher frequencies than the coincidence frequencies should be reduced [7]. The radiation condition can be extended to moving objects such as ships or air planes. When a moving object moves faster than the speed of the radiated wave, sound propagates into the surrounding medium.

10.5.4 Superposition of Vibration

The structural vibration can be expanded in terms of spectral components using Fourier synthesis between spatial and wavenumber regions. Suppose that the structural vibration is represented as a single sinusoidal wave

$$v = v_x \mathrm{e}^{\mathrm{i}\omega_0 t}, \quad (\mathrm{m/s}) \tag{10.68}$$

where $v(x)$ is not necessarily a real function. The function $v(x)$ signifying the spatial distribution of the vibration with respect to x can be represented via a Fourier transform as

$$v = \int_{-\infty}^{\infty} V(k_x)\mathrm{e}^{-\mathrm{i}k_x x} dk_x, \quad (\mathrm{m/s}) \tag{10.69}$$

where

$$V(k_x) = \frac{1}{2\pi}\int_{-\infty}^{\infty} v(x)\mathrm{e}^{\mathrm{i}k_x x} dx. \quad (\mathrm{m^2/s}) \tag{10.70}$$

Similarly, the sound pressure generated by the vibration can be expressed as

$$p(x, z)|_{z=0} = p(x) = \int_{-\infty}^{\infty} P(k_x)\mathrm{e}^{-\mathrm{i}k_x x} dk_x, \quad (\mathrm{Pa}) \tag{10.71}$$

where the sound pressure on the structure can be written as

$$p = p(x)\mathrm{e}^{\mathrm{i}\omega_0 t} \quad \text{(Pa)} \tag{10.72}$$

and

$$P(k_x) = \frac{1}{2\pi}\int_{-\infty}^{\infty} p(x)\mathrm{e}^{\mathrm{i}k_x x}dx. \quad (\text{Pa}\cdot\text{m}) \tag{10.73}$$

The sound pressure field, including its dependence in the z-direction, can be represented by

$$p(x, z) = \int_{-\infty}^{\infty} P(k_x)\mathrm{e}^{-\mathrm{i}k_x x}\mathrm{e}^{-\mathrm{i}\sqrt{k_0^2-k_x^2}z}dk_x, \quad \text{(Pa)} \tag{10.74}$$

corresponding to the radiation condition $k_0^2 - k_x^2 > 0$. Subject to the radiation condition, the wave with wavenumber k_z propagates toward the z-axis. In contrast, under the condition $k_0^2 - k_x^2 < 0$, the wave component moving toward the z-axis is attenuated with respect to z.

From the boundary condition using Eq. 10.59

$$\begin{aligned} P(k_x) &= V(k_x)\cdot S_0 \cdot Z_{rad_A}(k_x) \\ &= V(k_x)\rho_0 c_0 \frac{k_0}{\sqrt{k_0^2 - k_x^2}}, \quad (\text{Pa}\cdot\text{m}) \end{aligned} \tag{10.75}$$

which corresponds to Eq. 10.60. Consequently, the sound pressure distribution on the structure is

$$\begin{aligned} p(x) &= \int_{-\infty}^{\infty} P(k_x)\mathrm{e}^{-\mathrm{i}k_x x}dk_x \\ &= \int_{-\infty}^{\infty} V(k_x)S_0 Z_{rad_A}(k_x)\mathrm{e}^{-\mathrm{i}k_x x}dk_x \\ &= \int_{-\infty}^{\infty} V(k_x)\rho_0 c_0 \frac{k_0}{\sqrt{k_0^2 - k_x^2}}\mathrm{e}^{-\mathrm{i}k_x x}dk_x. \quad \text{(Pa)} \end{aligned} \tag{10.76}$$

A sinusoidal source that excites the structure at a single point ($x = 0$) with a single angular frequency of ω_0 can be formulated as

$$\begin{aligned} v &= v(x)\mathrm{e}^{\mathrm{i}\omega_0 t} \\ &= V_0\delta(x)\mathrm{e}^{\mathrm{i}\omega_0 t}. \quad (\text{m/s}) \end{aligned} \tag{10.77}$$

The Fourier transform of $v(x)$ is

$$V(k_x) = \frac{1}{2\pi}\int_{-\infty}^{\infty} V_0\delta(x)\mathrm{e}^{\mathrm{i}k_x x}dx = \frac{V_0}{2\pi}. \quad (\mathrm{m}^2/\mathrm{s}) \tag{10.78}$$

The sound pressure generated on the structure ($z = 0$) is

$$\begin{aligned} p(x) &= \int_{-\infty}^{\infty} P(k_x)\mathrm{e}^{-\mathrm{i}k_x x}dk_x \\ &= \int_{-\infty}^{\infty} V(k_x)S_0 Z_{rad_A}(k_x)\mathrm{e}^{-\mathrm{i}k_x x}dk_x \\ &= \int_{-\infty}^{\infty} \frac{V_0}{2\pi}\rho_0 c_0 \frac{k_0}{\sqrt{k_0^2 - k_x^2}}\mathrm{e}^{-\mathrm{i}k_x x}dk_x \\ &= \frac{V_0}{2\pi}\rho_0 c_0 \int_{-\infty}^{\infty} H_z(k_x)dk_x \quad (\mathrm{Pa}) \end{aligned} \tag{10.79}$$

$$H_z(k_x) = \frac{k_0}{\sqrt{k_0^2 - k_x^2}}\mathrm{e}^{-\mathrm{i}k_x x}. \tag{10.80}$$

The radiation impedance is a complex function, for which the integration is divided into two parts:

$$\int_{-\infty}^{\infty} H_z(k_x)dk_x = \left[\int_{k_0^2 > k_x^2} + \int_{k_0^2 < k_x^2}\right] H_z(k_x)dk_x. \quad (1/\mathrm{m}) \tag{10.81}$$

The first term of the integration can be rewritten as

$$\begin{aligned} I_1 &= \int_{k_0^2 > k_x^2} H_z(k_x)dk_x \\ &= \int_{k_0^2 > k_x^2} \frac{k_0}{\sqrt{k_0^2 - k_x^2}}\left[\cos k_x x - \mathrm{i}\sin(k_x x)\right]dk_x. \quad (1/\mathrm{m}) \end{aligned} \tag{10.82}$$

Evaluating the real part of I_1 gives

$$\begin{aligned} V_0\frac{\rho_0 c_0}{2\pi}\Re[I_1] &= V_0\frac{\rho_0 c_0}{2\pi}\cdot 2\int_0^{k_0} \frac{k_0}{\sqrt{k_0^2 - k_x^2}}\cos k_x x dk_x \\ &= V_0\frac{\rho_0 c_0 k_0}{2}\mathrm{J}_0(k_0 x), \qquad (\mathrm{Pa}) \end{aligned} \tag{10.83}$$

where

$$\frac{2}{\pi}\int_0^{\pi/2} \cos[k_0 x\cos\theta]d\theta = \mathrm{J}_0(k_0 x), \tag{10.84}$$

is the zeroth-order Bessel function [8]. For the imaginary part of the integration, I_1 becomes 0 as

$$-\mathrm{i}\int_{k_0^2>k_x^2}\frac{k_0}{\sqrt{k_0^2-k_x^2}}\sin k_x x dk_x = 0, \tag{10.85}$$

where the integrand is an odd function of k_x.

The second integration of Eq. 10.81 can be rewritten as

$$\begin{aligned} I_2 &= \int_{k_0^2<k_x^2}\frac{k_0}{\sqrt{k_0^2-k_x^2}}[\cos k_x x - \mathrm{i}\sin k_x x]\,dk_x \\ &= \left[\int_{-\infty}^{-k_0}+\int_{k_0}^{\infty}\right]\frac{k_0}{-\mathrm{i}\sqrt{k_x^2-k_0^2}}[\cos k_x x - \mathrm{i}\sin k_x x]\,dk_x \\ &= 2\int_{k_0}^{\infty}\frac{k_0}{-\mathrm{i}\sqrt{k_x^2-k_0^2}}\cos k_x x dk_x. \quad (1/\mathrm{m}) \end{aligned} \tag{10.86}$$

By taking the imaginary part

$$\begin{aligned} V_0\frac{\rho_0 c_0}{2\pi}\Im[I_2] &= V_0\frac{\rho_0 c_0}{2\pi}\cdot 2\int_{k_0}^{\infty}\frac{k_0}{\sqrt{k_x^2-k_0^2}}\cos k_x x dk_x \\ &= V_0\frac{\rho_0 c_0 k_0}{2}(-\mathrm{Y}_0(k_0 x)), \quad (\mathrm{Pa}) \end{aligned} \tag{10.87}$$

where

$$-\frac{2}{\pi}\int_0^{\infty}\cos[k_0 x\cosh\theta]d\theta = \mathrm{Y}_0(k_0 x), \tag{10.88}$$

called the zeroth-order Neumann function [8].

Consequently, the sound pressure on the structure is given by

$$\begin{aligned} p(x) &= V_0\frac{\rho_0 c_0 k_0}{2}[\mathrm{J}_0(k_0 x) - \mathrm{i}\mathrm{Y}_0(k_0 x)] \quad (\mathrm{Pa}) \\ &= V_0\left[\frac{\rho_0\omega_0}{2}\mathrm{J}_0(k_0 x) + \mathrm{i}\frac{\omega_0\rho_0}{2}(-\mathrm{Y}_0(k_0 x))\right] \end{aligned} \tag{10.89}$$

$$p(x,t) = p(x)\mathrm{e}^{\mathrm{i}\omega_0 t}, \tag{10.90}$$

where

$$v(x,t) = V_0\delta(x)\mathrm{e}^{\mathrm{i}\omega_0 t}. \quad (\mathrm{m/s}) \tag{10.91}$$

The real part of the Eq. 10.89 determines the radiated sound pressure on the structure at ($z = 0, x = 0$) that is in-phase with the vibration velocity and thus generates

Fig. 10.21 Zeroth-order Bessel and Neumann functions

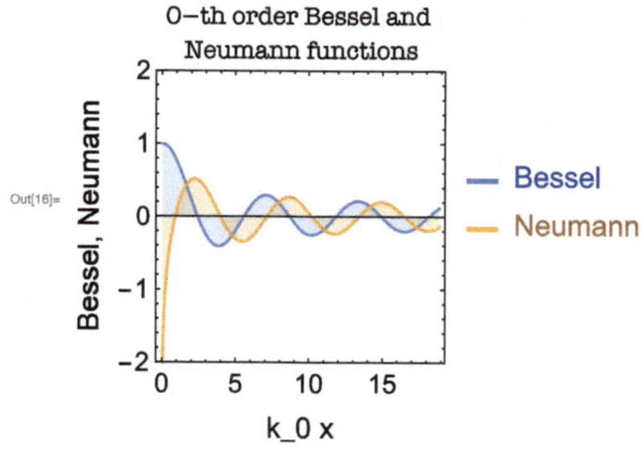

the sound field in the surrounding medium ($z > 0$). In contrast, the imaginary part represents the sound pressure that is out-of-phase with the vibration velocity, and no progressive waves are rendered in the medium. In other words, the imaginary part accounts for the additional inertial effect of the surrounding medium on the vibrating structure.

Figure 10.21 presents plots of the zeroth-order Bessel and Neumann functions. The Neumann function is not well defined at $x = 0$, because the sound pressure is not defined at the source position ($x = 0$) under ideal conditions such as a single-point excitation.

10.5.5 Sound Radiation from an Impulsive Vibration

The previous section formulated the sound radiation from a sinusoidal vibration using radiation impedance, where the external force was given at a single point. Under the same excitation conditions, the impulsive sound radiation can be formulated using a superposition of sinusoidal waves. Take the real part of the sound pressure above and consider the waves produced by the sound field in the surrounding medium. To obtain an analytical example, assume the spectral condition of the impulsive vibration velocity v_i (m/s) that can be expressed as [8]

$$V_{spec} \sim \frac{1}{\omega_0}. \tag{10.92}$$

The radiated sound due to this impulsive source velocity (Pa) can be written as

$$\begin{aligned} p_i &= \int_{-\infty}^{\infty} (1/\omega_0)\Re[p(x)]e^{i\omega_0 t} d\omega_0 \\ &= \rho_0 V_0 \int_0^{\infty} J_0(\omega_0 x/c_0) \cos \omega_0 t d\omega_0 \end{aligned} \tag{10.93}$$

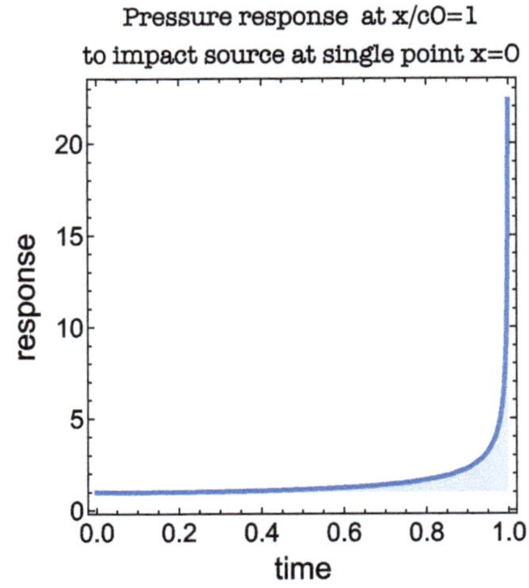

Fig. 10.22 Image of an impulsive pressure wave at $x/c_0 = 1$ on the structure following a pulse-like excitation at a single point with spectral characteristics of $1/\omega_0$

$$= \begin{cases} \rho_0 V_0 \frac{1}{\sqrt{(x/c_0)^2 - t^2}}, & t < |x|/c_0 \\ 0 & t > |x|/c_0. \quad \text{(Pa)} \end{cases}$$

Suppose that $x > 0$ is the observation point, and x/c_0 gives the time when the sound pressure wave reaches the observation point on the structure. The sound pressure response to an impulsive structural vibration can be observed before time x/c_0 when the pressure wave reaches the observation point. The pulse-like vibration concentrated at a single point on the structure can be Fourier transformed and represents the vibration as a superposition of plane waves propagating along the structure in wavenumber space. Of those superposed plane waves, only those components that have wavelengths longer than those in the sound pressure waves in the surrounding medium contribute to the sound pressure fields in the medium. Consequently, the sound pressure response from these wave components of longer wavelengths than those in the surrounding medium is observable when $t < x/c_0$. After $t > x/c_0$, the sound pressure wave generating the sound pressure field in the medium has passed the observation point, and no waves are able to excite the sound field at the observation point. Figure 10.22 presents a sample of the pressure response to the impulsive structural vibration velocity, where the observation point x/c_0 is set to unity.

10.6 Exercises

1. Solve the wave equation for an acoustic tube under the conditions that [3]

   (1) pressure source at $x = 0$ and open end $x = L$,

(2) velocity source at $x = 0$ and open end $x = L$,
(3) pressure source at $x = 0$ and close end $x = L$,
(4) velocity source at $x = 0$ and close end $x = L$.

2. Obtain the driving-point impedances for the above cases.
3. Confirm the eigenfrequencies (fundamental and harmonics) in the above cases.
4. Confirm that the driving-point impedances are pure imaginary in the above cases.
5. Derive the eigenfrequencies for mixed or generalized boundary conditions following the procedure given in this chapter.
6. Derive Eq. 10.24.
7. Obtain the expression in closed form for Eq. 10.24 when $|r| = 1$ and $z = e^{i\Omega}$.
8. Derive the expression giving the trend of the minimum phase in the above closed form [1–3].

References

1. R.H. Lyon, Progressive phase trends in multi-degree-of-freedom systems. J. Acoust. Soc. Am. **73**(4), 1223–1228 (1983)
2. R.H. Lyon, Range and frequency dependence of transfer function phase. J. Acoust. Soc. Am. **76**(5), 1435–1437 (1984)
3. M. Tohyama, *Sound and Signals* (Springer, Berlin, 2011)
4. T. Hasegawa, M. Tohyama, Analysis of spectral and temporal waveforms of piano-string vibration. J. Audio Eng. Soc. **60**(4), 237–245 (2012)
5. M. Tohyama, *Waveform Analysis of Sound* (Springer, Berlin, 2015)
6. R.H. Lyon, *Machinery Noise and Diagnostics* (Butterworth, Washington, 2000)
7. M. Tohyama, T. Itow, Theoretical interpretation of the mass law based on the wave theory. Acustica **30**(1), 1–11 (1974)
8. A. Jeffrey, *Handbook of Mathematical Formulas and Integrals*, 3rd edn. (Elsevier Academic Press, New York, 2004)

Chapter 11
Reverberation in Rooms

Abstract Reverberation is a typical auditory event in the time domain. Assuming a random superposition, the exponential and Rayleigh distributions are derived for the sound energy and amplitude distributions in a room indicating that the sound field is no longer uniform. Reverberation decay curves and modulation transfer functions can be derived from the impulse responses at observation points in a room assuming white noise as a source signal. Reverberation can be represented as a random process for the collisions of waves with the surrounding walls, assuming that the sound field is a superposition of sinusoidal waves with random magnitude and phase. Assuming a stochastic model for collisions and the energy balance equation in the sound field, various types of reverberation formulae are derived. Most of those formulae feature exponentially decay of reverberation in a room. In addition to the stochastic model, this chapter describes the non-exponentially decaying reverberation that can often be heard in a room, assuming that the sound field is a mixture composed of three-, two-, and one-dimensional sound fields. Interestingly, from a theoretical perspective, the energy decay does not take an exponential form, even if the sound field is purely one or two dimensional. Very steep initial decays show a $1/t$ or $1/t^2$ time dependence rather than the exponential form in the one- or two-dimensional field.

Keywords Random superposition of sinusoidal waves · Exponential distribution · Rayleigh distribution · Transmission of envelopes · Transmission of modulation · Reverberation process · Stochastic reverberation process · Reverberation time formula · Reverberation time · Binomial distribution · Poisson distribution · Multinominal distribution · Energy balance equation · Mean free path · Image sources · Three-Dimensional field · Two-Dimensional field · One-Dimensional field

M. Tohyama, *Sound in the Time Domain*, Signals and Communication Technology,
https://doi.org/10.1007/978-981-10-5889-9_11

11.1 Spatial Dynamics for Superposition of Sinusoidal Waves with Random Phase

11.1.1 Statistical Approach to Sound Field in an Enclosed Space

The sound field rendered by a source is deterministic in principle. It may be represented as a superposition of wave solutions of the governing wave equation. However, if the sound source radiates a sinusoidal wave, the sound distribution within an enclosed space is highly sensitive to the positions of both source and receiver. A problem is that in a room the eigenfrequencies of the sound waves are too densely distributed to be separately identified. The sound field is composed of responses that highly overlap in the eigenmodes defined by the eigenfrequencies. The conditions associated with overlapping are sensitive to the positions of all sources and the point of observation. This explains why sound transmission in a room is overly dependent on positions.

In addition, from a practical point of view, eigenmode superpositions make analyses and predictions of sound fields far from likely, even if the sound field is deterministic from a theoretical point of view. The difficulty in a deterministic treatise looks in essence to be random noise. Hence, rather than a deterministic approach based on theory, a statistical approach to randomness in sound fields would be a good place to begin an analysis from a practical standpoint [1, 2].

11.1.2 Superposition of Random Sinusoidal Waves

An example of a statistical approach to the superposition of sinusoidal waves would be the representation of Gaussian random noise. Suppose a wideband source of random noise is located in an enclosed space. If the sound field is observed through a very narrowband filter, the response at the observation point might be approximated by a sinusoidal wave but with a magnitude and phase that were both random such that

$$\begin{aligned} h(t) &= x(t)\cos\omega_c t + y(t)\sin\omega_c t \\ &= \sqrt{u(t)}\cos(\omega_c t + \phi(t)) \\ &= \sqrt{x^2 + y^2}\cos(\omega_c t + \phi), \end{aligned} \tag{11.1}$$

where ω_c denotes the central frequency of the narrowband filter. The random variable U representing the squared envelope is the superposition of the squares for the mutually independent Gaussian variables X^2 and Y^2.

A random variable U that is defined by a sum of squares of two independent Gaussian variables with zero mean and variance of σ^2 follows an exponential distribution with $\mu = \sqrt{2}\sigma$ and variance of $2\sigma^2$ such that [3, 4]

$$p_{\mathrm{U}}(u) = \frac{1}{\mu}\mathrm{e}^{-\frac{u}{\mu}}. \tag{11.2}$$

The probability density function above shows that $u = 0$ gives the most-probable squared envelope as a random variable. This may look at first sight nonintuitive; however, taking the envelope instead of the sum of squares, the probability density function is intuitively understandable.

11.1.3 Distribution of Envelopes

The random variables X and Y are mutually independent and follow a Gaussian distribution such that

$$p_{\mathrm{X}}(x) = \frac{1}{\sqrt{2\pi\sigma^2}}\mathrm{e}^{-\frac{x^2}{2\sigma^2}} \tag{11.3}$$

$$p_{\mathrm{Y}}(y) = \frac{1}{\sqrt{2\pi\sigma^2}}\mathrm{e}^{-\frac{y^2}{2\sigma^2}}. \tag{11.4}$$

Introducing a transformation of the variables such that

$$\mathrm{X} = \mathrm{Z}\cos\theta, \tag{11.5}$$

$$\mathrm{Y} = \mathrm{Z}\sin\theta, \tag{11.6}$$

the probability density function of X and Y is therefore

$$\begin{aligned} p_{\mathrm{X,Y}}(x, y)dxdy &= p_{\mathrm{Z},\Theta}(z\cos\theta, z\sin\theta)zdzd\theta \\ &= \frac{1}{2\pi\sigma^2}\mathrm{e}^{-\frac{z^2}{2\sigma^2}}zdzd\theta, \end{aligned} \tag{11.7}$$

where

$$\mathrm{X}^2 + \mathrm{Y}^2 = \mathrm{Z}^2. \tag{11.8}$$

Consequently, the probability density function of Z can be written as

$$p_{\mathrm{Z}}(z) = \frac{z}{\sigma^2}\mathrm{e}^{-\frac{z^2}{2\sigma^2}}, \tag{11.9}$$

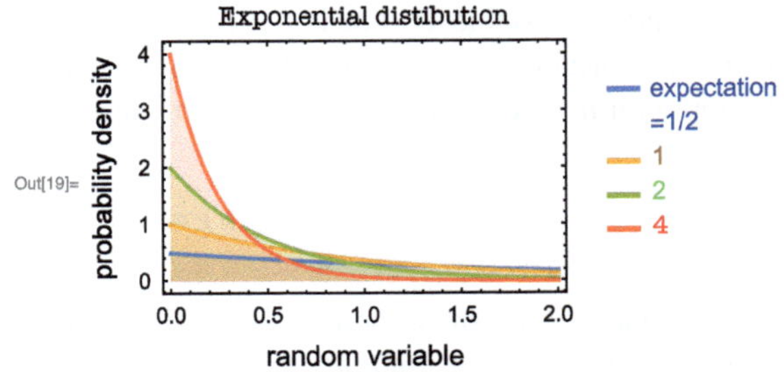

Fig. 11.1 Exponential distribution

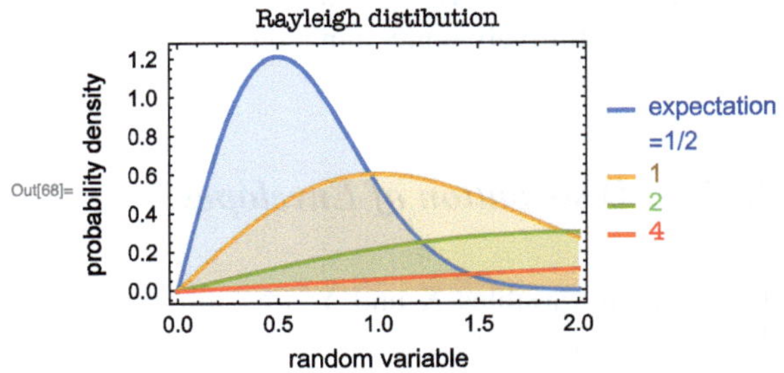

Fig. 11.2 Rayleigh distribution

where

$$p_Z dz = \frac{1}{2\pi\sigma^2}\int_0^{2\pi} \mathrm{e}^{-\frac{z^2}{2\sigma^2}} z dz d\theta \qquad (11.10)$$
$$= \frac{z}{\sigma^2}\mathrm{e}^{-\frac{z^2}{2\sigma^2}} dz.$$

The probability density function of Z is called the Rayleigh distribution. Figures 11.1 and 11.2 illustrate examples of the exponential and Rayleigh distributions. The sound field is no longer uniform in a reverberation room where a sinusoidal wave is traveling.

11.2 Transmission of Envelopes and Reverberation

11.2.1 Reverberation Response to Wideband Noise Source

Reverberation is a typical time-domain phenomenon encountered in an enclosed space where sound propagates. Suppose there is a wideband source of noise (such as white noise). Assuming the impulse response from the source to the receiving position is $h(n)$, then the response can be written as

$$y(n) = x * h(n), \qquad (11.11)$$

where $x(n)$ denotes the signal from the noise source. When the sound source stops at $n = 0$, the response $y(n)$ can be written as

$$y(n) = \sum_{m=n}^{N-1} x(n-m)h(m) \tag{11.12}$$

for $n \geq 0$, where N gives the length of the impulse response record. Taking the ensemble average of the squared response [5]

$$\begin{aligned} \mathrm{E}[\mathrm{Y}^2] &= \mathrm{E}\left[\sum_{m=n}^{N-1} \mathrm{X}(n-m)h(m)\right]\left[\sum_{l=n}^{N-1} \mathrm{X}(n-l)h(l)\right] \\ &= \sum_{m=n}^{N-1}\sum_{l=n}^{N-1} h(m)h(l)\mathrm{E}[\mathrm{X}(n-m)\mathrm{X}(n-l)] \\ &= U\sum_{m=n}^{N-1} h^2(m), \end{aligned} \tag{11.13}$$

where

$$\mathrm{E}[\mathrm{X}(n-m)\mathrm{X}(n-l)] = U\delta(m-l). \tag{11.14}$$

The expression above is called the reverberation decay curve corresponding to the setup of source and receiving positions in the room. The decay curve mostly follows an exponentially decay in a three-dimensional space [6].

Figure 11.3 illustrates an example of a reverberation decay curve derived from the impulse response, where the multiplying method for two impulse response records is applied instead of squaring the impulse response record [5–7]. Suppose that two impulse response records h_1 and h_2 are measured at the same position [7]. Multiplying the data pair in the same way as in taking the correlation coefficient gives

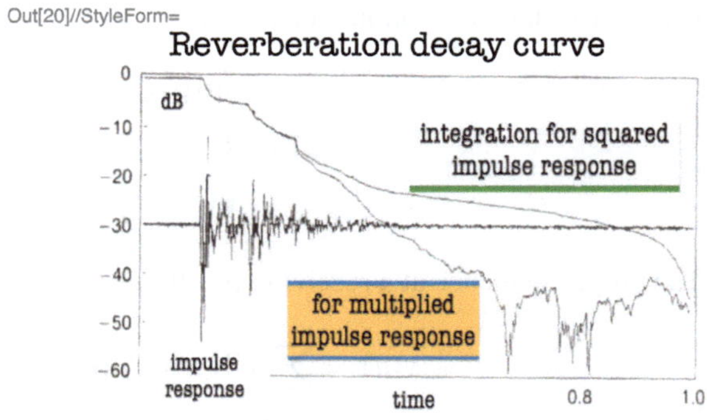

Fig. 11.3 Sample of a reverberation decay curve and the impulse response from Fig. 3.3.3. [6]

$$\int_t^\infty h_1(\tau)h_2(\tau)d\tau \quad (11.15)$$
$$= \int_t^\infty h^2(\tau) + h(\tau)n_1(\tau) + h(\tau)n_2(\tau) + n_1(\tau)n_2(\tau)d\tau$$
$$= \int_t^\infty h^2(\tau)d\tau + K(t)$$

where

$$h_1(t) = h(t) + n_1(t), \quad (11.16)$$
$$h_2(t) = h(t) + n_2(t), \quad (11.17)$$

and $n_1(t)$ and $n_2(t)$ are the background noise for the corresponding measurement. The first term in Eq. 11.15 is nonnegative and gives the reverberation decay curve. However, the second term $K(t)$ is either positive or negative and appears as a fluctuation to the first term. The figure shows a comparison of the results with the conventional squaring method. By taking the positive part of the result given by the multiplying method, the reverberation decay curves are obtained without the effects of background noise.

11.2.2 Reverberation Response to Modulated Wideband Noise Source

The distribution of the narrowband envelope was described in Subsect. 11.1.3 assuming a wideband noise source. The response of the envelope to modulated wideband noise is determined by the complex modulation transfer function [8, 9]. Suppose a modulated input sequence is given such that

$$x(n) = \cos \Omega_m n \cdot r(n) \quad (11.18)$$

where $r(n)$ denotes random noise, and

$$\mathrm{E}[\mathrm{R}(n)] = 0 \quad (11.19)$$
$$\mathrm{E}[\mathrm{R}(l)\mathrm{R}(l-n)] = N\delta(n) \quad (11.20)$$
$$\mathrm{E}[\mathrm{R}^2(n)] = N. \quad (11.21)$$

The output sequence at a receiving position in the space is written as

$$y(n) = \sum_l h(l) \cos \Omega_m (n-l) r(n-l), \quad (11.22)$$

where $h(n)$ denotes the impulse response between source and receiving positions in the space. Therefore, taking the ensemble average of the squared output sequence, then

$$
\begin{aligned}
\mathrm{E}[\mathrm{Y}^2(n)] &= N\sum_l h^2(l)\frac{1+\cos 2\Omega_m(n-l)}{2} \\
&= NI * h^2(n) \\
&= N\sum_l h^2(l)\left[\frac{1+\Re[M_c \mathrm{e}^{\mathrm{i}2\Omega_m n}]}{2}\right] \\
&= N\sum_l h^2(l)\left[\frac{1+|M_c|\cos(2\Omega_m n+\theta)}{2}\right]
\end{aligned}
\tag{11.23}
$$

is obtained where

$$
\begin{aligned}
M_c(2\Omega_m) &= \frac{\sum_l h^2(l)\mathrm{e}^{-\mathrm{i}2\Omega_m l}}{\sum_l h^2(l)} \\
&= |M_c(2\Omega_m)|\mathrm{e}^{\mathrm{i}\theta(2\Omega_m)}
\end{aligned}
\tag{11.24}
$$

$$
I = \frac{1+\cos 2\Omega_m n}{2}, \tag{11.25}
$$

Eq. 11.24 is called the complex modulation transfer function. Comparing the square of the input modulated signal gives

$$
\begin{aligned}
x^2(n) &= \cos^2\Omega_m n \cdot r^2(n) \\
&= \frac{1+\cos 2\Omega_m n}{2}\cdot r^2(n), \\
\mathrm{E}[\mathrm{X}^2] &= N\frac{1+\cos 2\Omega_m n}{2},
\end{aligned}
\tag{11.26}
$$

and therefore the magnitude of the complex modulation transfer function yields the modulation index [9], whereas the phase gives the change in the phase of the envelope.

The square of the impulse response renders the transfer function of the signal dynamics represented by the squared envelopes. The Fourier transform of the squared impulse response is expressed as the auto-convolution of the frequency or spectral characteristics of the impulse response in the frequency domain that can also be understood as the auto-correlation of the complex spectral function. Figure 11.4 illustrates the modulation indexes measured in a large reverberating room of volume 640(m^3); here r denotes the distance from the source [10]. The modulation index decreases with distance from the sound source.

Out[21]//StyleForm=

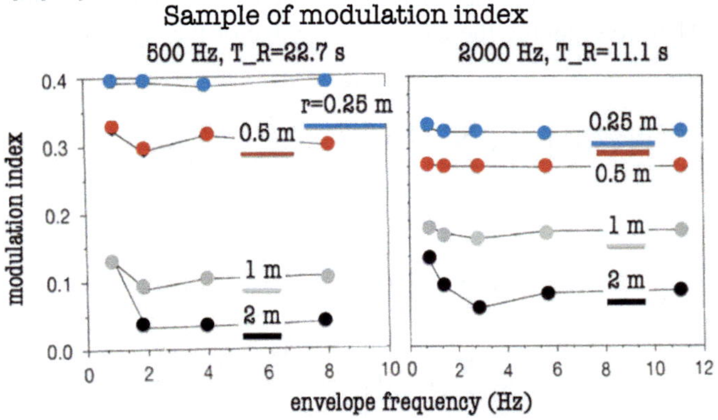

Fig. 11.4 Sample of modulation indexes in a large reverberation room from Fig. 3.7.2 [6]

11.2.3 Transmission of Modulated Sinusoidal Signal

The transmission of modulated sinusoidal signals can be formulated using the modulation index and envelope phase given the impulse response $h(t)$. Consider a modulated sinusoidal signal of the form

$$
\begin{aligned}
x(t) &= \frac{1+\cos\omega_m t}{2}\cos\omega_c t \\
&= \frac{1}{2}\left[\frac{1}{2}\cos(\omega_c - \Delta\omega)t \right. \\
&\quad \left. + \cos\omega_c t + \frac{1}{2}\cos(\omega_c + \Delta\omega)t\right],
\end{aligned} \tag{11.27}
$$

where $\omega_m = \Delta\omega$. If the signal radiates into the space, then the received signal can be expressed as

$$
\begin{aligned}
y(t) = \frac{1}{2}&\left[\frac{1}{2}|H_L|\cos(\omega_L t + \phi_c + \Delta\phi_L)\right. \\
&+ |H_c|\cos(\omega_c t + \phi_c) \\
&\left. + \frac{1}{2}|H_H|\cos(\omega_H t + \phi_c - \Delta\phi_H)\right],
\end{aligned} \tag{11.28}
$$

where H denotes the frequency characteristics for the impulse response between the source and receiving positions, and

$$
\omega_L = \omega_c - \Delta\omega \tag{11.29}
$$

$$
\omega_H = \omega_c + \Delta\omega \tag{11.30}
$$

$$H = |H|e^{i\phi} \tag{11.31}$$

$$\phi(\omega_L) = \phi(\omega_c) + \Delta\phi_L \tag{11.32}$$

$$\phi(\omega_H) = \phi(\omega_c) - \Delta\phi_H. \tag{11.33}$$

Setting $\phi_c = 0$ and $|H_c| = 1$, $\Delta\phi_L = \Delta\phi_H = \Delta\phi$, $|H_L| = |H_H| = |M||H_c|$ $(0 \leq |M| \leq 1)$

$$\begin{aligned} y(t) &= \frac{1}{2}\left[\frac{1}{2}|M|\cos(\omega_L t + \Delta\phi) \right. \\ &\quad \left. + \cos\omega_c t + \frac{1}{2}|M|\cos(\omega_H t - \Delta\phi)\right] \\ &= \frac{1 + |M|\cos(\Delta\omega t - \Delta\phi)}{2}\cos\omega_c t. \end{aligned} \tag{11.34}$$

Examining the envelope of the source signal, the modulation index depends on the local (or narrowband) property of the magnitude frequency characteristics for the impulse response, whereas the delay of the envelope is sensitive to the group delay of the impulse response. Extending the modulated sinusoidal signal as a compound envelope of signals with multiple frequencies,

$$x(t) = \sum_m \frac{1 + \cos\omega_m t}{2}\cos\omega_c t \tag{11.35}$$

then the receiving signal may be written as

$$y(t) \cong \sum_m \frac{1 + |M_m|\cos(\omega_m t - \Delta\phi_m)}{2}\cos\omega_c t, \tag{11.36}$$

where the modulation index represents the magnitude spectral characteristics in the envelope, whereas the group delay determines the delay of the individual envelope.

An example of transmission is afforded by a single resonator. Setting $|H_c| = 1$, $|H_L| = |H_H| = 1/\sqrt{2}$, and $\Delta\phi = \pi/4$; then the receiving signal from the single resonator becomes

$$y(t) = \frac{1 + \frac{1}{\sqrt{2}}\cos(\Delta\omega t - \pi/4)}{2}\cos\omega_c t, \tag{11.37}$$

and for the source signal

$$x(t) = \frac{1 + \cos\omega_m t}{2}\cos\omega_c t. \tag{11.38}$$

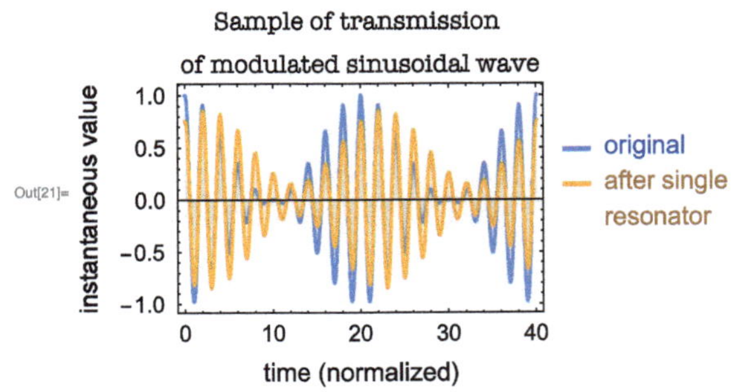

Fig. 11.5 Sample of a modulated sinusoidal signal transmitted by a single resonator where the carrier frequency is set to be that for the resonator

Figure 11.5 presents an example of a modulated sinusoidal signal observed through the resonator with a resonance frequency setting the frequency of the carrier; the half-power bandwidth is equal to $\Delta\omega$.

The signal dynamics represented by the envelope in the time domain is deformed by the local characteristics of the magnitude and phase in the frequency domain for the transmission space. The local behavior of the magnitude for the transmission system changes the modulation index of the signal dynamics, and the phase renders the time delay (or advance) of the envelope.

11.3 Stochastic Reverberation Process and Reverberation Time Formula in an Enclosed Space

The exponential function provides a mathematical means to represent sound waves traveling through some space or medium and undergoing collisions with wall surfaces surrounding the space. Consider a sound source radiating sound waves in an enclosed space. The path the sound takes from source to receiving position is, in principle, deterministically predictable, provided wave propagation in the enclosed space is governed by the linear wave equation. The time history of the reflections associated with reverberant sound is expressed by the impulse response between the sound source and the receiving position. However, the impulse response is highly sensitive to the location of both source and receiver. It may be appropriate to sample the impulse responses at different locations creating an ensemble of outcomes through stochastic processes. Probability distributions are introduced for the number of reflections [11].

11.3.1 Ensemble Average of Reverberation Energy

Suppose the boundaries of the enclosed space are composed of K different surfaces, indexed by $k = 1, \ldots, K$. For the kth surface, α_k is the coefficient of absorption of sound and is given by $1 - r_k$, where r_k denotes the coefficient of reflection for sound energy. If n_k is the number of collisions into the kth surface in the time interval t, then the energy density in the collision process may be formulated as

$$I(t) = I_0 \prod_{k=1}^{K} (1 - \alpha_k)^{n_k}, \quad (\text{J/m}^3) \tag{11.39}$$

where I_0 is the energy density in the room at the steady state, specifically, the time when the sound source stops at $t = 0$. Assume that the energy density samples are recorded at many different locations in the space; then a new ensemble is created with respect to the locations of the source and receiver. Introducing a probability distribution for n_k, the ensemble average for reverberation with respect to the ensemble of locations is expressed by

$$\begin{aligned} \text{E}[\text{I}(t)] &= \overline{I}(t) \\ &= \overline{I}_0 \sum_{n_1, n_2, \ldots, n_K} P(n_1, n_2, \ldots, n_K) \prod_{k=1}^{K} (1 - \alpha_k)^{n_k}, \quad (\text{J/m}^3) \end{aligned} \tag{11.40}$$

where $P(n_1, n_2, \ldots, n_K)$ is the joint probability distribution for n_k.

The expectation value for the number of collisions in a unit interval of time or equivalently the collision frequency can be estimated as [12–14]

$$\text{E}[\text{n}_\text{k}] = \frac{cS_k}{4V} = \mu_k, \qquad (1/\text{s}) \tag{11.41}$$

where c, V, and S_k denote the speed of sound (m/s), volume of the space (m^3), and the area of the kth surface (m^2). Hence, the expected number of collisions into the kth surface in a time interval t is given by

$$N_k(t) = \text{E}[\text{n}_\text{k}] \cdot t = \frac{cS_k}{4V} \cdot t = \mu_k \cdot t, \tag{11.42}$$

which increases linearly with the traveling time of sound. The collision process can be interpreted as a series of random events within the time interval t. Here, $4V/S$(m) is called the mean free path of the sound field [12–14].

11.3.2 Binomial Distribution for the Collision Process

Consider an enclosed space with walls of surface areas $S_1(\text{m}^2)$ and $S_2(\text{m}^2)$. Assume there are $n_1(t)$ collisions with wall 1 in a time interval t, and simultaneously $N(t) - n_1(t) = n_2(t)$ collisions with wall 2. If the collision process is assumed to obey a binomial distribution, then the energy density can be written as [11]

$$I(t) = I_0(1-\alpha_1)^{n_1(t)}(1-\alpha_2)^{n_2(t)}, \qquad (\mathrm{J/m^3}) \tag{11.43}$$

where α_1 and α_2 are the coefficients of absorption of sound from walls 1 and 2, respectively, and

$$N(t) = \mu \cdot t = cS/4V \cdot t \tag{11.44}$$

$$\begin{aligned}\mu &= \mu_1 + \mu_2 \\ &= \frac{c}{4V}(S_1 + S_2) \\ &= \frac{cS}{4V}. \qquad (1/\mathrm{s})\end{aligned} \tag{11.45}$$

Taking the ensemble average according to this binomial distribution yields

$$\mathrm{E[I(}t\mathrm{)]} = \overline{I}_0 \sum_{n_1=0}^{\mu\cdot t} P(\mu\cdot t, n_1)(1-\alpha_1)^{n_1(t)}(1-\alpha_2)^{\mu\cdot t - n_1(t)}. \qquad (\mathrm{J/m^3}) \tag{11.46}$$

Substituting

$$P(\mu t, n_1) = \mathrm{C}(\mu t, n_1)\left(\frac{S_1}{S}\right)^{n_1}\left(\frac{S_2}{S}\right)^{\mu t - n_1} \tag{11.47}$$

then

$$\begin{aligned}\mathrm{E[I(}t\mathrm{)]} &= \overline{I}_0\left(\frac{S_1}{S}(1-\alpha_1) + \frac{S_2}{S}(1-\alpha_2)\right)^{\mu t} \\ &= \overline{I}_0(1-\overline{\alpha})^{\mu t} \\ &= \overline{I}_0 \mathrm{e}^{\mu t \log_e(1-\overline{\alpha})} \qquad (\mathrm{J/m^3})\end{aligned} \tag{11.48}$$

is derived where

$$\overline{\alpha} = \frac{\alpha_1 S_1 + \alpha_2 S_2}{S} \tag{11.49}$$

and C denotes combination. The reverberation time T_R(s) is customary defined by the condition

$$\overline{I}(t)|_{t=T_R} = \overline{I}_0 \cdot 10^{-6}, \qquad (\mathrm{J/m^3}) \tag{11.50}$$

where $I_0(\mathrm{J/m^3})$ denotes the energy density at the steady state when the sound source stops. The reverberation time is found to be

$$T_R = \frac{4V/cS}{-\log_e(1-\overline{\alpha})}\log_e 10^6 \tag{11.51}$$

$$\cong 0.162\frac{V}{-\log_e(1-\overline{\alpha})S}, \qquad \text{(s)}$$

which approaches 0 in the limit as $\overline{\alpha} \to 1$, where $c = 340\,(\text{m/s})$.

11.3.3 Multi-nominal Distribution in the Collision Process

The results above can be extended to instances when a multinomial distribution is assumed for the distribution of collisions [11]. Suppose, for example, three different surfaces for the enclosing boundary. Similar to Eq. 11.46, the ensemble averaged energy density $I(t)$ can be expressed as

$$\mathrm{E}[\mathrm{I}(t)] = \overline{I}_0 \sum_{n_1,n_2}^{N} P(N,n_1,n_2,n_3)(1-\alpha_1)^{n_1}(1-\alpha_2)^{n_2}(1-\alpha_3)^{n_3}, \qquad (\text{J/m}^3) \tag{11.52}$$

where $N(t) = n_1(t) + n_2(t) + n_3(t)$, S_k $(k = 1, 2, 3)$ are the areas (m^2) of the boundary surfaces of total $S = S_1 + S_2 + S_3$. Recalling that the multinomial distribution

$$P(N,n_1,n_2,n_3) = \frac{N!}{n_1!n_2!n_3!}\left(\frac{S_1}{S}\right)^{n_1}\left(\frac{S_2}{S}\right)^{n_2}\left(\frac{S_3}{S}\right)^{n_3} \tag{11.53}$$

is generated by expanding the polynomial

$$\left(\frac{S_1}{S}+\frac{S_2}{S}+\frac{S_3}{S}\right)^{n(t)} \tag{11.54}$$

then

$$\mathrm{E}[\mathrm{I}(t)] = \overline{I}_0\left(\frac{S_1}{S}(1-\alpha_1)+\frac{S_2}{S}(1-\alpha_2)+\frac{S_3}{S}(1-\alpha_3)\right)^{N(t)} \tag{11.55}$$

is obtained. Consequently,

$$\mathrm{E}[\mathrm{I}(t)] = \overline{I}_0(1-\overline{\alpha})^{\mu t} \qquad (\text{J/m}^3) \tag{11.56}$$

$$T_R = \frac{4V/cS}{-\log_e(1-\overline{\alpha})}\log_e 10^6 \tag{11.57}$$

$$\cong 0.162\frac{V}{-\log_e(1-\overline{\alpha})S} \qquad \text{(s)}$$

is derived where

$$\overline{\alpha} = \frac{\alpha_1 S_1 + \alpha_2 S_2 + \alpha_3 S_3}{S}. \tag{11.58}$$

11.3.4 Poisson Distribution as a Limiting Case

A Poisson distribution is derived as the limit of a binomial distribution when $N(t) = \mu t$ approaches infinity [4]. Introducing the binomial distribution,

$$P(N, n_1) = \frac{N(N-1)\cdots(N-n_1+1)}{n_1!}\left(\frac{\lambda}{N}\right)^{n_1}\left(1-\frac{\lambda}{N}\right)^{N-n_1}, \tag{11.59}$$

and taking the limit when N approaches infinity,

$$P(N, n_1) = \frac{\lambda^{n_1}}{n_1!}\cdot\frac{N(N-1)\cdots(N-n_1+1)}{N^{n_1}}\left(1-\frac{\lambda}{N}\right)^{N}\left(1-\frac{\lambda}{N}\right)^{-n_1}$$
$$\longrightarrow \frac{\lambda^{n_1}}{n_1!}\mathrm{e}^{-\lambda} \tag{11.60}$$

is derived; here the infinite product expansion of the exponential function has been exploited. The result above reduces to a Poisson distribution with expectation λ. Figure 11.6 illustrates examples of Poisson distributions for different values of λ.

By setting

$$\lambda_k = \mu_k t \tag{11.61}$$

and introducing the independent Poisson distributions into Eq. 11.40

$$\overline{I}(t) = \overline{I}_0\mathrm{e}^{-\frac{c\overline{\alpha}S}{4V}t} \quad (\mathrm{J/m^3}) \tag{11.62}$$

is derived [11, 15], where $4V/S(\mathrm{m})$ is called the mean free path of sound waves in rooms [12–14]. Thus the reverberation formula

$$T_R \cong 0.162\frac{V}{A} \quad (\mathrm{s}) \tag{11.63}$$

is obtained where

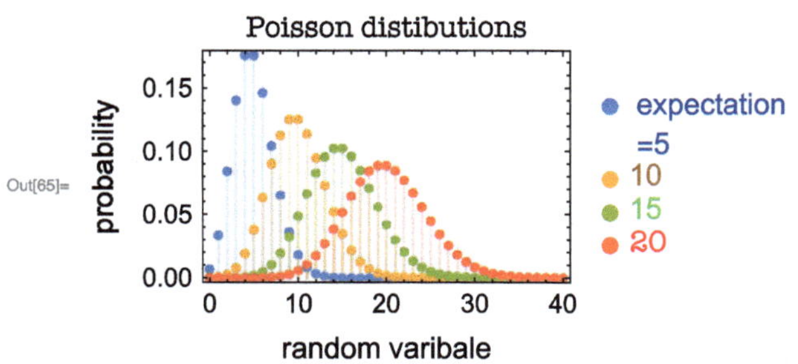

Fig. 11.6 Sample of Poisson distributions

$$\sum_k \alpha_k S_k = \overline{\alpha} S. \tag{11.64}$$

The reverberation formula above is applicable in instances when the expectation of the collision frequency is large; in other words, the formula does not work as well when the coefficient of absorption approaches unity. Under high absorption conditions, the reverberation time formula Eq. 11.51, in which binomial (or a multinomial) probabilistic process is assumed, provides a reasonable result.

11.4 Energy Balance Equation and Poisson Process

11.4.1 Energy Balance Equation

The reverberation formula stated above can be interpreted given the energy balance equation in an enclosed space. Suppose a sound source radiates a stationary wave in the enclosed space. Increasing the sound power density $\Delta I_0(\mathrm{W/m^3})$ in the space in a unit time [14] gives

$$V \Delta I_0 = W_x - W_b, \qquad (\mathrm{W}) \tag{11.65}$$

where V denotes the volume of the space ($\mathrm{m^3}$), $V\Delta I_0(\mathrm{W})$ represents the power increase in the space per unit time, $\Delta I_0(\mathrm{W/m^3})$ gives the averaged increase in the power density in the space, $W_x(\mathrm{W})$ denotes the power output of the sound source, and $W_b(\mathrm{W})$ denotes the absorbed power per unit time interval. The equation above describes the energy balance for the energy accumulation in the room between the supplied energy from the source and the absorbed energy by the wall. In particular, the energy increase can be estimated from the difference between the energy coming from the source and the energy lost by absorption.

The power absorbed over a unit time interval is given by

$$W_b = \overline{\alpha} I_{in} V = \overline{\alpha} \mu E V \qquad (\mathrm{W}) \tag{11.66}$$

$$I_{in} = \mu E, \qquad (\mathrm{W/m^3}) \tag{11.67}$$

where $I_{in}(\mathrm{W/m^3})$ indicates of the sound power density coming into the boundaries, $\overline{\alpha}$ denotes the averaged coefficient of absorption of the walls, $E(\mathrm{J/m^3})$ gives the energy density in the room, and $\mu(1/\mathrm{s})$ denotes the average frequency of collisions with the walls. The energy density at the steady state becomes

$$E_0 = \frac{W_x}{V\overline{\alpha}\mu} = \frac{4W_x}{c\overline{\alpha}S}, \qquad (\mathrm{J/m^3}) \tag{11.68}$$

where $\mu = c/MFP(1/\mathrm{s})$, $MFP = 4V/S(\mathrm{m})$, and $S(\mathrm{m^2})$ is the total area of the walls. According to the definition of the coefficient of absorption for the boundaries or for the room, μ determines the ratio of the energy in the room $E_0 V$ and the incident energy going into the boundaries $\mu E_0 V$ through collisions with the walls. That is, μ gives the number of collisions with the walls in unit time interval.

11.4.2 Differential Equation for Reverberation Formula

According to the energy balance equation, the energy decay of sound during reverberation after the sound source stops can be formulated by a differential equation. Setting $W_x = 0$ as the power output from the sound source during reverberation, the reverberant energy density $E(t)$ then obeys

$$\frac{dE(t)}{dt} = \frac{-W_b(t)}{V} = -\overline{\alpha}\mu E(t) = -\frac{c\overline{\alpha}S}{4V}E(t). \quad (\mathrm{W/m^3}) \tag{11.69}$$

Consequently

$$E(t) = \frac{4W_x}{c\overline{\alpha}S}\mathrm{e}^{-\frac{c\overline{\alpha}S}{4V}t} \quad (\mathrm{J/m^3}) \tag{11.70}$$

subject to the initial condition

$$E(0) = E_0 = \frac{4W_x}{c\overline{\alpha}S}. \quad (\mathrm{J/m^3}) \tag{11.71}$$

This reverberation decay formula is identical to that assuming a Poisson process for randomly occurring collisions [11]. In contrast, Eq. 11.70 above assumes no probabilistic distribution but only the average of the collision frequency and the energy balance equation.

The coincidence of the two reverberation formulas is understandable by interpreting the energy balance equation as describing the randomness in the probabilistic estimate for the time interval between successively occurring random events that obey a Poisson process. Consider a Poisson process for a random variable N over a time interval t,

$$P(N = n, t) = P(n, t) = \frac{(\mu t)^n}{n!}\mathrm{e}^{-\mu t}, \tag{11.72}$$

where $\mu t = \mathrm{E}[N, t]$, which gives a probabilistic estimate during the time interval of successively occurring random events [4]. The probability $P(0, \Delta t)$ that there is no event in the small time interval Δt can be expressed as

$$P(0, \Delta t) = 1 - P(1, \Delta t) \cong 1 - \mu\Delta t, \tag{11.73}$$

where $P(1, \Delta t)$ denotes the probability for a single event occurring in the small time interval, discarding the probabilities when multiple events occur. The probability $P(n, t)$ that n events occur in the time interval t is then

$$P(n, t + \Delta t) = P(n, t)P(0, \Delta t) + P(n-1, t)P(1, \Delta t) \\ = P(n, t)(1 - \mu\Delta t) + P(n-1, t)\mu\Delta t, \tag{11.74}$$

from which

$$\frac{P(n, t + \Delta t) - P(n, t)}{\Delta t} = \mu\left(P(n-1, t) - P(n, t)\right) \tag{11.75}$$

is derived. Taking the limit when $\Delta t \to 0$, the equation above can be rewritten as

$$\frac{\partial P}{\partial t}(n, t) = \mu\left(P(n-1, t) - P(n, t)\right) \tag{11.76}$$

and

$$\frac{\partial P}{\partial t}(0, t) = -\mu P(0, t) \tag{11.77}$$

by setting $P(-1, t) = 0$. The solution for probability $P(0, t)$ in the above differential equation is

$$P(0, t) = \mathrm{e}^{-\mu t} \tag{11.78}$$

where $P(0, 0) = 1$.

This solution determines the probability when no event occurs in the time interval t. It estimates the probable time interval (in which there is no event) between successively occurring random events obeying a Poisson process. The differential equation that gives the exponential distribution is identical to Eq. 11.69 representing the energy balance equation from which the energy decay is derived. In particular, provided that the collision process is assumed Poissonian with expectation μt, the function for the decay in energy can be interpreted as a probability that no collision occurs and decreases exponentially. Here, the exponential function representing the decay in energy can also be interpreted as the probability for random collisions in sound.

11.5 Reverberation and the Geometric Dimension of the Space

11.5.1 Mirror Image in Three-Dimensional Space

A rectangular room is a typical three-dimensional setting for sound fields. Figure 11.7 shows an array of mirror-image sources for a rectangular room. The number of reflected sound waves returning to the sound source position after the source stops can be estimated by the number of mirror-image sources that are contained in a sphere centered at the source position of radius ct, i.e., [14]

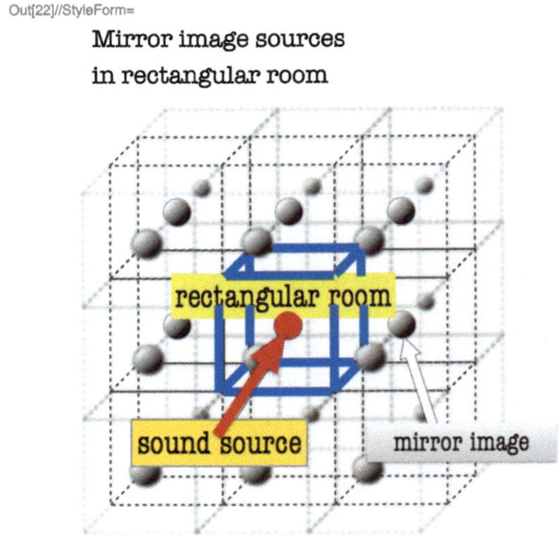

Fig. 11.7 Mirror images of a sound source in a rectangular room from Fig. 11.1 [14]

$$N_{R_3} \cong \frac{4\pi(ct)^3}{3V}, \tag{11.79}$$

where t denotes the time after the sound from the source has stopped. The density of the reflection sound coming back to the source position within a unit time interval can be estimated from

$$n_{R_3} = \frac{d}{dt}N_{R_3} = \frac{4\pi c^3 t^2}{V}, \qquad (1/\mathrm{s}) \tag{11.80}$$

which decreases inversely proportional to the room volume $V(\mathrm{m}^3)$, but increases proportional to the square of the time. The number of collisions with walls from reflecting sound returning to the source position can be estimated using the mean free path (MFP) [12–14] of the propagating sound,

$$N_{c_3} = \frac{ct}{MFP_3}, \tag{11.81}$$

where

$$MFP_3 = \frac{4V}{S_3} \qquad (\mathrm{m}) \tag{11.82}$$

and $S_3(\mathrm{m}^2)$ denotes the area of the walls.

If no collisions occur at the walls, the intensity of sound at the source position returning from a single mirror-image source is obtained from

$$J_s = \frac{W_0}{4\pi(ct)^2}, \qquad (\mathrm{W/m^2}) \tag{11.83}$$

where W_0(W) denotes the sound power output of the source. Each collision removes certain intensity $\overline{\alpha_3} J_s$ in reflecting sound. The sound intensity in the room at the source position (which is assumed to be the center of the room) may be expressed as

$$\begin{aligned} J_3(t) &\cong \frac{W_0}{4\pi(ct)^2}(1-\overline{\alpha}_3)^{N_{c3}(t)} n_{R_3} \Delta t \\ &= \frac{W_0 \Delta t c}{V}(1-\overline{\alpha}_3)^{N_{c3}(t)}. \qquad (\mathrm{W/m^2}) \end{aligned} \tag{11.84}$$

The reverberation decay curve is given by

$$\begin{aligned} \hat{J}_3(t) &= \int_t^\infty \frac{W_0 \Delta \tau c}{V}(1-\overline{\alpha}_3)^{N_{c3}(\tau)} \\ &= \frac{W_0 c}{V}\int_t^\infty (1-\overline{\alpha}_3)^{N_{c3}(\tau)} d\tau \\ &= \frac{4W_0}{A_3} e^{-\frac{cA_3}{4V}t} \\ &= \frac{4W_0}{A_3} e^{\frac{c \log_e(1-\overline{\alpha}_3)}{MFP_3}t}, \end{aligned} \tag{11.85}$$

where

$$A_3 = -\log_e(1-\overline{\alpha}_3) S_3 \qquad (\mathrm{m^2}) \tag{11.86}$$

and S_3 denotes the area of the walls in the three-dimensional space. According to the definition of the reverberation time

$$\begin{aligned} T_{R_3} &= \frac{\log_e 10^6}{c} \cdot \frac{4V/S_3}{-\log_e(1-\overline{\alpha}_3)} \\ &\cong 0.162 \frac{V}{A_3}. \qquad (\mathrm{s}) \end{aligned} \tag{11.87}$$

In a three-dimensional array of mirror-image sources, the decrease in energy due to the distance between source and mirror-image is canceled by the increase in density of the mirror-image sources as distance increases. This explains why a smooth energy decay curve is expected. A two-dimensional array of mirror-image sources is the exception [16] as described in the next subsection.

11.5.2 Mirror-Image Sources in a Two-Dimensional Space

Figure 11.8 illustrates a two-dimensional array of the mirror-image sources [14, 16, 17]. A two-dimensional field may be constructed from so-called tangential waves that travel within the $x - y$ plane, for example, in the sound field of a rectangular room in which the floor (or ceiling) is covered by sound-absorbing materials as shown in Fig. 11.9 [10, 14, 17–19].

The reverberation decay curve is obtained using the multiplying method [6, 7, 14] so that its non-exponential behavior can be confirmed. The initial part of the decay curve mainly follows the three-dimensional reverberation formula developed in the previous subsection; however, it follows another decay law corresponding to that for the two-dimensional reverberation theory being described.

The density of the mirror-image sources (reflection sounds coming back to the source position) in the two-dimensional array can be written as

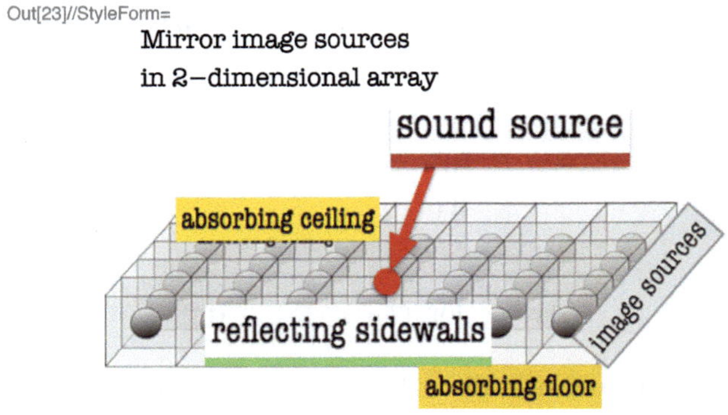

Fig. 11.8 Mirror-image sources in a two-dimensional space from Fig. 11.2 [14]

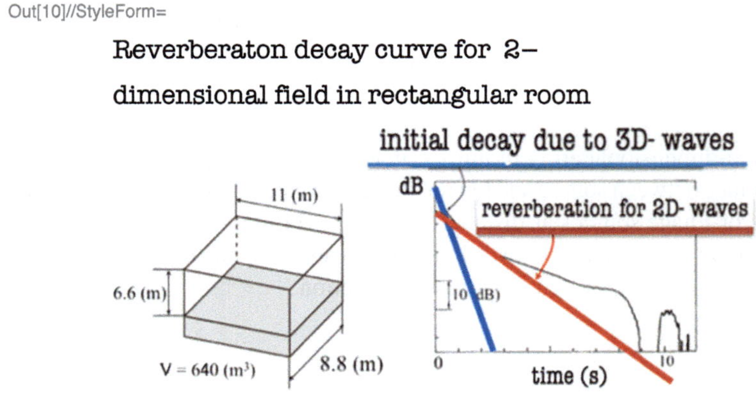

Fig. 11.9 Reverberation decay curve in a two-dimensional field constructed in a rectangular room with carpeted floor to absorb sound from Fig. 11.8 [14]

$$n_{R_2} = \frac{\pi c^2 2t}{S_2}, \qquad (1/\mathrm{s}) \tag{11.88}$$

where $S_2(\mathrm{m}^2)$ denotes the area of the two-dimensional field. The sound intensity in the two-dimensional field becomes

$$\begin{aligned} J_2(t) &\cong \frac{W_0}{4\pi(ct)^2}(1-\overline{\alpha}_2)^{N_{c2}(t)} n_{R_2} \Delta t \\ &= \frac{W_0 \Delta t}{2S_2 t}(1-\overline{\alpha}_2)^{N_{c2}(t)}. \qquad (\mathrm{W/m^2}) \end{aligned} \tag{11.89}$$

The collision number $N_{c_2}(t)$ is given by [12–14, 17]

$$N_{c_2} = \frac{ct}{MFP_2} = \frac{L_2 ct}{\pi S_2}, \tag{11.90}$$

where

$$MFP_2 = \frac{\pi S_2}{L_2} \qquad (\mathrm{m}) \tag{11.91}$$

$$S_2 = L_x L_y \qquad (\mathrm{m}^2) \tag{11.92}$$

$$L_2 = 2(L_x + L_y). \qquad (\mathrm{m}) \tag{11.93}$$

The intensity and decay curves of reverberation are not well defined at $t = 0$ because of the term $1/t$ [16]. This singularity arises because the decrease in sound intensity with distance from the mirror-image source is not compensated by the increase in density of the mirror-image sources. It differs from the three-dimensional case in which the reverberation decay curve is well defined.

The formal representation of the reverberation decay curve is written as

$$\begin{aligned} \hat{J}_2(t) &= \int_t^{\infty} \frac{W_0}{2S_2}\frac{1}{\tau}(1-\overline{\alpha}_2)^{N_{c2}(\tau)} d\tau \\ &= \frac{W_0}{2S_2}\int_t^{\infty} \frac{1}{\tau} \mathrm{e}^{-\delta_2 \tau} d\tau, \qquad (\mathrm{W/m^2}) \end{aligned} \tag{11.94}$$

where

$$\delta_2 = -\frac{L_2 c}{\pi S_2}\log_{\mathrm{e}}(1-\overline{\alpha_2}). \qquad (1/\mathrm{s}) \tag{11.95}$$

Excluding $t = 0$, a steep decay is seen in the initial stage in Fig. 11.10.

The decay rate after the sound source stops, however, may be approximated by $J_2(t)$ given by Eq. 11.89 instead of the reverberation decay curve. Taking the logarithm of $J_2(t)$, then

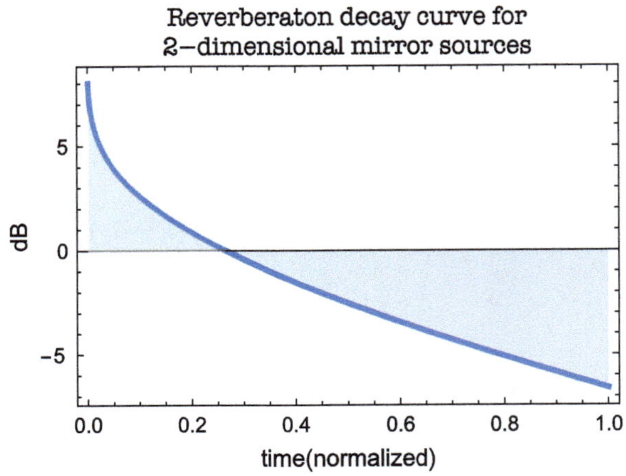

Fig. 11.10 Reverberation decay curve in a purely two-dimensional field constructed using a two-dimensional array of mirror-image sources

$$\log_e J_2(t) = \log_e \frac{W_0 \Delta t}{2S_2} + N_{c_2}(t)\log_e(1-\overline{\alpha}_2) - \log_e t. \tag{11.96}$$

The energy decay in the two-dimensional field is the result of acoustic conditions at the boundaries and the purely dimensional field-independent nature of these conditions. Assuming the $1/t$ decay rate arising from the dimensional condition can be compensated, the energy decay resulting from the acoustic conditions can be estimated from

$$\log_e J_{d_2} \cong N_{c_2}(t)\log(1-\overline{\alpha}_2), \tag{11.97}$$

which over a logarithmic scale becomes

$$J_{d_2} \cong e^{-\delta_2 t}, \tag{11.98}$$

where

$$\begin{aligned} \delta_2 t &= -N_{c_2}(t)\log_e(1-\overline{\alpha_2}) \\ &= -\frac{L_2 ct}{\pi S_2}\log_e(1-\overline{\alpha_2}). \end{aligned} \tag{11.99}$$

The reverberation time is defined as

$$\begin{aligned} T_{R_2} &= \frac{\log_e 10^6}{c} \cdot \frac{\pi S_2/L_2}{-\log_e(1-\overline{\alpha}_2)} \\ &\cong 0.128\frac{S_2}{A_2}. \quad \text{(s)} \end{aligned} \tag{11.100}$$

The energy decay characteristics for two-dimensional traveling waves in a sound field rendered in the three-dimensional rectangular room are shown in Fig. 11.9 and obey the decay exhibited by tangential (two-dimensional) waves. The decay curve in

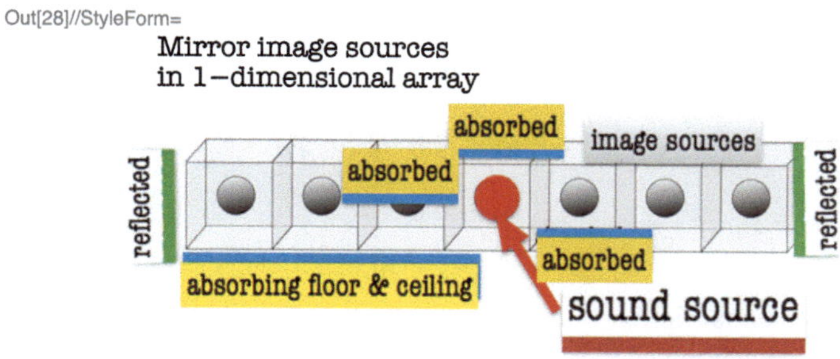

Fig. 11.11 One-dimensional array of mirror-image sources [20]

the two-dimensional field also obeys another form that is exhibited by an axial wave field—traveling sound occurring between two parallel walls. The decay formula will be introduced in the next subsection.

11.5.3 Mirror Image in One-Dimensional Space

Figure 11.11 shows an array of mirror-image sources in which a one-dimensional reverberation field is constructed [14, 17]. A one-dimensional reverberation field is possible in a space partitioned by parallel walls perpendicular to the floor. Similar to the previous subsections, the decay rate after the sound source stops can be approximated by taking the logarithm of $J_{d_1}(t)$; hence,

$$\log_e J_{d_1}(t) = \log_e \frac{W_0 dt}{4\pi L_1 c} + \frac{ct}{L_1}\log_e(1-\overline{\alpha}_1) - \log_e t^2. \tag{11.101}$$

Subject to the surrounding acoustic conditions and excluding the $1/t^2$ singularity due to the one-dimensional field, the reverberation time can be estimated from

$$\begin{aligned} T_{R_1} &= \frac{\log_e 10^6}{c} \cdot \frac{L_1}{-\log_e(1-\overline{\alpha}_1)} \\ &\cong 0.041 \frac{L_1}{-\log_e(1-\overline{\alpha}_1)}, \qquad \text{(s)} \end{aligned} \tag{11.102}$$

where the mean free path is $MFP_1 = L_1$, and L_1 denotes a typical length of the field such as the dimensional length. Like the two-dimensional field described previously, a very steep decay can be seen in the initial decay stage because of the $1/t^2$ dependence.

The reverberation formulas and decay constants described above can be rewritten for three-dimensional reverberation field as

$$T_{R_3} \cong 0.162 \frac{V}{-\log_e(1-\overline{\alpha}_3)S_3} \quad \text{(s)} \tag{11.103}$$

$$\delta_3 = \frac{c}{MFP_3}(-\log_e(1-\overline{\alpha}_3)) = \frac{cA_3}{4V} \quad \text{(1/s)} \tag{11.104}$$

$$MFP_3 = \frac{4V}{S_3}. \quad \text{(m)} \tag{11.105}$$

For two-dimensional field,

$$T_{R_{xy}} \cong 0.128 \frac{V}{-\log_e(1-\overline{\alpha}_{xy})S_{xy}} \quad \text{(s)} \tag{11.106}$$

$$S_{xy} = 2(L_x + L_y)L_z \quad (\text{m}^2) \tag{11.107}$$

$$\delta_{xy} = \frac{c}{MFP_{xy}}(-\log_e(1-\overline{\alpha}_{xy})) = \frac{A_{xy}c}{\pi V} \quad \text{(1/s)} \tag{11.108}$$

$$MFP_{xy} = \frac{\pi V}{S_{xy}}. \quad \text{(m)} \tag{11.109}$$

Similarly, for one-dimensional field,

$$T_{R_x} \cong 0.082 \frac{V}{-\log_e(1-\overline{\alpha}_x)S_x} \quad \text{(s)} \tag{11.110}$$

$$S_x = 2L_yL_z \quad (\text{m}^2) \tag{11.111}$$

$$\delta_x = \frac{c}{MFP_x}(-\log_e(1-\overline{\alpha}_x)) = \frac{cA_x}{2V} \quad \text{(1/s)} \tag{11.112}$$

$$MFP_x = L_x. \quad \text{(m)} \tag{11.113}$$

The summation above is over the superposition of three-, two-, and one-dimensional wave motions. The mixing ratio of these three modal types is frequency dependent. The mixture renders a frequency-dependent non-exponential decay curve in general [14, 17–19, 21, 22].

11.6 Exercises

1. Derive the expectation and variance for (1) the exponential distribution and (2) the Poisson distribution.
2. Derive the complex modulation transfer function given in Eq. 11.24.
3. Consider a modulated input sequence

$$x(n) = \mathrm{e}^{-an} r(n),$$

where $r(n)$ is white noise and the impulse response $h(n)$ at a receiving position is

$$h(n) = \mathrm{e}^{-bn}.$$

Obtain the ensemble average of the squared receiving signal.

4. Obtain the series expansion for an exponential function $\mathrm{e}^{-\lambda t}$.
5. Derive the ensemble average of the energy density of the reverberation when the collision process of sound is Poissonian.
6. Verify that the exponential function $\mathrm{e}^{\lambda t}$ satisfies the differential equation

$$\frac{d}{dt}x(t) = \lambda x(t).$$

7. Derive Eqs. 11.80, 11.88, 11.89, 11.94, and 11.101.
8. Suppose the impulse response for a sound recording condition is

$$h_O(t) = \mathrm{e}^{-at},$$

and similarly that for the sound reproducing condition

$$h_R(t) = \mathrm{e}^{-bt}.$$

Obtain the impulse response for the sound recording and reproducing system [23].

9. Verify that the transient time response to an input signal

$$x(t) = \mathrm{e}^{-at}$$

for a system whose impulse response is given by

$$h(t) = \mathrm{e}^{-bt}.$$

Consider good reproducing conditions for the input signal.

References

1. M.R. Schroeder, Statistical parameters of the frequency response curves in large rooms. J. Audio Eng. Soc. **35**(5), 299–306 (1987)
2. K.J. Ebeling, *Physical acoustics, Properties of Random Wave Fields*, vol. XVII (Academic Press, London, 1984), pp. 233–310
3. R. Nelson, *Probability, Stochastic Processes, and Queueing Theory* (Springer, Berlin, 1995)
4. R. Durrett, *Essentials of Stochastic Processes* (Springer, Berlin, 1999)
5. M.R. Schroeder, New method of measuring reverberation time. J. Acoust. Soc. Am. **37**(3), 409–412 (1965)

6. M. Tohyama, H. Suzuki, Y. Ando, *The Nature and Technology of Acoustic Space* (Academic Press Limited, London, 1995)
7. Y. Hirata, A method of eliminating noise in power responses. J. Sound Vib. **84**, 593–595 (1982)
8. T. Houtgast, H.J.M. Steeneken, R. Plomp, A review of the MTF concept in room acoustics and its use for estimating speech intelligibility in auditoria. J. Acoust. Soc. Am. **77**(3), 1069–1077 (1985)
9. M.R. Schroeder, Modulation transfer functions: definition and measurement. Acustica **49**(3), 179–182 (1981)
10. H. Nomura, H. Miyata, T. Houtgast, Speech intelligibility and modulation transfer function in non-exponential decay fields. Acustica **69**(4), 151–155 (1989)
11. M.R. Schroeder, D. Hackman, Iterative calculation of reverberation time. Acta Acustica **45**(4), 269–273 (1980)
12. C.W. Kosten, The mean free path in room acoustics Acta. Acustica **10**, 245–260 (1960)
13. L. Batchelder, Reciprocal of the mean free path. J. Acoust. Soc. Am. **36**(3), 551–555 (1964)
14. M. Tohyama, *Sound and Signals* (Springer, Berlin, 2011)
15. A. Jeffrey, *Handbook of Mathematical Formulas and Integrals*, 3rd edn. (Elsevier Academic Press, London, 2004)
16. T. Houtgast, *Private communication*
17. Y. Hirata, Geometrical acoustics for rectangular rooms. Acustica **43**(4), 245–252 (1979)
18. M. Tohyama, S. Yoshikawa, Approximate formula of the averaged sound energy decay curve in a rectangular reverberant room. J. Acoust. Soc. Am. **70**(6), 1674–1678 (1981)
19. P.M. Morse, R.H. Bolt, Sound waves in rooms. Rev. Mod. Phys. **16**, 69–150 (1944)
20. Y. Hara, *Private communication*
21. Y. Hirata, Dependence of the curvature of sound decay curves and absorption distribution on room shapes. J. Sound Vib. **84**, 509–517 (1982)
22. M. Tohyama, A. Suzuki, Reverberation time in an almost-two-dimensional diffuse field. J. Sound Vib. **111**(3), 391–398 (1986)
23. Y. Hirata, Reverberation time of listening room and the Definition of reproduced sound. Acustica **41**(3), 222–224 (1978)

Chapter 12
Signal Dynamics and Sound Source Distance

Abstract The sound field in a room is composed of direct and reflection or reverberation sounds. Direct sounds mostly represent the source characteristics, whereas reverberation sounds convey the sound-path information in a room. The energy ratio of the two components is a function of the distance between the source and receiving positions, and it is an important parameter in establishing auditory events in room acoustics. This chapter develops the characteristics in a sound field close to the sound source in terms of the relationship between direct sound and reverberation. In accordance with the energy ratio, the reverberation time is apparently dependent on the distance in the region close to the source. The magnitude and phase of the transfer function depend on the distance, although the propagation phase may be dominant in the region close to the source. The dependency of the transfer function on distance produces variations in the envelope and spectral characteristics arising from the distance from the source. Curiously, the positive effects of the variation in the envelope on speech intelligibility can be seen at some distance from the source, because the time envelope (or signal dynamics) depends on the fine structure of the spectral characteristics of the transfer function. Similarly, the magnitude spectral variation enhances loudness in the direct sounds of music at the same distance as that for intelligibility. The envelope analysis for speech signals and the auto-correlation functions for musical sounds partly explain the positive effects on intelligibility and loudness, respectively.

Keywords Ratio of direct and reverberation sounds · Critical distance · Coherent length · Standard deviation of magnitude spectral characteristics · Propagation phase · Center of impulse response · Random noise · Modulated noise · Phase correlation · Speech envelope · Intelligibility · Loudness · Coherent region · Auto-correlation

M. Tohyama, *Sound in the Time Domain*, Signals and Communication Technology,
https://doi.org/10.1007/978-981-10-5889-9_12

12.1 Direct Sound and Distance from Source

12.1.1 Energy Ratio of Direct Sound and Reverberation

The ratio of energy in direct sound to that in reverberation is a function of distance from the source. Specifically,

$$K = E_d/E_{rev} = \frac{W_0/4\pi r^2}{4W_0/A} = A/16\pi r^2, \tag{12.1}$$

where

$$A = -\log_e(1 - \overline{\alpha}) \cdot S \quad (\mathrm{m}^2), \tag{12.2}$$

S (m^2) denotes the area of the boundary, W_0 (W) gives the sound power output of the source, and r (m) is the distance from the source. The distance r_c for which $K = 1$,

$$r_c = \sqrt{\frac{A}{16\pi}}, \quad (\mathrm{m}) \tag{12.3}$$

is called the critical distance.

The notion of direct sound, however, can be extended to sound that includes early sound reflections within 30–50 (ms) after the initial direct sound. Therefore the perceptual energy ratio is defined as [1, 2]

$$D_{30} = \frac{\int_0^{30\ \mathrm{ms}} h^2(t)dt}{\int_0^{\infty} h^2(t)dt} \tag{12.4}$$

where $h(t)$ denotes the impulse response between the source and receiving positions. The ratio D_{30} above is called deutlichkeit or *definition*. As well as the critical distance, another parameter, the coherent length [3, 4], is also defined,

$$L_c \cong \sqrt{\frac{A}{64}}. \quad (\mathrm{m}) \tag{12.5}$$

Within the coherent length, a sound wave is expected approximately to travel from the source as a spherical wave.

Figure 12.1 shows the impulse response records in the reverberation space under the conditions seen in the photos of Fig. 12.2. The impulse responses were windowed with an exponential function to match the reverberation time of about 1 second. Each impulse response record is normalized by its maximum.

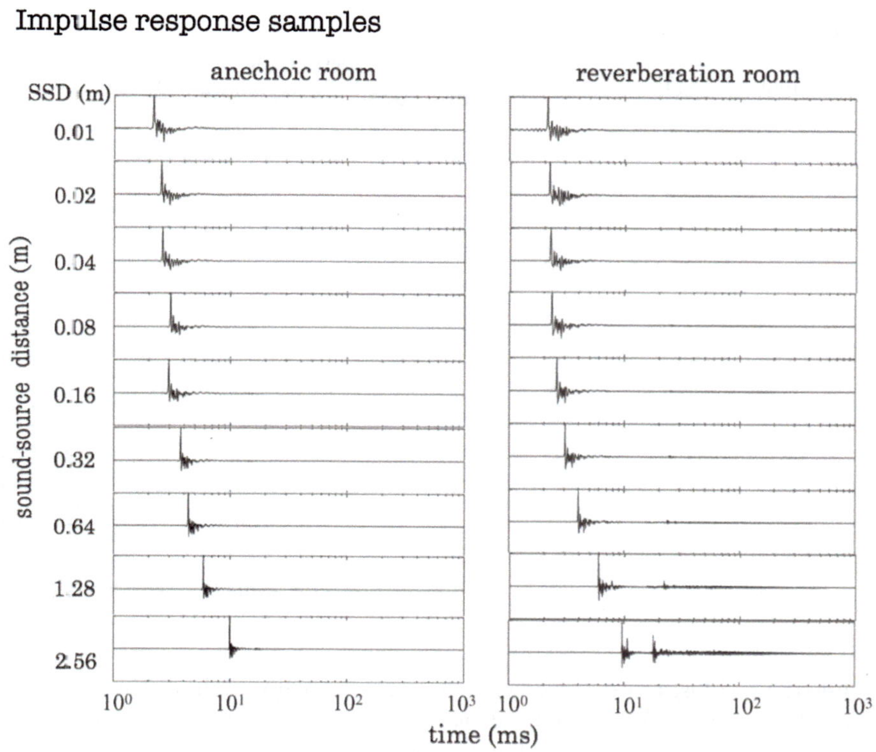

Fig. 12.1 Samples of the impulse response records in a reverberation room and an anechoic room from Figs. 2.3 and 2.5 [5]

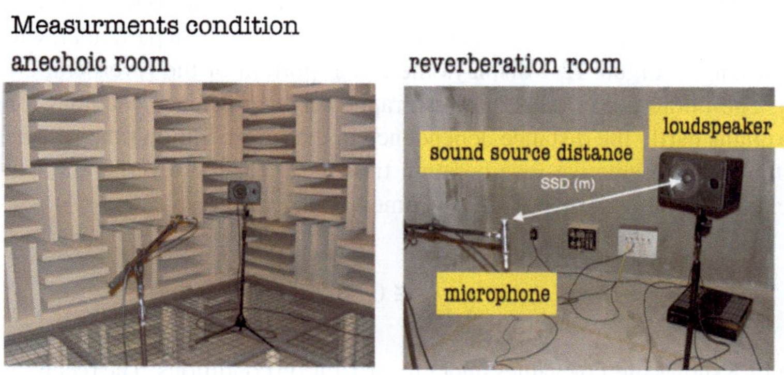

Fig. 12.2 Measurements recorded in a reverberation room and an anechoic room from Fig. 2.1 [5]

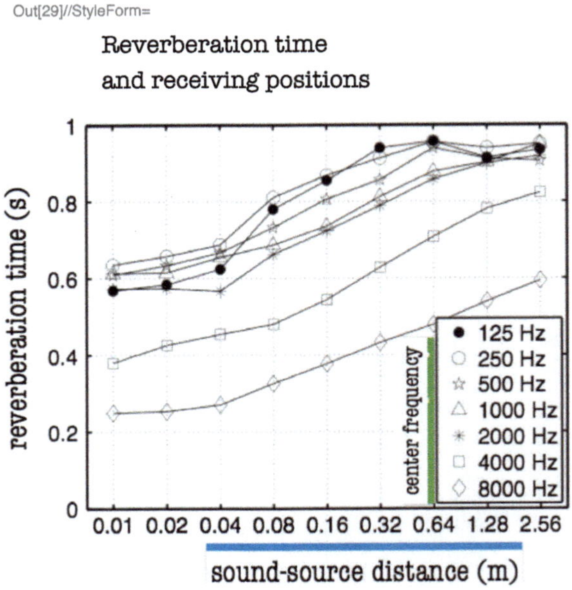

Fig. 12.3 Reverberation time at each receiving position from Fig. 2.8 [5]

Figure 12.3 illustrates the reverberation time as a function of the distance from the source following a 1/3−octave-band analysis. Although from a theoretical point of view, the reverberation time is defined conventionally based on averaged distances between the source and receiving points, it may however apparently depend on the source and receiving locations as seen in Fig. 12.3. The reverberation time becomes shorter as the receiving position approaches the source of the sound, because direct sound contributes most to the sound field close to the source, in particular, within the coherent length. Figure 12.4 displays the initial portion of the reverberation decay curves. The initial decay rates diminish rapidly as the distance becomes longer. It also explains partly the distance dependence in the reverberation time records.

Setting a representative reverberation time 1 (s) for the reverberant space, the coherent length defined by Eq. 12.5 becomes [3, 4]

$$L_c \cong \sqrt{\frac{A}{64}} \cong 0.68. \qquad \text{(m)} \tag{12.6}$$

Figure 12.5 shows the D_{30} values under reverberation conditions. The receiving positions at which the impulse responses were measured are widely distributed from inside to outside the coherent region. In addition, sound receiving positions were also located inside and outside the direct sound dominant area as shown in the plot of D_{30} in Fig. 12.5. As the distance increases, D_{30} decreases; in particular, it rapidly decreases outside the coherent region. The sound quality changes significantly (positively enhanced or negatively degraded) when the receiving position is inside or outside the coherent region [5, 7, 8].

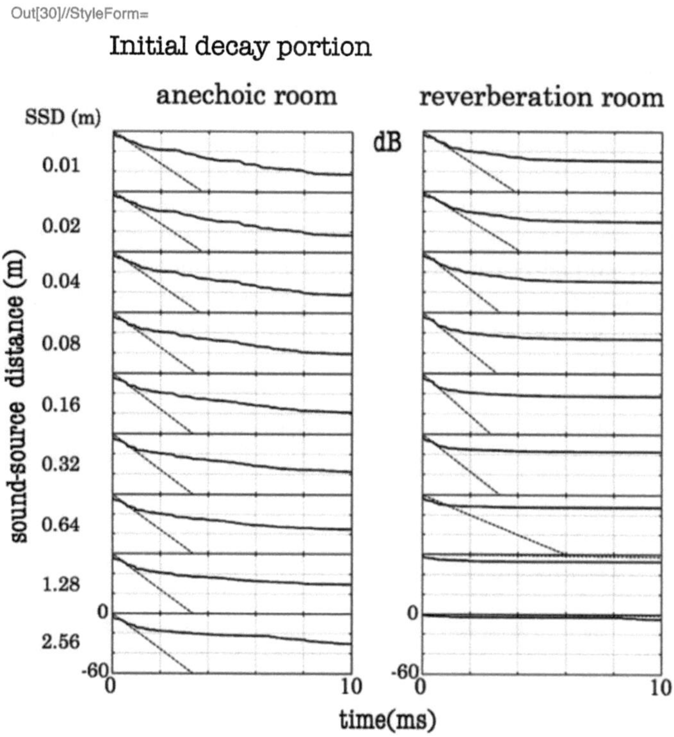

Fig. 12.4 Initial decay portion at receiving positions from Fig. 4.18 [6]

Fig. 12.5 D_{30} in a reverberation room from Fig. 2.9 [5]

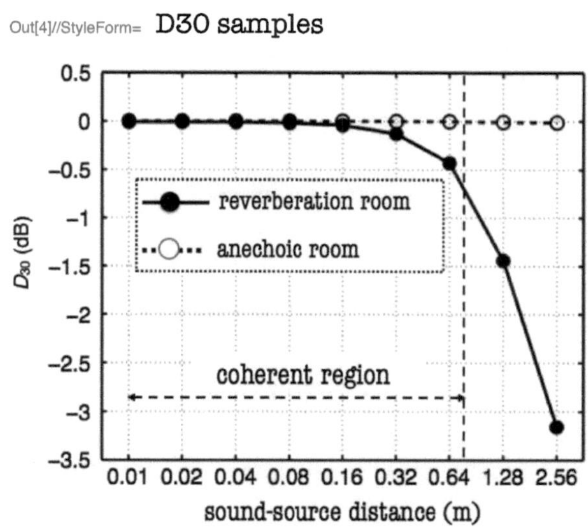

12.1.2 Standard Deviation of Magnitude Frequency Response

Along with the parameter D_{30}, the spectral characteristics of the magnitude of the frequency response may be sensitive to the distance from the source. Figure 12.6 compares the magnitudes of the frequency responses, defined as the Fourier transforms of the impulse responses, for the reverberation and anechoic rooms. These magnitudes are normalized by their maximum as determined on the dB scale. There is no significant difference between the reverberation and anechoic rooms in the coherent region at distances less than 0.32 (m). In contrast, many peaks and dips can be seen in the responses for the reverberation room as the distance from the source increases.

According to the reverberation theory [3, 9], the standard deviation in the dB scale with respect to the frequency is expected to be around 5.5 (dB) for the reverberation

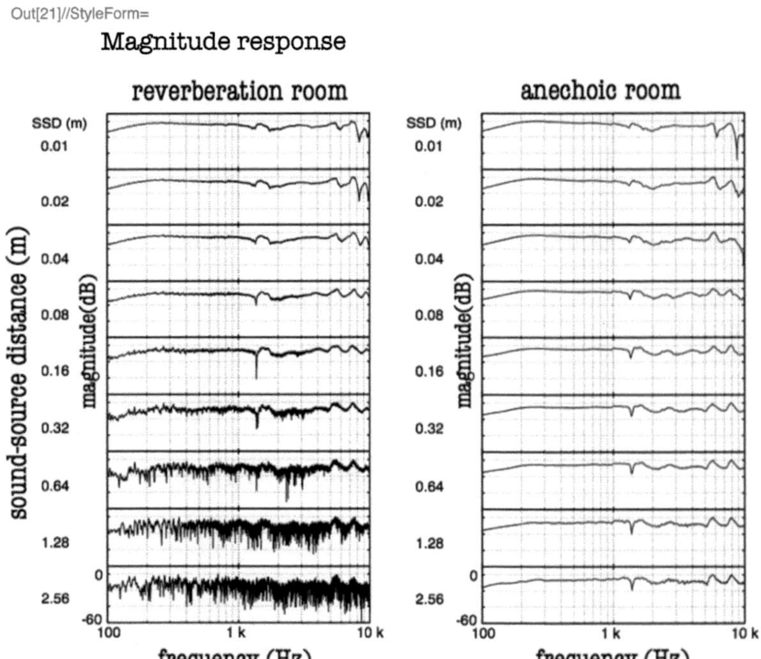

Fig. 12.6 Spectral characteristics of the magnitude of sound at various receiving positions in both anechoic and reverberation rooms from Figs. 2.4 and 2.6 [5]

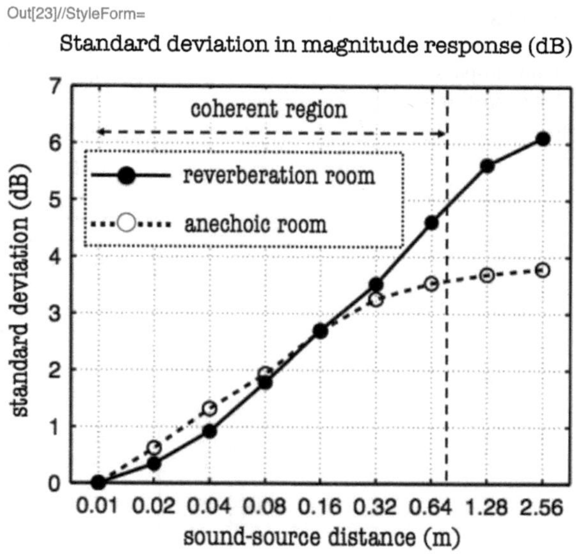

Fig. 12.7 Standard deviation for magnitude response in dB from Fig. 2.11 [5]

field. Setting the magnitude response at $r = 0.01$ as a reference, Fig. 12.7 displays the standard deviation in the dB scale. The standard deviation reaches around 5–6 (dB) in the reverberation room when distance exceeds the coherent length, whereas that for the anechoic room it is within almost 3–4 (dB). The increase in the standard deviation for the anechoic room exhibits the characteristics of the sound source (the loudspeaker displayed in Fig. 12.2), whereas that for the reverberation room results from reverberation in the transmission path from the source to the receiving positions. The spectral characteristics for the transmission path is sensitive to distance in the coherent region, even in the anechoic room, although like D_{30} the energy ratio indicates no such spectral difference. No prominent differences from the anechoic room are seen within the coherent region for the reverberation room. At distances beyond 0.16 (m), marked differences in the standard deviation between the reverberation and anechoic rooms are seen in Fig. 12.7.

12.1.3 Propagation Phase and Distance

Like the magnitude spectral response, the phase spectral response is a function of distance in the reverberation room. As described in Sect. 10.4, the phase response follows the propagation phase given by kr in a one-dimensional system such as an acoustic tube [10, 11]. In the coherent region, a propagation phase may be expected even in a three-dimensional reverberant space [3, 10–12]. Figure 12.8 presents examples of a linear regression analysis for the narrowband minimum-phase characteristics normalized by the accumulated phase at $r = 0.01$ (m).

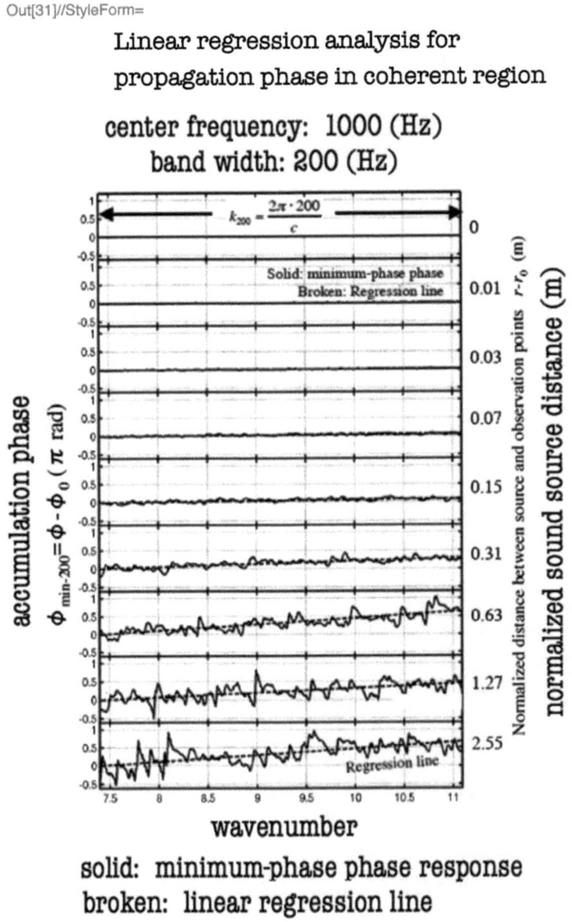

Fig. 12.8 Linear regression analysis for narrowband minimum-phase characteristics from Fig. 5 [12]

Figure 12.9 illustrates the slopes of the regression lines derived from the regression analysis in Fig. 12.8. The phase response mostly follows the propagation phase as long as the distance is within the coherent region $r < 0.68$ (m). The details of the analysis may be found in Refs. [3, 12].

The propagation phase indicates that the field in the coherent region is mostly dominated by direct sound. Intuitively that understanding is conveyed by the centers of the waveforms of the impulse responses. The center is defined by

$$T_c = \frac{\sum_{n=0}^{N-1} nh^2(n)}{\sum_{n=0}^{N-1} h^2(n)}, \tag{12.7}$$

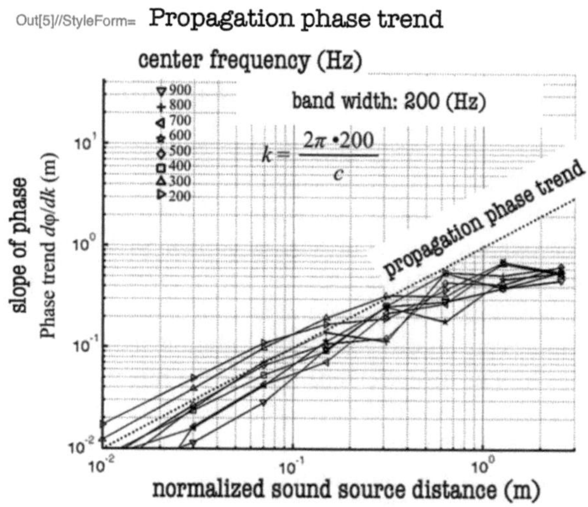

Fig. 12.9 Slopes of linear regression lines (phase trend) for narrowband minimum-phase characteristics from Fig. 6 [12]

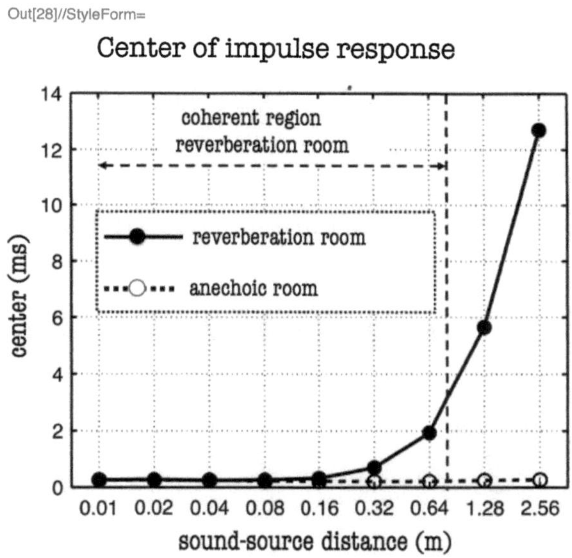

Fig. 12.10 Center of the impulse response records in reverberation space from Fig. 2.10 [5]

where $h(n)$ denotes the impulse response after removing the initial delay portion, and N the length of the impulse response. Figure 12.10 displays the center of the impulse responses in the reverberation room. After removing the initial delay portion, the center is almost 0 in the coherent region. This indicates that the phase response mostly follows the propagation phase in the coherent region.

12.2 Envelope Spectrum and Sound Source Distance

12.2.1 Random Noise and Modulated Noise

The feature of the magnitude frequency response introduced in the previous subsection illustrates the characteristics of the transmission path in the frequency domain. The signal dynamics in the time domain such as envelopes are quite important signal features, in particular, speech intelligibility [13, 14]. An intriguing question would be how the envelopes are related to speech quality in the coherent region. Speech samples are almost perfectly intelligible in the coherent region; however, the quality may still be distance dependent in the region close to the sound source.

Figure 12.11 illustrates examples of waveforms of random and modulated noise along with their spectral characteristics. The difference in the pair of waveforms is quite obvious because their difference is clearly understandable in the time domain. In contrast, the difference in the magnitude and phase spectral characteristics in the frequency domain is not explicitly observable. One way to represent the difference in the frequency plane is to take phase correlations defined formally as [3, 6, 14]

$$R_{ph} = \mathrm{E}[\mathrm{e}^{\mathrm{i}(\phi(\Omega') - \phi(\Omega' - \Omega))}] \tag{12.8}$$

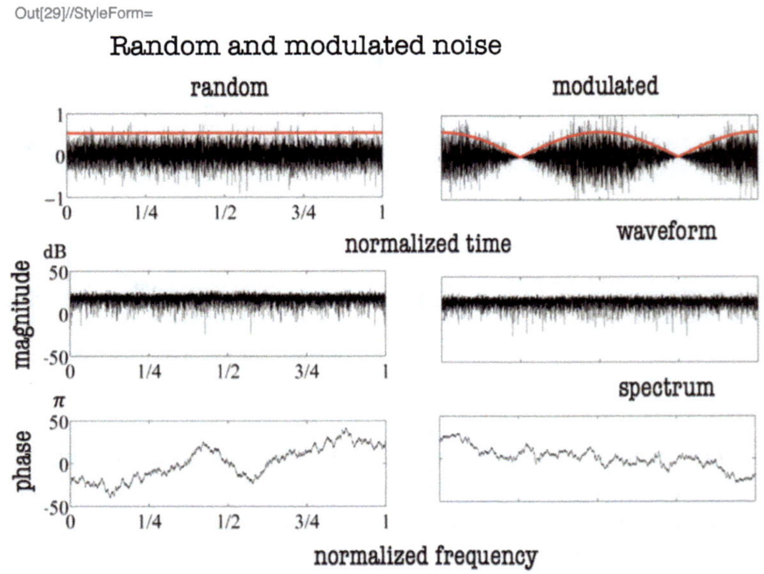

Fig. 12.11 Waveforms and their spectral characteristics for random and modulated noise from Fig. 5 [14]

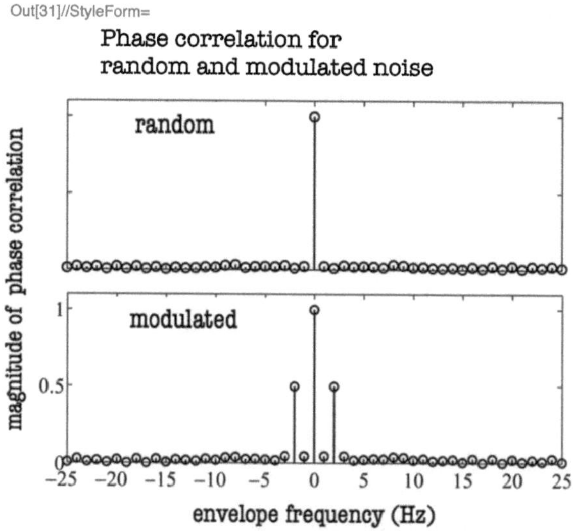

Fig. 12.12 Absolute of the phase correlation for random and modulated noise from Fig. 6 [14]

where $\phi(\Omega)$ denotes the phase spectral function of Ω. The phase correlation estimates the closely located frequency components in the envelopes representing the signal dynamics in the time domain that arise from the fine structure in the spectral characteristics. Figure 12.12 shows the magnitude of the phase correlation for random noise and modulated noise where the envelope frequency may be read. The frequencies of the envelope indicate that fine structure is required in the spectral domain so that the envelopes might be expressed.

12.2.2 Speech Envelopes and Distance from Source

Even in the coherent region, where spectral characteristics for the transmission path is significant and some spectral deviation can be seen, the envelopes of a speech sample may be modified because of the fine structure in the spectral features for the transmission path. Figure 12.13 gives an example of the spectral analyses of two narrowband speech envelopes in which two representative frequency bands (in 1/4 octave band) are observed. The main frequency components of the envelopes are mostly distributed below 20 (Hz)[13].

It would be intriguing to see how the envelopes are modified as the distance from the sound source increases in the coherent region. Figures 12.14 and 12.15 are presented to compare the speech envelopes with respect to the envelope frequencies when source distance is changed in the anechoic and reverberation rooms, respectively. Looking at the details in these figures, the envelope around the center frequency of 700 (Hz) is prominent for source distances up to 0.16 (m) for both the anechoic and reverberation rooms. The difference in the spectral energy over the

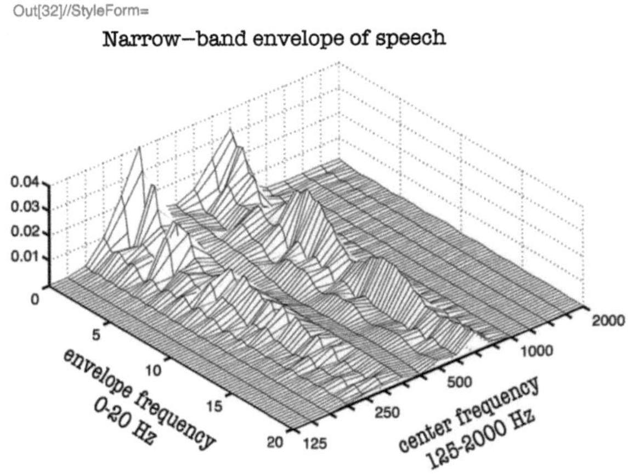

Fig. 12.13 Samples of narrowband speech envelopes with their frequencies from Fig. 3 [15]

envelope frequencies (up to 20 (Hz)) from the reference at $r = 0.01$ (m) are shown in Fig. 12.16 [5, 15].

For the two representative frequency bands, no prominent differences can be seen between the reverberation and anechoic rooms as long as the distance is within 0.16 (m). In addition, for 707 (Hz) frequency band, the envelope spectral energy increases within the area for both the anechoic and reverberation rooms. The increase or enhancement in the envelope energy may arise from source characteristics (or the loudspeaker) rather than reflected sounds. However, from reflections and reverberation in the room, the envelope energy changes considerably as distance increases and approaches the coherent length. If enhancement is also confirmed in audio signal tests, then the results may be informative for microphone settings for sound recordings of direct sound from musical instruments.

12.2.3 *Intelligibility Test*

Enhancements in the envelope energy may be confirmed by audio signal tests [5, 15]. Intelligibility, however, is almost perfect in the coherent region. In the Ref. [15], the results of paired-comparison tests were reported in which speech samples were found to be more intelligible. Figure 12.17 gives the experimental conditions for the paired-comparison tests of speech intelligibility [5, 7, 8, 15]. A masking noise made up of a convolution of white noise and an impulse response at $r = 2.56$ (m) in a reverberation room is added to speech samples. Figure 12.18 displays the results of paired-comparison tests for intelligibility heard in both a reverberation room and an anechoic room.

Out[33]//StyleForm=

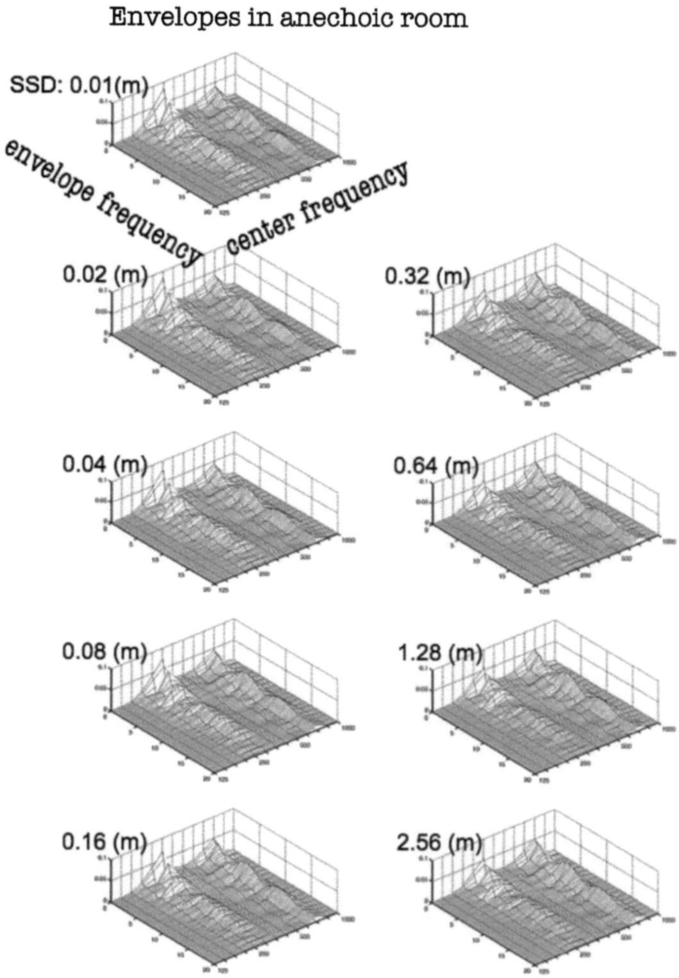

Fig. 12.14 Variation of the speech envelopes with source distance in an anechoic room from Fig. 5.4 [5]

Comparing these intelligibility test results and Fig. 12.5, the intelligibility results mostly follow D_{30}. Speech intelligibility rapidly decreases outside the coherent region because of the effects of reflections and reverberation diminishing the quality of the signal in the time domain. Similar results in an anechoic room would be intriguing. The increase in the envelope spectral energy at 707 (Hz) seen in Fig. 12.16 partly explains the enhancement in intelligibility in the anechoic room as distances become longer. This may be an example of envelope enhancement in the time domain through the fine structure of the spectral characteristics.

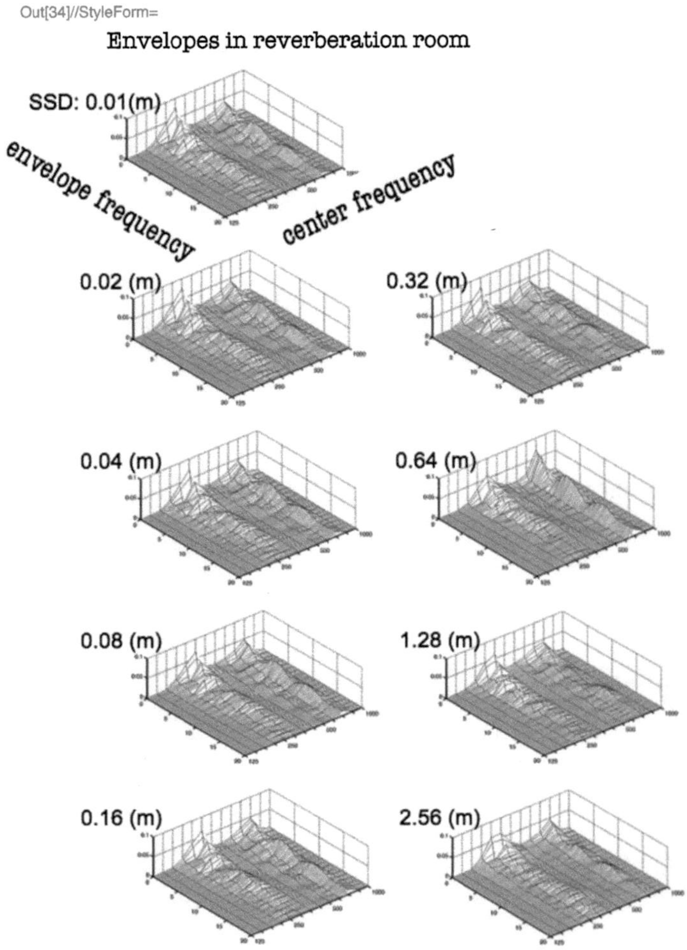

Fig. 12.15 As for Fig. 12.14 but recorded in a reverberation room from Fig. 5.3 [5]

The results in the reverberation room also follow the increase in the anechoic room, as long as the distance is within 0.16 (m) in the coherent region. However, intelligibility rapidly decreases as distance moves beyond this region. The envelope energy at 707 (Hz) in Fig. 12.16 may explain the results in the reverberation room (Fig. 12.18).

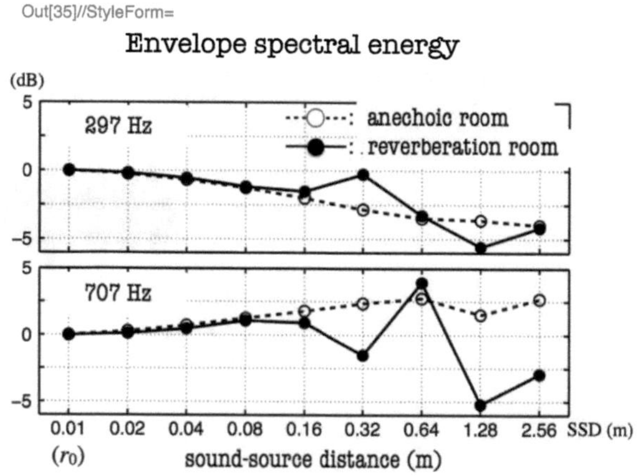

Fig. 12.16 Differences in the envelope spectral energy for representative frequency bands as a function of source distance from Fig. 4 [15]

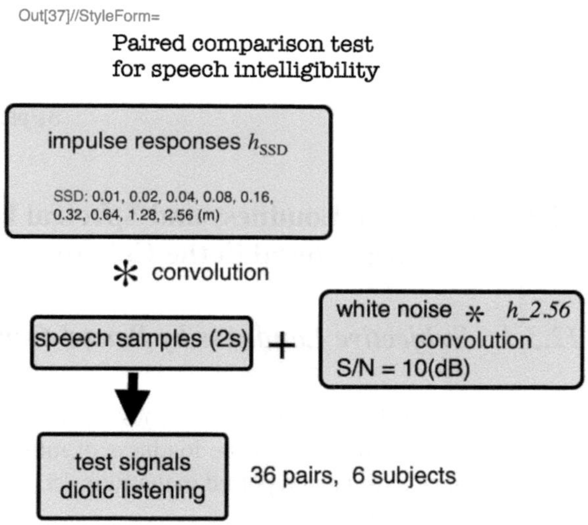

Fig. 12.17 Experimental conditions for paired-comparison tests of speech intelligibility

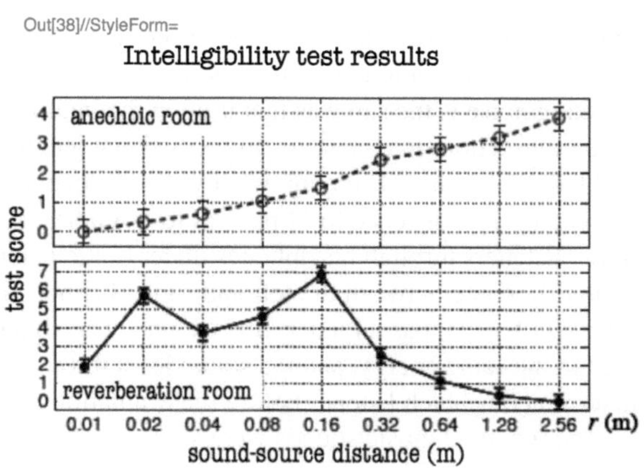

Fig. 12.18 Results of intelligibility for paired-comparison tests heard in a reverberation and in an anechoic room from Fig. 1 [15]

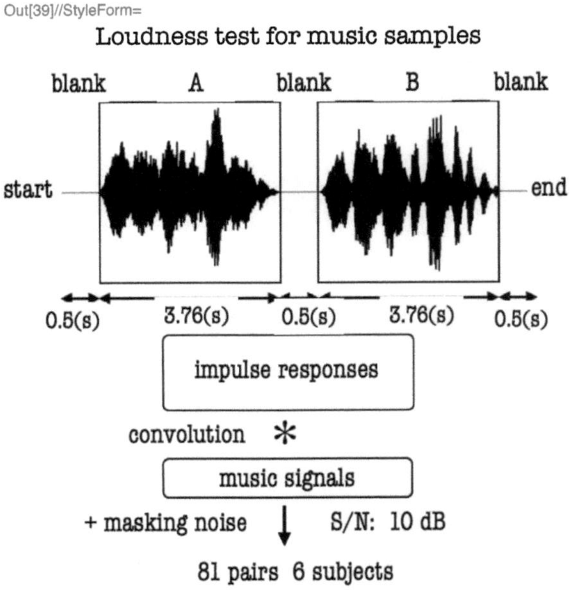

Fig. 12.19 Musical samples analyzed in paired-comparison tests from Figs. 3.4 and 3.5 [5]

12.3 Subjective Loudness and Spectral Deformation of Musical Sound in the Coherent Region

12.3.1 Subjective Loudness by Paired-Comparison Tests

Along with speech intelligibility, loudness is a fundamental sound quality. Some results of audio signal tests of the loudness of musical sounds as a function of the source distance have been reported in the Refs. [5, 8]. The test signals generated are shown in Fig. 12.19.

A short musical passage (G.F. Haendel, "Water Music Suite" [16]) that was originally recorded in an anechoic room, is convolved with the impulse responses under the same reverberation condition as that described in the previous subsection. The masking noise is the same as that used in the intelligibility test described in the previous subsection. Each test signal has a signal length of 3.76 (s) including a fade-in and fade-out using a Hanning window. Six listeners participated in the listening test and each person listened to 81 pairs of test samples. All audio signals were presented to the listeners through headphones under diotic listening conditions in which there is no difference between the left and right channels. Each listener was asked which interval (A or B) has a louder direct sound. The signal in each interval was normalized to the same energy, and the sound level through the headphones was arbitrarily determined by each listener before the evaluation test.

The recorded music sounds were subjectively evaluated with respect to loudness of the direct sound based on paired-comparison tests. Figure 12.20 shows the results

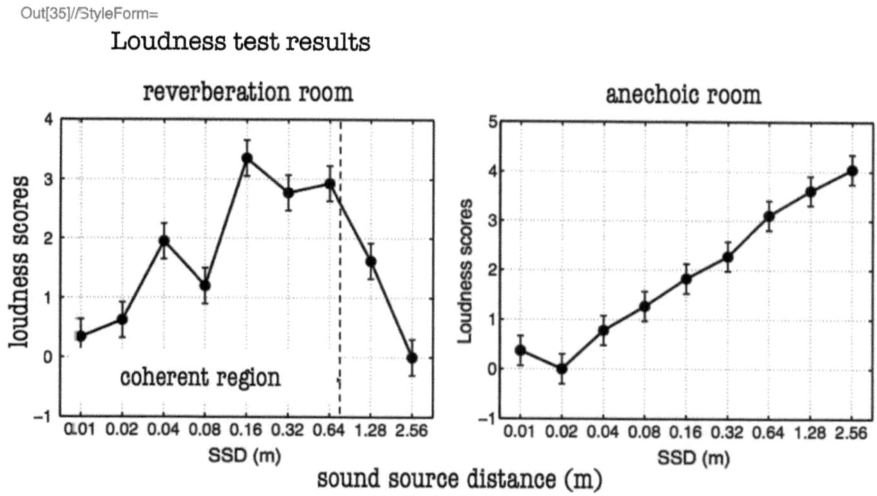

Fig. 12.20 Subjective evaluation scores of loudness from Fig. 3.6 [5]

from the paired-comparison tests for loudness as a function of the source distance with error bars [5, 8, 17].

The decrease in the loudness mostly follows the sound definition D_{30} shown in Fig. 12.5. Loudness rapidly decreases outside the coherent region because of the effects on direct sound from reflection or reverberation. In contrast, in the anechoic room, loudness rises almost uniformly as distance increases. The increase in the anechoic room may be from the effect of the source (loudspeaker used in the experiments) on loudness.

As in the anechoic room, loudness may even be enhanced in the reverberation room as long as the distance is within 0.16 (m). If the sound is decomposed into direct and other reflected sounds, then the loudness of the direct sound would decrease. The correspondence between the results for the anechoic and reverberation rooms implies that direct and other reflected sounds may have fused into a single sound from a perceptual point of view, and hence the subjective loudness increases for both rooms within the distance 0.16 (m). Such a correspondence or increase in loudness stops at 0.16 (m), and thereafter the loudness decreases because a decomposition of the sound into direct and other commences.

The fusion of direct and other reflected sounds into a single sound may require spectral similarity with the direct sound even after the other sounds are mixed in. An auto-correlation analysis of test samples would be informative in interpreting the results of enhancement or deterioration in loudness [18].

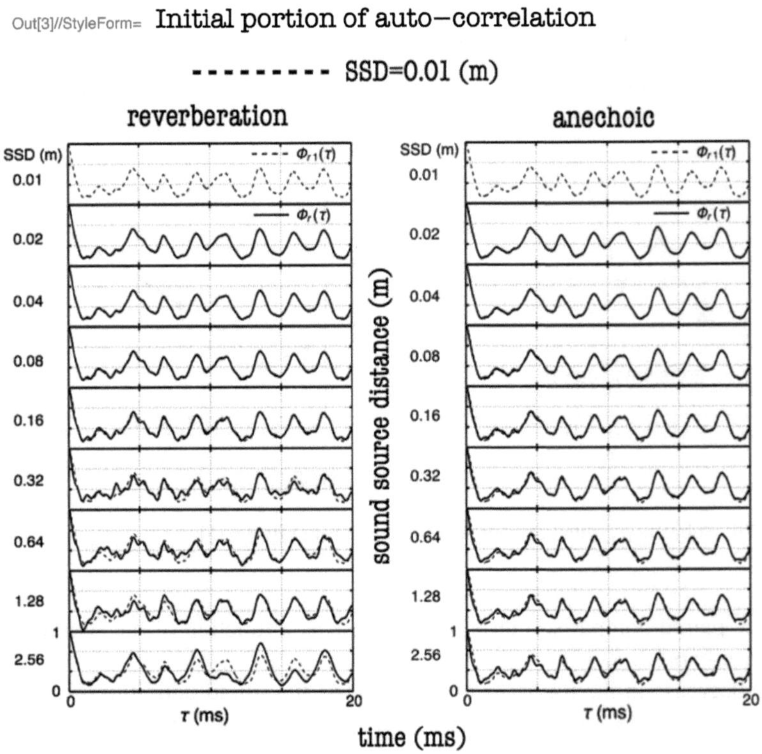

Fig. 12.21 Example of auto-correlation sequences for musical sounds in reverberation and anechoic rooms [19]

12.3.2 Auto-correlation Functions for Early Echoes

Spectral deformation as a consequence of source distance may be represented using auto-correlation functions [5]. Figure 12.21 illustrates examples of the auto-correlation sequence for the short-duration test signals with a length of 30 (ms) and a time lag of up to 20 (ms) for both reverberation and anechoic rooms.

The auto-correlation varies depending on the source distance even at the coherent length. Taking the auto-correlation sequence at $r = 0.01$ (m) as reference, the absolute of the difference from the reference is shown in Fig. 12.22.

The difference in the dependence on source distance is notable between the reverberation and anechoic rooms. The total square of the difference from the reference representing the spectral dissimilarity from the reference (mostly direct sound) is presented in Fig. 12.23.

There is no prominent difference in the spectral dissimilarity between the results for the reverberation and the anechoic rooms as long as the distance is within 0.16 (m). However, the difference becomes marked when the distance exceeds 0.16 (m), that

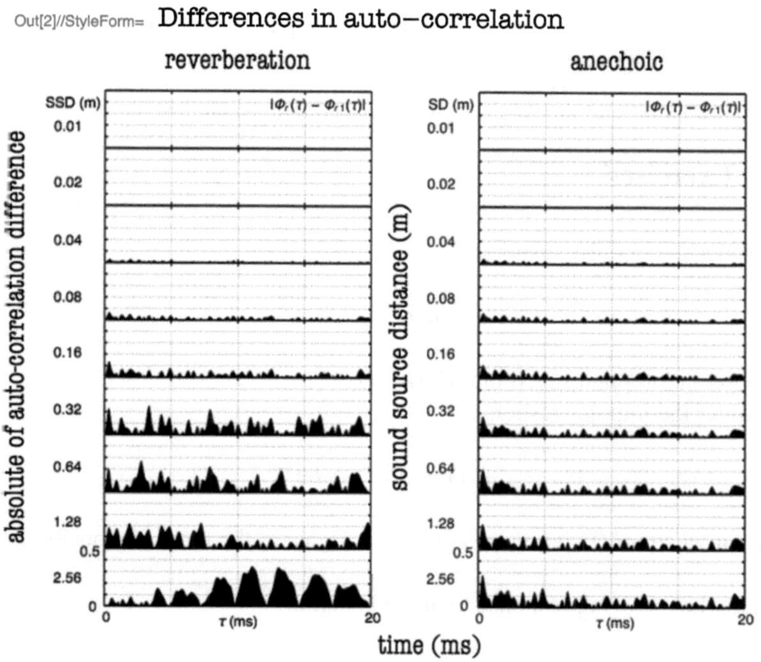

Fig. 12.22 Absolute of the differences in the auto-correlation sequences as a function of source distance where $r = 0.01$ is taken as Ref. [19]

Fig. 12.23 Total square of the differences in the auto-correlation sequences as a function of source distance where $r = 0.01$ is [19]

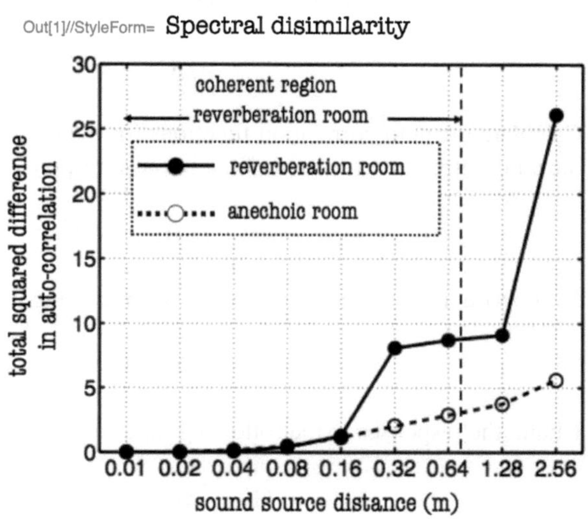

is, going outside the coherent region. This trend partly explains the results seen in the loudness of signals shown in Fig. 12.20.

12.4 Exercises

1. Consider the impulse response

$$h(t) = \mathrm{e}^{-at}$$

at an observation point in a room. Derive D_{30}.

2. Consider the impulse response

$$h(t) = a\delta(t) + b\mathrm{e}^{-at}$$

in a room where $a\delta(t)$ represents the direct signal at the receiving position. Derive D_{30}.

3. Obtain the receiving response to an input signal

$$x(t) = \mathrm{e}^{-\lambda t}$$

for the above receiving point.

4. Suppose the auto-correlation function of an input signal is

$$r_{in}(t) = \mathrm{e}^{-at}.$$

Calculate the auto-correlation functions for the output response in a system where the auto-correlation function of the impulse response is given by

$$r_{sys}(t) = \mathrm{e}^{-bt}.$$

5. Suppose that the impulse response at a receiving position is

$$h(t) = \delta(t) + b\delta(t - \tau).$$

Obtain the responses to the following input signals,

$$\begin{aligned} &(1) \quad x(t) = \cos \omega t \\ &(2) \quad x(t) = \delta(t) \\ &(3) \quad x(t) = \mathrm{e}^{-\lambda t} \\ &(4) \quad x(t) = \mathrm{e}^{-\lambda t} \cos \omega t. \end{aligned} \tag{12.9}$$

Consider what D_{30} implies.

References

1. R. Thiele, Richitungsverteilung und Zeitfolge der Schalrueckwuerfe in Rauemen. Acustica **3**, 291–302 (1953)
2. T.J. Schultz, Acoustics of the concert hall. IEEE Spect. **2**, 56–67 (1965)
3. M. Tohyama, *Sound and Signals* (Springer, Berlin, 2011)
4. P.M. Morse, R.H. Bolt, Sound waves in rooms. Rev. Mod. Phys. **16**, 69–150 (1944)
5. Y. Hara, *Sound perception and temporal-spectral characteristics in sound field near sound source (in Japanese with English abstract)* Ph.D. Thesis, Kogakuin University (2014)
6. M. Tohyama, *Waveform Analysis of Sound* (Springer, Berlin, 2015)
7. Y.Hara, Y.Takahashi, H.Nomura, M.Tohyama, and K.Miyoshi *Perception of sound source distance and loudness in a coherent field* Acoustics'08, 2196, 4pPPd6 (2008)
8. Y. Hara, H. Nomura, M. Tohyama, K. Miyoshi, Subjective evaluation for music recording positions in a coherent region of a reverberant field, in *Proceedings of AES 2008 124th Convention* (2008), p. 7446
9. M.R. Schroeder, Statistical parameters of the frequency response curves in large rooms. J. Audio Eng. Soc. **35**(5), 299–306 (1987)
10. R.H. Lyon, Progressive phase trends in multi-degree-of-freedom systems. J. Acoust. Soc. Am. **73**(4), 1223–1228 (1983)
11. R.H. Lyon, Range and frequency dependence of transfer function phase. J. Acoust. Soc. Am. **76**(5), 1435–1437 (1984)
12. Y. Takahashi, M. Tohyama, and Y. Yamasaki Phase response of transfer functions and coherent field in a reverberation room. Electron. Commun. Jpn. Part 3 **90**(4), 1–8 (2007)
13. T. Houtgast, H.J.M. Steeneken, R. Plomp, A review of the MTF concept in room acoustics and its use for estimating speech intelligibility in auditoria. J. Acoust. Soc. Am. **77**(3), 1069–1077 (1985)
14. M. Kazama, S. Gotoh, M. Tohyama, T. Houtgast, On the significance of phase in the short term Fourier spectrum for speech intelligibility. J. Acoust. Soc. Am. **127**(3), 1432–1439 (2010)
15. Y. Hara, Y. Takahashi, K. Miyoshi, Narrow-band envelope spectral analysis close to sound source (in Japanese). J. Inst. Elect. Inf. Comm. Eng. Jpn. **J97–A**(3), 221–223 (2014)
16. *Water music suite, anechoic-recorded samples for evaluation* Denon professional test CD, Track 9 II(6) Bars 1-11
17. E.D. Montag, Empirical formula for creating error bars for the method of paired comparison. J. Electron. Imaging **15**(1) 010502.1–0.0502.3 (2006)
18. Y. Ando, *Auditory and Visual Sensation* (Springer, Berlin, 2009)
19. Y. Hara, *Private communication* (2017)

References

1. K. Lincke, Richt...
   29, 302 (1953)
2. T. Schiller, Ann...
3. M. Tobiyama, Ro...
4. M. Meng, R. H...
5. Y. Hara, Nucl. ...
   tors for fundamen...
6. M. Tobiyama, ...
7. J. Urata, Y. Tan...
   rators and dimensions...
8. M. Hara, E. Kong...
   ...
   (2011), p. ...
9. ... N. Neeb...
   Nucl. Phys. Sci. ...
10. R.H. Lyon, Progr...
    Mon. 1274–1...
11. R.H. Lyon, Res...
    Nucl. 1185–1...
12. Y. Takahashi, M. ...
    ... in a ...
13. T. Kanazawa, M. ...
    ... use for estim...
    ...
14. M. Kamimura, ...
    Photon spectrum ...
15. Y. Hara, Y. Kak...
    source. In Symp...
16. ...
    Lack, HEP Ree...
17. J. D. Morgan, Ben...
    ... Interaction, ...
18. Y. Ando, Nucl. ...
19. Y. Hara, Proc. ...

Index

M. Tohyama, *Sound in the Time Domain*, Signals and Communication Technology,
https://doi.org/10.1007/978-981-10-5889-9

M

N

O

P

MIX
Papier aus verantwortungsvollen Quellen
Paper from responsible sources
FSC® C105338

If you have any concerns about our products,
you can contact us on
ProductSafety@springernature.com

In case Publisher is established outside the EU,
the EU authorized representative is:
Springer Nature Customer Service Center GmbH
Europaplatz 3, 69115 Heidelberg, Germany

Printed by Libri Plureos GmbH
in Hamburg, Germany